RUSSIA FROM 1812 TO 1945

RUSSIA
from 1812 to 1945

A HISTORY

Graham Stephenson

PRAEGER PUBLISHERS
New York • Washington

BOOKS THAT MATTER

Published in the United States of America in 1970
by Praeger Publishers, Inc.
111 Fourth Avenue, New York, N.Y. 10003

Library of Congress Catalog Card Number: 77–96294

Printed in the United States of America

Contents

CONTENTS

List of Plates

Between pages 224 and 225

All photographs are reproduced by kind permission of the Novosti Press Agency

Maps

Acknowledgements

Lyashchenko, *A History of the National Economy of Russia* for the tables on pp. 366, 369; Mrs J. E. Thorpe and Collier-Macmillan International, for extracts from *The Russian Peasant and other Stories*, by John Maynard.

Preface

THIS book is intended to give the reader an introduction to some of the most characteristic features of Russian history between Napoleon and Hitler. It deals with a period in which some Europeans hoped to be able to eliminate Russian power; it is written in a period in which that hope is still alive but is less likely to be realised than in the period 1812–1945. A Frenchman or a German might feel regret that their respective nations had failed to extinguish Russia; an Englishman could not share this sentiment, for both in 1812 and 1945 the survival of Britain was bound up with that of Russia. This book is based upon the assumption that during the period treated Britain and Russia were 'natural' allies; no opinion is expressed as to whether this is still the case.

The reader will not find here a chronological summary posing as a history book. In the period before 1917 much use has been made of thematic treatment. The idea is to suggest to the reader that such topics as the peasantry and the intelligentsia are both crucial and characteristic. Both groups might well be omitted from a study of nineteenth-century Britain; in this respect the English-speaking reader needs to have his imagination jogged. This is what the topical chapter is intended to do. If the reader wishes to accumulate all the material on any single personality—say Stolypin—the rather elaborate table of contents should enable him to find his way round the book. After 1917 the thematic treatment is used much less. This change in organisation is intended to reflect a substantial change in the history of Russia. Up to 1917 the book analyses the decline and fall of the Tsarist political system. The forces of discontent and dissolution were strong and the political structure which tried to contain them comparatively weak. The structure of the book reflects the conflict between official and unofficial Russia, between the Emperor and his bureaucrats on the one hand and the revolutionary forces on the other. But after 1917 the focus of interest is concentrated more exclusively upon the new political masters of Russia. This does not mean to imply the absence of discontent in the post-revolutionary period but merely that discontent has (so far) been less well organised, less articulate and less successful.

Some readers may find that the treatment of Russian authoritarianism is not to their taste. There are many different reasons for hating Stalin; the author has tried to avoid them all. He has not aligned himself with the Khrushchev revisionists, the Trotskyites, the Maoists or the Cold War agitators. He has not been unduly influenced by the lurid tales of Stalin's daughter. He has found it impossible to believe that Stalin was mad. He has been horrified by the descriptions of camp life and death; but he has considered that the duty of the historian is to refrain from loud cries of outraged virtue. It is no doubt regrettable that the Russians have not shown much inclination to follow the decent example of western liberalism; but it is more interesting to understand what actually happened than to regret what failed to materialise. Stalinist Russia presents the historian with almost intolerable problems. Dreadful human suffering was so closely intermingled with outstanding public triumphs. Again and again the question has to be faced: do the triumphs justify the suffering? But this is not a question which the historian is called to answer. It can be left to a group of people whom history has frequently called into existence to shed light in dark places – the Russian intelligentsia. The western reader may – if he is sufficiently humble – learn from writers like Pasternak and Solzhenitsyn some clues to the riddle. But the western reader (and writer) who has not himself been touched by the contradictions of recent Russian history will do well to refrain from striking virtuous attitudes.

Dates: In the nineteenth century the Russian calendar was twelve, and in the twentieth, thirteen days behind the Gregorian calendar in use in Western Europe. This peculiar custom was abolished by the Bolsheviks in 1918, and they brought Russia into line with the West. In this book, dates are given according to the old (or Julian) calendar up to and including the February revolution of 1917. The second or Bolshevik revolution is dated according to the new or Gregorian calendar. Thus, in this book the fall of Nicholas II is described as 'the February revolution' and the victory of the Bolsheviks is described as 'the November revolution'.

Measures: Pood = 36 lbs
Arshin = 28 inches
Verst = 3,500 feet
Desyatin = 2·7 acres.

I

Russia on the Eve of the French Invasion

THE LEGACY OF PETER THE GREAT

The invasion of 1812 was the most serious challenge to the Russian State since the invasion of Charles XII of Sweden in 1709. The development of Russia since that time is well symbolised by the immense bronze statue of Peter the Great erected by the most famous of his successors, Catherine II. It shows the maker of modern Russia mounted on a horse which even he controls only with difficulty; he points towards the river Neva and the city which he had built. Inscribed upon the huge granite block which supports the rider are the simple words, in Russian and in Latin: 'Petro Primo Catherina Secunda'. In this laconic fashion did the two illustrious statebuilders greet one another. This statue inspired Pushkin to write one of his greatest poems, 'The Bronze Horseman'. In it he imagines Peter's thoughts as he surveys the city of his creation:

> 'From here we shall threaten the Swede; here a city shall be founded to spite our haughty neighbour. By nature we are fated here to cut a window through to Europe, to stand with firm foothold on the sea. Here, on waves unknown to them, ships of every flag will come to visit us, and we shall revel on the open sea.' A hundred years have passed, and the young city, the adornment and marvel of the northern lands, has risen splendidly and proudly from the gloom of the forests and the swamp of the marshes; where once the Finnish fisherman . . . cast his time-worn net into the unknown waters, there now huge harmonious palaces and towers crowd on the bustling banks; ships in their throngs speed from all ends of the earth to the rich quays; the Neva is clad in granite; bridges hang

poised over her waters; her islands are covered with dark-green gardens. And before the younger capital ancient Moscow has paled. . . .

St Petersburg had been intended by its founder to be a fortress and a port; it became, during the eighteenth century, one of the most beautiful cities in Europe. Imperial palaces, the houses of the nobility and the merchants, theatres and schools, arose on the inhospitable marshes. They were the work of foreign architects. St Petersburg always had a large foreign population including a substantial English and Scottish colony. From this exotic city, the Venice of the North, the rulers of Russia surrounded by their courtiers, soldiers and bureaucrats, looked outward at a Europe which seemed similar to their own capital and inward at a land which seemed strange and foreign. Yet the inhabitants of that land accepted with little criticism their German-born autocrats and French-speaking court nobility. The alternative to autocracy was thought to be anarchy. St Petersburg might be a burden upon the people of Russia but it was also a centre of authority, a protection against foreign invasion and internal disorder.

Autocracy was Russia's peculiar institution. It was far older than Peter the Great. He had made it work more efficiently; Stalin made it work better than the Romanovs. During the nineteenth century many of the most intelligent subjects of the Tsar criticised the principles of autocracy, but for most Russians it was an acceptable form of government. The reasons for this go deep into the Russian past. The absence of clearly marked geographical frontiers tended to encourage dispersion; autocracy countered this tendency. Autocracy expressed the religious and cultural unity of a people who continued to believe that the Orthodox form of Christianity was closer to the spirit of Jesus Christ than any other. Moscow became the heir of the authoritarianism of Byzantium.

Russia had neither a reformation, nor a religious war, nor a capitalist class. Consequently her history lacked the shape of English development – a shape which England transmitted to the United States. English history of the seventeenth and eighteenth centuries is in essence the record of how cohesive social classes eroded the power of the monarchy. As each class won freedom for itself it gave an example to another class. This struggle was fought within the framework of a sovereign legal system. The idea of Law was more important than the idea of the State. The laws were the rules by which the political game was played. In the end, English development tended to reduce the State to little more than a referee. Society was everything and government was nothing: liberty was widespread but so was inequality. The English development was hateful to most Russians in the nineteenth century. They saw it as nothing but an excuse for the application of jungle law, and thought little of the boasted English freedom. They thought that it was merely a freedom for the strong to oppress the weak. They thought that freedom was dangerous because it gave free rein to the worst instincts of the worst men. They thought that the English system was both naive and irreligious: it failed to take into account the great fact of original sin.

Enlightened Russians realised that their country needed change. In particular, they understood the need to adapt it to the Industrial Revolution, but they wanted to do this without adopting an English type of parliamentary democracy. They thought that change should be initiated and pushed through by an enlightened autocrat. Peter the Great had modernised Russia: there seemed to be no reason why one of his nineteenth-century successors should not imitate his achievement. Russian political thought was always more attracted by the idea of equality than by that of liberty. Autocracy was alleged to create equality. All were equally the subjects of the most high Tsar. His unfettered will was more humane than any impersonal legal system. The belief in autocracy has been one of the constant features of Russian history.

The eighteenth-century rulers of Russia had been outstandingly successful in their conduct of foreign policy. Peter the Great had established Russia upon the shores of the Gulf of Finland; his successors pushed the frontiers westward and southward. Catherine the Great occupied the whole of eastern Poland (1772–1795) and gave Russia a common frontier with Prussia and Austria. She also completed the work of Peter the Great by defeating the Turks and annexing the northern coast of the Black Sea. This task was finished in 1783. The new fortress of Sevastopol, constructed for Catherine by Sir Samuel Bentham, symbolised the fact that Russia was not confined to the Baltic alone. The Baltic proved to be no more than a window into

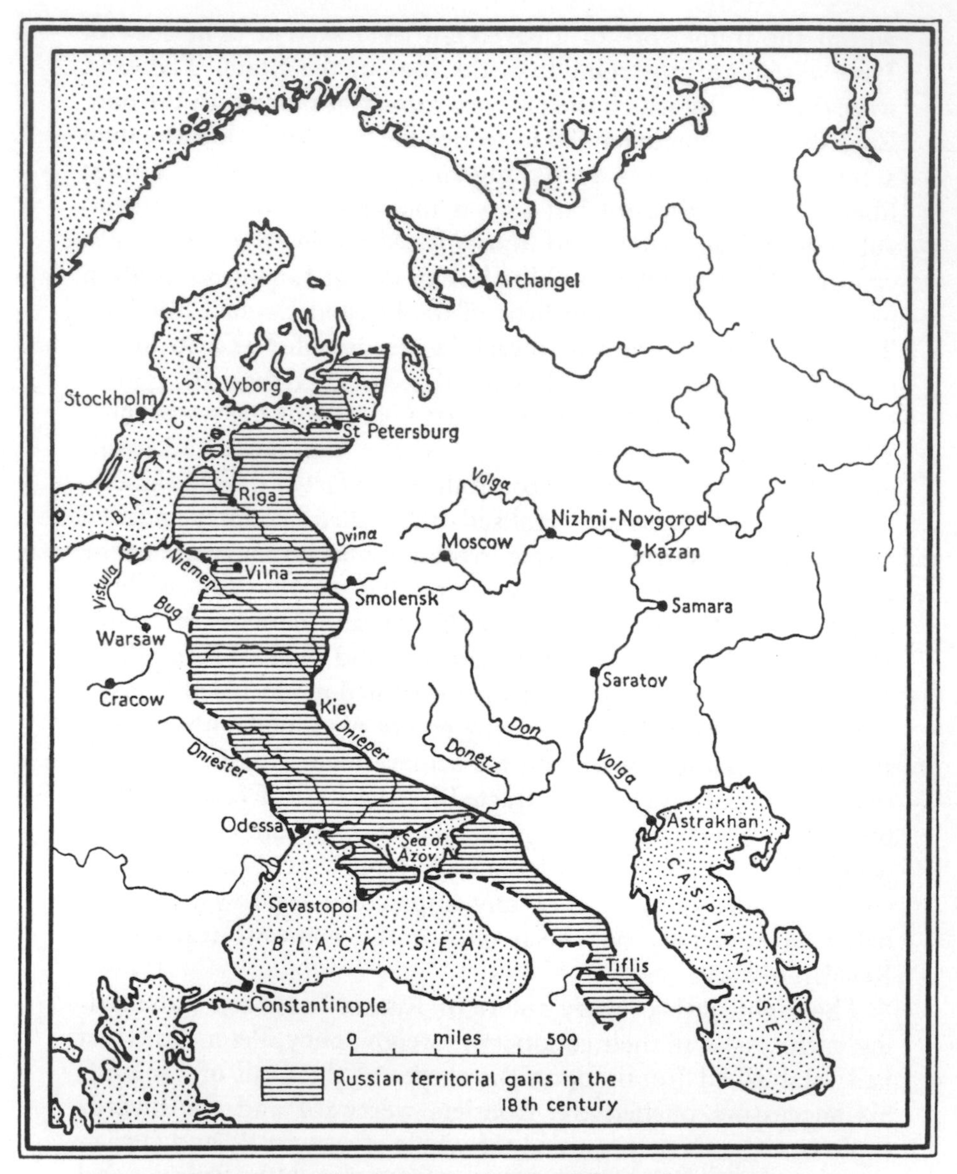

Russian territorial gains in the eighteenth century

Europe, but the Black Sea became a door through which nineteenth-century Russia hoped to dominate the eastern Mediterranean. Its ports eclipsed those of the Baltic in their commercial activities. The movement south became the most important in Russian nineteenth-century history. Turkey alone could not have halted it. Behind Turkey stood Britain, that power which was in most respects the exact antithesis of autocratic Russia. Here was one of the tragic paradoxes of the nineteenth century. Britain successfully resisted the Russian drive towards the Mediterranean; she maintained the integrity of Turkey. Then, in 1915, when by a strange reversal of fortunes Russia and Britain fought on the same side, Britain was unable to get through to her ally because Turkey stood in the way. British hostility was already foreshadowed when Catherine the Great died (1796). But the map at that date showed nothing but success for the Petrine State in the field of foreign policy.

A glance at the internal history of Russia during the eighteenth century would not seem to confirm this verdict. There were several facts which indicated internal instability. Disputed successions were frequent in eighteenth-century Russia, disputes which were often settled by the intervention of the Guards. Catherine II herself helped to depose her husband Peter III in 1762: with great coolness she subsequently announced to the world that Peter had died (he had in fact been murdered) and that she intended to set aside the right of his son to occupy the throne. Catherine managed to carry off this *coup d'état* triumphantly. It was a shocking action by the legalistic standards of the Western monarchies but it did have excellent practical results; the most able available candidate became monarch. The same rather cavalier attitude towards the laws of succession was shown in 1801. Then a group of Guards officers, with the knowledge and approval of the heir to the throne, later Alexander I, murdered Paul I in his bedroom. It was afterwards given out that Paul was mad and the opinion supported by a number of dramatic stories. Modern research, however, inclines to the view that Paul merely had a rather morbid sense of humour. There were no more palace revolutions until March 1917. None of the nineteenth-century Romanovs was as able as either Peter I or Catherine II. A palace revolution which placed a person of that calibre upon the Russian throne might well have saved the *ancien régime*. Legitimacy has its disadvantages.

A more serious weakness at the end of the eighteenth century lay in the machinery of government. Outwardly it retained the imprint which had been placed upon it by Peter I, but profound modifications had been made by Catherine II. Peter had been a centraliser. He had placed the entire administrative machinery under the control of the Senate and its chief, the Procurator-General. This institution acted as a legislature, the supreme judiciary, and as the supervisor of the executive departments which Peter organised on a collegiate basis. This was a device quite commonly used in the eighteenth century – in the British Board of Admiralty for example – and meant that any given department was not under the control of a single individual but was run by a group of persons regarded as equals. The system prevented any one person from assuming too much power and was particularly suitable where public honesty and competence were at a low level. But it did not make for speed of decision or consistency of policy.

Catherine II decentralised the structure of government both by transforming the Senate into little more than a supreme court and by creating a new provincial administration (Decree for the Establishment of the Gubernii, 1775). The Empire was divided into some fifty provinces on the basis of population: little care was taken to observe natural or historical boundaries. The provinces were further subdivided into counties (*uezd*), each of which was supposed to have a population of 20–30,000. Several provinces were grouped together under a Governor-General, who was appointed directly by the Empress and reported to her without reference to the administrative colleges. The local nobility was permitted to elect certain officials, especially those concerned with petty justice and minor administrative affairs. If Catherine intended this to give the nobles experience of government which would prepare them to participate in an elected legislature she must have been sadly disillusioned, for the ineptitude and servility of the elected noble officials became a scandal. The richer and better educated landowners would not serve, and those who did were merely the unpaid tools of the all-powerful Governor and his bureaucratic staff. Participation in local administration never carried the social prestige which gave local government so much vigour in eighteenth-century England. By the beginning of the nineteenth century the abuses of Catherine's system demanded reform. The Governors exercised tyrannical authority, executive, judicial and even legislative; the local bureaucracy was corrupt and poorly educated, for men of ability would only serve in St Petersburg. The central bureaucratic institutions had been mostly destroyed by Catherine after 1775. Her grandsons had little choice but to build them up again.

The most serious legacy of Catherine's reign, however, was financial insolvency. Catherine's conquests involved many wars and she failed either to open new sources of taxation or to create public credit. Instead she turned to the printing press, at first as a means of facilitating exchange but later to pay for her wars. By the end of her reign the inevitable depreciation had occurred; the paper rouble was worth only 68 per cent of its face value. The government was bound to accept payment of taxes in paper but had to pay silver for its own purchases. Catherine's successor had to correct this cancerous legacy at the very time when Russia was threatened by Napoleon. The question which faced Alexander I in 1801 was: would the confused administrative and financial condition of Russia permit her to retain the brilliant foreign conquests of the eighteenth century?

THE NOBILITY

Even an autocracy, such as Russia was in the eighteenth century, required a governing class through which the will of the autocrat could be transmitted. This was provided by the nobility, whose status and organisation can be a puzzle to the Western student whose ideas have been shaped by the 'estate' or caste systems of England and France. Russian autocrats were faced by a fundamental dilemma, never resolved under the *ancien régime*: should they make the State more efficient by encouraging the aspirations towards independence of their most able and educated subjects, or should they avoid the risk of having to share political power by keeping the nobility chained to state service. Peter I had adopted the latter course; in 1722 he issued his Table of Ranks (which remained unaltered until 1917), and compelled all noblemen to enter state service in either a military or a bureaucratic capacity. Those who did not would lose the most prized privilege of the nobleman – the right to keep serfs; even those who did enter the state service would not receive a title of nobility until they had passed Grade 8 in the fourteen-rung ladder of promotion. In practice, however, the sons of the older and richer families had no difficulty in passing this grade while the ambitious and able sons of non-noble parents found many impediments in the way of their promotion. But compulsory state service was very unpopular with the nobility. In 1762 it was abolished by Peter III, and never restored.

This relaxation in the system established by Peter I had profound results. A proportion of the nobility retired from the capital altogether, some to reside upon their rural estates busying themselves as best they could, others to live in or about Moscow, the old capital, whose noble population nursed grievances against the 'modernising' St Petersburg government. But even though it had ceased to be compulsory, state service had become both a tradition and an economic necessity with a large section of the nobility. The principle of primogeniture had never been established in spite of Peter I's efforts. Estates were subdivided among sons and daughters until there was hardly anything left. Trade and industry were so little developed and carried so little prestige that they could not employ the sons of noblemen. In consequence, there remained only state service, ususally in the army. Thus from 1762 there developed in the nobility a split between the provincial nobility who never visited the capital, who were uneducated, who had little to occupy them apart from field sports, who were unable to adopt new ideas about agriculture, who clung to the one thing which gave them status, the ownership of serfs; and the metropolitan nobility who spent their lives in state service and who went to their estates only to collect the rents and, possibly, in old age. During the nineteenth century the rural nobility were to give birth to the revolutionary intelligentsia, and the metropolitan nobles produced the 'liberal' reforming bureaucracy. The tradition of state service prevented the nobility from developing a powerful corporate sentiment, from becoming an 'estate' strong enough to curb the autocracy, for the autocracy constantly drew away the most able of the nobles into the army and bureaucracy. 'The noble is just a bureaucrat in a dressing-gown while the bureaucrat is a noble in uniform.'

Even at the height of their power during the reign of Catherine II, the nobles were unable to force the crown into an admission of their political rights. The Charter of the Nobility (1785) confirmed their freedom from taxation and physical chastisement, and their right to dispose of their property as they liked. But the provincial noble assemblies were strictly confined to local issues and placed under the tutelage of the provincial Governors. It was as if, in eighteenth-century England, the House of Lords did not exist but the nobles were allowed to discuss matters concerning roads, bridges and drains with the permission of the Lord-Lieutenant of the county. Such an occupation was unlikely to appeal to the owner of a thousand serfs; small wonder that the more intelligent nobles sought employment in St Petersburg.

The nobility, if not the only, was by far the largest educated class. Under Peter I the government had tried to force its noble conscripts to learn military tactics, engineering and related technical subjects. But with the increasing freedom of the nobility these irksome disciplines were neglected and the more enlightened members of the class fell with enthusiasm upon the literary and philosophical works of French, English, Scottish and German writers. These tastes were encouraged by Catherine who indefatigably consumed the latest authors and the newest theories of the Enlightenment. French became the language of educated society – a fact which Tolstoy brings out in the opening pages of *War and Peace*. The metropolitan nobility despised the Russian language and many could not speak it fluently, thus opening another chasm between themselves and the rural nobles. Much of this education was the work of foreign tutors, men who frequently had to leave their native countries because of their political radicalism and who found a ready market for their talents in autocratic Russia. Their noble pupils travelled much abroad, especially after 1762, and there excited the wonder of Western observers by the vigour with which they absorbed new ideas and by their energetic dissipations. At home, Catherine failed to increase the number of universities (two) or to inspire them with fresh vigour. She did, however, encourage the creation of secondary schools in provincial cities; about thirty such schools existed by the end of the century. But their effectiveness was greatly hampered by the general lack of money already described.

Towards the end of the eighteenth century a significant change in the cultural bias of the Russian nobility began to make itself felt. Some influential writers began to turn away in disgust from the foreign models which had hitherto dominated the scene. In the same spirit they turned towards the Russian language which had not hitherto been thought capable of serious literary use. Such writers were the historians Shcherbatov and Karamzin, who not only wrote in Russian but had the effrontery to attack the whole tendency of the post-Petrine state and to claim shining virtues for the institutions and the morals of Old Muscovy. These writers, patriotic, conservative and xenophobic, were to have a numerous spiritual progeny during the nineteenth century. Other significant figures were the writers Radishchev and Novikov. The former published in 1790 his *Journey from St Petersburg to Moscow* in which he gave a bitter account of serfdom and spoke slightingly of the autocratic form of government. His book made the aged Empress very angry in spite of the fact that Radishchev merely repeated what Catherine herself had said in earlier years. Radishchev was sent to Siberia – the first of a long line of writers who suffered because they wrote what everyone knew to be true. Novikov was a Freemason who was responsible for the publication of a large number of educational books and the creation of scholarships for foreign study. But in 1792, suspected of propagating radical ideas, he was sentenced to fifteen years imprisonment.

These were the opening shots of a long war between the State and the intelligentsia, a war which was to grow in intensity during the nineteenth century. This struggle is a unique feature of modern Russian history; its tensions helped to transform the stilted and imitative literature of the eighteenth century into the most original and lively of the national literatures of the nineteenth. For many of the individual writers, conflict with authority brought exile and personal tragedy, but for mankind at large it brought the refreshment of a new literary horizon. This was the dying gift of the Russian nobility.

THE PEASANTRY

The economic basis of the nobility was the land. By the beginning of the nineteenth century private estates were everywhere worked by peasant serfs, who were by far the most numerous class in the State and entirely deprived of political and personal rights. The bonds of serfdom had been strengthened during the eighteenth century with the intention of compensating the nobility for their lack of political power. But the direction of policy changed with the accession of Alexander I in 1801. From that date the autocrats, at first timidly and then with desperate courage, attacked serfdom because it was inhumane, because it was inefficient and because it gave too much social influence to the nobility. Emancipation at length came in 1861; the long delay is perhaps a testimony to the obstinacy of the nobility even in Russia. The peasant question festered for too long. It helped to set the intelligentsia against the State and prevented Russia from taking steps towards industrialisation at an early date. Defeat in 1812 might even have been an advantage; the total victory over Napoleon strengthened the arguments of those who claimed that Russian institutions were not in need of reform. There were widespread fears that Napoleon would provoke social war in Russia by declaring emancipation during his march to Moscow. He failed to do so and the peasants fought with courageous patriotism. In spite of the condition to which he had been reduced – in practice little short of slavery – the Russian peasant retained a superstititious veneration for the person of the Tsar, seeing in him the personal representative of God. Evils were usually blamed upon the landowner and even when the peasant did revolt – an event which had occurred frequently during the eighteenth century and which was to occur still more frequently during the nineteenth – he frequently 'discovered' a 'true' Tsar and claimed that the actual occupant of the throne was a usurper. The peasant lived in a world which was touched hardly at all by the concerns and interests of the educated world. His life was ruled by his labour, by ancient myths and customs and by the liturgical presentation of Christianity. One of the most important developments of the nineteenth century was the 'discovery' of the peasantry by the intellectuals. For centuries the peasant had been ignored when servile and punished when disobedient, but during the early nineteenth century the Slavophil writers[1] suddenly perceived that the peasant was, in fact, the true Russian, the sole receptacle of national virtue. No one was more, astonished than was the peasant by this belated appreciation of his worth.

THE CHURCH

The greatest formative influence, apart from the autocracy, had been the Church. It had taken its doctrine from Byzantium,

[1] See Chapter 5, p. 129-34

it had helped to nurse Russia through the Mongol invasions, it had preserved and transmitted a culture in many respects fundamentally different from that of the Latin West. From Byzantium the Church had inherited the idea of Caesaropapism – that is to say, the idea that the autocrat should be supreme head of both Church and State. But having given so much, the Church had fallen upon evil days in the eighteenth century. Its independent head, the patriarchate, had been abolished by Peter I. In its place he created a Holy Synod which was composed of the three metropolitan bishops (Moscow, Kiev and St Petersburg). To complete the subjection of the Church to the State, Peter created the office of Chief Procurator to the Holy Synod. This post was occupied by a secular official who involved the Church hierarchy in the state bureaucratic machine, a fate which tended to deprive the bishops of spiritual authority. The Church lost the independent control of its lands and its clergy were reduced to a bureaucratic caste. Individuals of great personal sanctity still managed to exist among the celibate 'black' clergy in the monasteries, but the official Church became to an increasing extent an extra arm of the State, forcing uniformity of belief upon the many millions within the borders of Russia who did not easily accept the Orthodox tradition. This side of the Church's activity became more pronounced as the nineteenth century progressed, and gave rise to a fierce religious persecution at the time when, in the rest of Europe, the tolerance born of indifference was falling upon the nations. In a country in which education was hard to get the church seminaries were of some importance. They helped to form the minds of many poor boys who later (like Speransky) became useful bureaucrats or who (like Stalin) served the State in other ways. The Church, like the nobility, was deprived of the strength which came from being an independent 'estate'; by the beginning of the nineteenth century it had lost the will and courage to restrain the autocracy. Nevertheless, by way of a majestic liturgy, it mediated a powerful spiritual message to the masses.

THE TOWNS AND INDUSTRY

Urban life was little developed outside the old and the new capitals. The total number of town dwellers in 1800 was about two million in a total population of 36 million. It is true that the proportion of urban dwellers had increased during the eighteenth century, yet the conditions did not exist in Russia for the rapid urbanisation which took place throughout Western Europe during the early nineteenth century. This sluggish development was one of the decisive features of Russian history both before and after 1917. It was welcomed by the Slavophils, who saw in it a barrier against the domination of society by an urban middle class, but it was regretted by those who saw in urbanisation the only means by which Russia could maintain her position in the world. Compared with the nineteenth-century development of Western societies, Russia lacked a bourgeoisie, the class which almost everywhere else in Europe created constitutional states out of conflict with ancient monarchies and aristocracies. In other words, Russia remained notably uninfluenced by the French Revolution and its aftermath.

Urbanisation occurred in the West because the economic forces connected with the Industrial Revolution compelled previously scattered manufacturing enterprises to congregate in larger units in the towns. The same development was prevented in Russia by serfdom. During the latter part of the eighteenth century the nobility used their control over the only large supply of labour – the bonded peasantry – to prevent the merchant class from setting up factories which made use of free labour. Consequently a merchant with capital to invest had only two courses open to him: he could either use the very small supply of labour in the towns or he could purchase one of the 'possessionary' factories which entitled him to use serf labour but which also forced him to supply his products exclusively to the State. 'Possessionary' enterprises were particularly common in the mining and metallurgical industries and in anything to do with military supplies. Such industries were well developed in eighteenth-century Russia which for a time had been the world's largest exporter of pig-iron, but the existence of unlimited supplies of cheap labour had its inevitable result; it discouraged the application of capital and of new knowledge. By 1800 the metal industries had become antiquated by British standards.

Outside these 'possessionary' works, industrial production was under the jealous control of the nobility. In some cases landowners merely used their serfs to supply the needs of a closed manorial economy. Everything required on the estate was manufactured in the most primitive way by its inhabitants. Such a system was strictly 'feudal' and prevented the growth of a free market, inhibited the division of labour and ensured a very low level of production. But more often the landlords encouraged their serfs to produce for the market, and of course took care to absorb the profits of their serf-workmen. Such industrial enterprises frequently employed thousands of serfs but they were rarely grouped into factories. The serfs worked in cottages and attics, often with great skill but without the aid of machinery. The income from this cottage industry was very important to most landowners; it was against their interests to permit the growth of urban industrial life. As long as serfdom lasted industry would remain chained to this picturesque but inefficient system.

ALEXANDER I AND NAPOLEON: TILSIT AND AFTER

The conflict between Napoleon and Alexander I was already eleven years old when the French troops crossed the Niemen in June 1812. During this decade Russia, alone among the continental powers, retained her independence and the integrity of her frontiers. Alone among the eighteenth-century autocracies she proved capable of withstanding the French Revolution in its Napoleonic form. This success had many causes – distance, climate, the patriotism of the Russian people – but not least the diplomatic skill of Alexander himself. He was the most intelligent of the monarchs against whom the French Emperor exerted himself; in him the simple terror felt by the rulers of Prussia and Austria was tempered by an admiration for Napoleon, an admiration which gave him some understanding of the mind of that cynical genius.

It must be admitted, however, that Alexander's earliest efforts did not reveal the finesse which he later acquired. The renewal of the war between France and England in 1803 did not in itself damage Russia's interests; on the contrary it gave Russia an opportunity to pursue her ambitions in the area of the Black Sea. Alexander, however, allowed himself to be influenced by the anti-French sentiment in court circles which had been inflamed by the murder of the Duc d'Enghien and by Napoleon's assumption of the imperial title. In 1805 Alexander signed a treaty of alliance with England under which Pitt agreed to play the paymaster. Austria unwillingly joined this alliance but Prussia remained neutral. The Russian army marched to the Danube only to find that the Austrians had already been routed at Ulm and that Napoleon was in Vienna. Kutuzov, the Russian commander, wanted to withdraw his army; but Alexander insisted upon advance and the result was the total defeat of the Russian army at Austerlitz (December 1805). Alexander himself fled ignominiously from the battlefield and rode through the night accompanied by only three cossacks. His reception in St Petersburg was frigid; the salons, encouraged by the outspoken criticisms of his mother, openly attacked his policy. The historian Karamzin rather more moderately wrote:

> Russia had set in motion all her forces to help England and Vienna, that is, to serve them as a tool in the quarrel with France, without any particular advantage to herself. . . . What did we intend to do if we had won the war? [To establish] the greatness, the primacy of Austria, which from gratitude would have relegated Russia to a secondary position. . . .

But worse was to follow. In 1806 Sebastiani, the new French ambassador in Constantinople, persuaded the Turks to declare war on Russia. Soon after, Prussia, which had come under the control of its Queen, the romantic Louise, belatedly declared war on Napoleon. Alexander had already promised aid to Prussia in spite of Prussia's neutrality in 1805. The Russian army marched again and was again too late. The Prussian ally had been rapidly defeated and the results were far more serious than those of Austerlitz, for the French army poured into Poland and thus threatened a vital Russian interest. The fighting was bitter and indecisive during the winter of 1806–7. Russia was isolated but Napoleon's army was not prepared for a winter campaign. At length in June 1807 the French won, at Friedland, the victory they so badly required. Alexander could have continued the war on Russian territory – Karamzin thought that he should – but the state of the army, the lack of a plan for defending Russia, the quarrels of his generals and the confusion of his finances were all arguments in favour of peace. Immediately after Friedland the two Emperors met at Tilsit on a raft roughly constructed by army engineers.

There is no record of what passed between the sovereigns at Tilsit but there is no doubt that Napoleon set out to dazzle Alexander. He talked vaguely but attractively about the division of the world between France and Russia and claimed that there was no necessary clash of interests between the two powers. He flattered Alexander by treating him as an equal. But the terms which Napoleon exacted were onerous. A French-controlled Duchy of Warsaw brought Napoleon's empire up to the Russian frontier; Prussia was dismembered, in spite of Alexander's protests; Russia was forced to join the continental blockade against English goods and had to promise to declare war against England if that power did not agree to peace terms within a specified time. The fact that Russia had now become a French ally was most clearly demonstrated by a secret clause which bound Russia to fight on the side of France in any European war. In return for these concessions Russia gained little enough. Alexander hoped for a free hand in Turkey but he did not get it; Napoleon expressly forbade him to seize Constantinople and gave only vague approval to other territorial ambitions. Only with regard to Sweden did Napoleon encourage Alexander to compensate himself for his humiliations. Russia consequently seized Finland from Sweden (1808–9), thus greatly strengthening herself by the removal of a foreign frontier from the close proximity of St Petersburg.

But, most precious of all, Russia had gained time. In 1807 Napoleon was irresistible: Alexander had to become his ally; mere neutrality was not sufficient to satisfy the master of Europe. But when the two Emperors met for the second time at Erfurt in 1808, the situation had already changed. The Spanish war had started, Napoleon had to withdraw some troops from Germany, and in consequence he loosened some of the bonds knotted at Tilsit. He permitted Alexander more freedom of action in Turkey by recognising Russia's annexation of the Danubian provinces of Moldavia and Wallachia. Napoleon tried to stiffen the loyalty of his wavering ally by suggesting a grandiose plan for the division of the Turkish Empire. At Erfurt Alexander found support in an unexpected quarter. Talleyrand encouraged him to recreate an anti-French alliance; surprising advice when it came from Napoleon's former foreign minister. Alexander's almost complete freedom in practice from the Tilsit bonds was strikingly illustrated in 1809. France was again forced into war with Austria, and Russia was bound to supply aid. But the Russian army advanced with such masterly delay that it arrived too late for the campaign. Shortly afterwards Alexander politely refused to permit Napoleon to marry his sister Catherine and the disappointed Emperor had to make do with Marie-Louise, an Austrian princess.

By 1810 it had become clear that Alexander had escaped from his Tilsit obligations. Napoleon hoped to bring him to heel again by threatening Russia in a particularly tender spot – Poland. After the defeat of Austria in 1809 he enlarged the Duchy of Warsaw by adding to it a slice of Austrian Galicia. But instead of reacting meekly to this threat of revived Polish nationalism, Alexander in his turn sought out a tender spot. At the end of 1810 he refused to impose the terms of the Trianon tariff, the latest of Napoleon's measures against English trade. This amounted to a public rebellion; it could not be ignored by Napoleon. Both sides now prepared for war. French troops poured into the Duchy of Warsaw, and Prussia and Austria were forced to fulfil their treaty obligations towards their conqueror. Alexander hastened to make peace with Turkey and Sweden and to prepare for a fighting retreat into Russia. He was now going to fight upon better terms than in 1807, but the previous performances of the Russian army gave the dispassionate observer little hope that Napoleon would be defeated. Alexander himself was more determined than optimistic; the nobles were nervous because they feared the effects of a war in Russia upon the serfs. It seemed likely that Napoleon would emancipate them and thus gain their support. But from his London embassy Count Vorontsov wrote:

> I fear only the diplomatic and political events; I do not fear the military events at all. Even if the operations should prove unfavourable to us in the beginning, we can win by stubbornly waging a defensive war and continuing to fight as we retreat. If the enemy pursues us . . . he will be surrounded by an army of Cossacks and in the end will be destroyed by our winter, which has always been our loyal ally.

In other words Vorontsov, like many Russian noblemen, most feared that Alexander would capitulate after the first defeat and negotiate another Tilsit. In order to reduce the likelihood of this, the nobles forced Alexander to dismiss and exile his allegedly pro-French minister Speransky early in 1812. After the French army had crossed the frontier Alexander was advised to leave head-quarters. He spent these critical months in St Petersburg, far from the battles and marches.

What had Alexander gained from the five-year breathing space? The same question has to be asked about the 'Tilsit peace' negotiated by Stalin in 1939. In 1807 Alexander could have fought Napoleon with allies: in 1812 he was alone; Austria and Prussia were under contract to Napoleon. Both powers sent their contingents in 1812 although both were less than willing allies. Alexander's Balkan and Polish ambitions had frightened the Austrians. Was there any advantage for them in a French defeat which might merely hand eastern Europe over to Russia? Relations with Britain were restored but the British navy, even British money, could be of little immediate aid in the great trial of strength on the Russian plains. The advantage gained in the extreme north and south (Finland and Turkey) seemed to be small compensation for the loss of the Austrian and Prussian alliances. Yet Alexander had been wise to seek this delay. Russians had learned to understand the implications of living in a Europe dominated by Napoleon. It meant the loss of the important British trade; it meant the possible revival of Polish nationalism; it meant that Russia could not be free to follow her ambition in the Balkans; it meant that the western frontier was constantly threatened: in brief, it meant an intolerable loss of independence. By 1812 there was a mood of national defiance which had not been universal in 1807. Then many educated Russians (like Rumyantsev, the foreign minister) had felt that Napoleon could be appeased. Now it was clear that he would have to be fought, and the fight would have to take place on Russian soil. National unity was essential and Tilsit allowed time for this sentiment to develop.

1812

Napoleon's army crossed the Niemen on 24 June 1812. Including reinforcements, he brought about 600,000 men to Russia, most of whom were of little military value. Napoleon was such a liar that it is impossible to be certain about what exactly he intended to accomplish by his invasion of Russia. At one moment he spoke of winning a single battle, making peace and forcing Alexander to accept once more the Tilsit terms; at others he spoke of marching to Moscow and then on to India where he reckoned to deal English commerce its death blow. It seems most likely, however, that his objective was the more limited one; he did not expect to get drawn on as far as Moscow. He hoped merely to force Alexander (or his successor if a palace revolution took place) to join the servile troop of European monarchs who had crept out to wish the Grand Army good fortune. This is why he did not emancipate the serfs; the Russian nobility would have become irreconcilable.

The Russian armies numbered about 225,000 and were divided. A small section had to be left to watch the Austrians; the two main armies under Barclay de Tolly and Prince Bagration had, during the first weeks of the campaign, to reckon with the possibility that Napoleon would advance either to Moscow or St Petersburg. Napoleon aimed to fight and defeat the northern Russian army under Barclay de Tolly before it united with the southern army at Smolensk. This he was nearly able to do, thanks to the plan devised by the Prussian general Phull. This strategist, who was bitterly criticised by his great compatriot von Clausewitz who was also serving with the Russian army, advised the construction of an armed camp at Drissa into which the Russian army should retreat and threaten the enemy line of communications. The fortifications of Drissa seemed so weak that Barclay wisely continued to retreat. His numerical inferiority made any other course impractical, but his plan was bitterly assailed both by Bagration and by court circles. Shishkov, the Tsar's secretary, wrote: 'To run away, to abandon so many towns and territories to the enemy, and with all that, to boast of such a beginning! What more could the enemy wish for? Nothing, except perhaps to advance unhindered to the very gates of our two capitals! Oh merciful Lord! My words are washed in bitter tears!' These words suggest that Alexander and his advisers had contemplated merely a short retreat from the frontiers. Already the war was imposing its own pattern upon the combatants. Neither side was fighting the campaign which it originally intended to fight.

Throughout July Barclay moved obstinately back. His rearguards inflicted much damage upon the French, but the heat, the devastation of the land and the lack of supplies were even more damaging. Even at this stage of the campaign the loss of horses was crippling the cavalry – a most important arm in a war of movement. At Vitebsk, at the end of July, Napoleon paused for several days. He had outmarched his supplies: should he go on? He decided to do so because he expected that the Russians would stand and fight at Smolensk when their two armies joined. This is what Barclay did; he defended the burning city with great ferocity and then slipped away during the night. Bagration accused his colleague of being 'irresolute, cowardly, senseless and slow' but it seems likely that Barclay saved Russia by his withdrawal. It was almost his last independent action. In the middle of August Kutuzov was made commander-in-chief of all the Russian armies.

Kutuzov has been so vividly depicted by Tolstoy that it is almost impossible to see him except through the eyes of the novelist. At the time of his appointment he was sixty-seven years of age and had hardly another year to live. He had been picked out for promotion by the great Suvorov and his popularity in the army was second only to that of Suvorov. At court he was not liked; and Alexander hated him because he had advised against fighting at Austerlitz, advice which Alexander should have followed. At this crisis there was no other general to turn to; popular and bureaucratic opinion demanded Kutuzov. He would have liked to continue Barclay's policy of retreat, but he knew that the morale of the Russian people would not survive the abandonment of the old capital without a battle. His task was to stand and fight and extricate the remnant of his army; victory was out of the question. This is what he did at Borodino on 7 September; he sacrificed half his army (58,000 casualties out of 112,000 combatants) in order to prove to Russia that the State was serious in its opposition to Napoleon. At the village of Fili on 13 September Kutuzov decided to abandon Moscow; he was opposed by most of his generals, by court opinion and by Sir Henry Wilson, the English commissioner at headquarters. Kutuzov explained himself in these words: 'As long as the army exists and is in condition to oppose the enemy, we preserve the hope of winning the war; but if the army is destroyed, Moscow and Russia will perish.' But the preservation of the army was not the only plan cogitated by Kutuzov. He realised that Moscow itself would act as a barrier to the French army. At Fili he said: 'Napoleon is like a stormy torrent which we are as yet unable to stop. Moscow will be the sponge that will suck him in.' On 14 September, Napoleon entered Moscow and complacently awaited the deputation which he imagined was going to bring him peace offers.

For thirty-three days he continued to wait for the deputation of boyars, but nothing happened. Fires broke out all over the city and the Emperor himself was driven out of the Kremlin. Nothing could stop the conflagration, which was possibly started by Rostopchin, Governor of Moscow; he later both confirmed and denied his responsibility. Disorder broke out among the French troops, exhausted by efforts to extinguish the blaze. Plunder was the only solace for such a hard campaign and the collection of plunder helped to turn an army into a rabble. On three occasions Napoleon sent word to his enemy offering to entertain peace conversations. One of his messengers was Herzen's father, Count Yakovlev, an absent-minded nobleman who had spent so long packing up his things that he was caught in Moscow by the French. After helping to put out some fires, he was led into the presence of the conqueror and later described his experience as follows:

> Napoleon began with those customary phrases, abrupt remarks, and laconic aphorisms to which it was the custom for thirty-five years to attribute some profound significance, until it was discovered that they generally meant very little. He then abused Rostopchin for the fires, and said it was mere vandalism; he declared, as always, that he loved peace above all things and that he was fighting England not Russia; he claimed credit for having placed a guard over the Foundling Hospital . . . and he complained of the Emperor Alexander. 'My desire for peace is kept from his Majesty by the people round him', he said.

In fact many of the courtiers, including Alexander's brother Constantine, were pressing the Tsar to open negotiations. But Alexander refused: he could not afford another Tilsit.

In the Russian camp the loss of Moscow caused widespread despair. But closer observers saw signs that Kutuzov's reading of the situation might be correct. The peasants had given up any hope they might have had that the coming of the French would bring liberation. Napoleon had made it clear that he did not intend to affront the one class upon which his future relations with Russia would lean, the landowners. Consequently, the peasants could give their natural hatred of the foreigners full scope; this sentiment became stronger after the burning of 'Holy' Moscow, and was daily fortified by the pillaging of crops and livestock which the hungry French had to undertake. During the autumn, the war assumed a more national character. In some cases the peasants spontaneously took up arms and murdered any isolated foreigners; in others, the peasants flocked to join the militia – a remarkable difference from their normal attitude towards compulsory military service; and in some cases peasants joined up with small bands of regulars and raided deep into the enemy lines of communications. Some of these partisan leaders became national heroes; the type has been incomparably described by Tolstoy in his character Dolohov. The changing character of the war is exemplified by Herzen in the following incident related by family tradition:

> The day before the enemy came, Count Rostopchin distributed arms of all kinds to the people at the Arsenal, and Platon [one of the family servants] had provided himself with a sword. Towards evening, a dragoon rode into the court-yard and tried to take a horse that was standing near the stable; but Platon flew at him, caught hold of the bridle, and said: 'The horse is ours; you shan't have it.' The dragoon pointed a pistol at him, but it can't have been loaded. . . . Platon pulled out his sword and struck the soldier over the head; the man reeled under the blow, and Platon struck him again and again. We thought we were doomed now; for, if his comrades saw him, they would soon kill us. When the dragoon fell off, Platon caught hold of his legs and threw him into a lime-pit, though the poor wretch was still breathing.

On 19 October Napoleon ordered retreat. His Russian enemy was not behaving according to the form which his other victories had taught him to expect. He made for Smolensk where supplies were collected but intended to use a more southerly road than the one along which he had advanced. This plan was prevented by

Kutuzov who, by a brilliant stroke, had already placed himself south of Moscow and barred the Kaluga Road. Soon the French army was trudging through the unburied corpses which still littered the battlefield of Borodino. Kutuzov slowly followed a parallel line to the south; he refused to bring the retreating French to battle, in spite of powerful pressure from his own generals and from the court. The most intemperate language was used by the Tsar; the old general's conduct was ascribed to the basest motives. But Kutuzov made no secret of the fact that his only objective was to see the French army off Russian soil. Since Napoleon was going there was no need to fight another battle. He had no wish to capture Napoleon, he did not want to take the pursuit beyond the Russian frontier, he was convinced that Russian interests would not be served by an invasion of Western Europe. Besides, he retained a healthy respect for Napoleon and the Grand Army. The Guard was still intact for Napoleon had held it in reserve at the battle of Borodino; its strength and discipline was noted far down the line of retreat by the partisan leader Davydov:

> The Old Guard approached with Napoleon himself among them. . . . We jumped on our horses and again appeared on the highway. The enemy, noticing our noisy crew, cocked their guns and went on proudly marching without accelerating their steps. No matter how hard we tried to detach at least one private from these closed columns, they remained unharmed, as though made of granite. . . . Napoleon and the Guard passed through our Cossacks like a hundred-gun warship through fishing boats.

The retreat did not become a rout until 11 November when the Grand Army left Smolensk. Then intense cold set in, with temperatures falling to 45°F below zero. The line of retreat was littered with frozen corpses and all vestiges of discipline vanished. With a flash of his old brilliance – and aided by Kutuzov's planned lethargy – Napoleon got 80,000 men back across the Berezina; but only 30,000 reached the Niemen. Before the return of the last frozen stragglers, Napoleon set off for Paris. He seemed gay and optimistic: 'I left Paris intending not to carry the war beyond the Polish frontiers. I was carried away by events. Perhaps I made a mistake in going to Moscow, perhaps I should not have stayed there long; but there is only a step from the sublime to the ridiculous and it is up to posterity to judge.' It is difficult to understand how anyone, even Napoleon, found the agony of December 1812 'ridiculous'.

Alexander appeared in Vilna on 22 December. Outwardly he treated Kutuzov as a hero, but behind the scenes he vilified him. 'In half an hour', he said to Wilson, 'I must . . . decorate that man with the great order of St George. . . . But I will not ask you to be present. I should feel too much humiliated if you were.' Alexander was now ready to make use of the great victory which his people had given him. By the effusion of much more Russian blood he was going to secure for his dynasty a position in Europe which it would never again achieve. The memory of 1812 lived on; but it was gradually forgotten that the victory was due only in part to the State and its army, that much was owed to the people, the land, the weather and the sacrifice of Moscow. Yet Alexander's part should not be underestimated. He had stood firm during the September days when all around him had counselled negotiation; he had been worthy of the Petrine tradition. Alone, without deflection from her autocratic ways, old Russia had stood up to the united forces of the West. National pride could be linked with conservatism. There was little need to change a system which worked so well. The French Revolution could be ignored, the Industrial Revolution was not needed. In the end, the Romanovs were to pay a heavy penalty for their victory in 1812.

2

Official Russia under the Autocratic Brothers

THE TSAR ALEXANDER I (1801–25)

During the night of 23 March 1801, a crowd of drunken noblemen broke into the apartments of the Emperor Paul and murdered him. The Semyonovsky regiment was on guard but its officers had been informed of the conspiracy and had decided to make no resistance. Even after several blows which cracked the skull of the victim, the body showed some signs of life. One of the murderers jumped upon Paul's stomach 'in order to expel the soul'. The apartments of the Tsarevich Alexander were in the same place; but he later claimed that he was asleep and that he heard nothing. It is certain, however, that he knew about the plot and that he had done nothing to bring it to his father's notice. The most that can be said in extenuation of Alexander's conduct is that he thought that the conspirators would merely depose and imprison Paul. Alexander was awoken by the murderers at once; in a state of nervous collapse, supported by the wife to whom he was openly unfaithful, he received the enthusiastic acclaim of the soldiers of the Guard and thus began the reign during which, by a quaint reversal, he became the chief supporter of the principle of legitimacy not only in Russia but throughout Europe. It was given out that Paul had died suddenly as the result of a stroke. Such a tale had been heard too often about the various victims of eighteenth-century palace revolutions in Russia. Talleyrand drily remarked: 'The Russians ought to invent another sickness to explain the death of their Emperors'. None of the conspirators was punished and several of them reached and maintained powerful positions in the court of the new ruler. But Alexander himself never forgot the night of 23 March: its memory helps to explain his later conversion to a form of mystical Christianity, as well as his nervous apprehension that he might be served in the same way as his father had been. These events, reminiscent of *Macbeth* or of some of the more sensational chapters of Gibbon, also help to explain the bouts of melancholy to which Alexander was subject.

Catherine had hated her son Paul, but she adored her grandsons and she probably intended Alexander (born 1777) to be her successor. She had supervised his education and had done her best to see that he was grounded in the most fashionable eighteenth-century doctrines. His tutor, Laharpe, had given the young man a tincture of 'liberalism'; Alexander spoke freely about his intention to give Russia a constitution, a settled body of law and to break with the barbarous Petrine autocratic system. But Laharpe's pupil soon showed that his philosophy went no deeper than phrases; he proved to be as obstinate in the defence of autocracy as any of his predecessors. Alexander became a good linguist – he spoke French and English rather more fluently than he spoke Russian. The necessity of keeping on good terms with his father and his grandmother gave him much early practice in dissimulation, a quality which was to help him to become one of the best diplomats of his era. Some of his contemporaries claimed that he showed a degree of mental instability which amounted to madness; but he had a power of consistent and logical effort which makes it difficult to accept this verdict. He knew how to use and then discard men of great ability; he never fell for any length of time under the spell of a single favourite. Alexander was reckoned a handsome man and was much influenced by women, especially German women. His mother was Prussian and his wife (whom he married in 1793) was a princess of Baden. During the earlier part of his reign his official mistress was a Pole, Maria Naryshkin, but neither the Emperor nor the mistress confined themselves to the embraces of the other. Alexander was much attracted to the Queen of Prussia, Louise, and it is not unlikely that his attempt to save Prussia from Napoleon in 1806 was in part prompted by his desire to play knight-errant to its Queen. Later on under the influence of religion, Alexander, like many another monarch, regretted the excesses of his youth. He began to speak to Maria Naryshkin about 'divine love', a subject which she found unattractive. During the last five years of his life he returned to the wife whom he had long neglected and managed to regain that patient woman's affections. But perhaps the most powerful feminine influence in Alexander's life was that of his sister Catherine. Until her death in 1818 he corresponded continuously with her; he sought her advice on all matters and she freely told him some of the unpleasant truths which were concealed from him by his courtiers. Much of the correspondence was destroyed by later Emperors; but enough survives to indicate that relations between Alexander and Catherine were more intimate than convention permits.

Alexander's religious conversion probably dates from 1812. He passed from an avowed atheism to the propagation of basic Christian notions. These had a profound effect upon his foreign policy. He tended to invest his own aims and objectives with divine approval – whoever resisted them resisted not only Alexander but also God. Alexander's strange views aroused the mirth of European statesmen, and it is true that in his later years he surrounded himself with a comic collection of pious women. Within Russia too Alexander's tastes aroused the indignation of the leaders of the Orthodox Church, only too aware that Russia was full of fundamentalist sects whose strange beliefs would be encouraged by the Emperor's own behaviour. Photius, an ecclesiastic of some vigour, undertook the dangerous task of combating the Emperor's religious mania. He at length persuaded Alexander to dismiss Golitsyn, his chief companion in mysticism, in 1824. But not until after Alexander's death was the Bible Society, the favourite instrument of the courtier-mystics, dissolved. It is tempting to divide Alexander's reign into two periods and to suppose that he became a reactionary after his conversion in 1812. But this is to oversimplify, for Alexander retained an interest in his earlier 'liberal' notions after 1812 and before 1812 he had already shown many of the 'reactionary' tendencies which were to become more obvious later. Alexander's conversion to mysticism merely lent his later utterances and attitudes a nauseating flavour of naive and egotistical 'holiness'.

SPERANSKY AND THE REFORM OF THE BUREAUCRACY

The assassination of Paul I had received the enthusiastic approval of the nobility. They hoped that under Alexander they would again have the powers and privileges which they had enjoyed under Catherine and which Paul had contemptuously swept aside. They hoped that the new ruler would grant them the right to share in the legislative work of the Empire by turning the Senate into an elected aristocratic body. Their hopes of an aristocratic constitution were not fulfilled, although Alexander did restore all the rights which the nobles had enjoyed in the eighteenth century. But the result of Alexander's constitutional work was the exact opposite of what the nobles had expected. Under him their power was finally subordinated to that of the imperial bureaucracy. For the remainder of the old régime the bureaucrat was the real ruler of Russia.

The most important architect of this development was Michael Speransky (1772–1839). He was a man of exceptional intellectual ability who had forced his way into the centre of power from humble origins – his father was a village priest. During the short period (1808–12) in which he enjoyed Alexander's confidence, his amazing industry enabled him to influence and reshape almost all the important organs of central government. His first major task was to bring some order into the jungle of the central ministries which had been allowed to decay during the decentralising period of Catherine's reign. In 1802 Alexander had ordered that the old collegiate system of control should be replaced by a ministerial system; that is to say, each department of state should be run by a single minister, appointed by and responsible to the Emperor alone. But the reform had been thoroughly carried out only in the Ministry of the Interior, in which Speransky himself held an important position. In his decree of July 1810, Speransky made sure that the 1802 order was carried out and, in addition, completely overhauled the existing relations between the departments. He found that many tasks were being duplicated and that there was no clear idea about the frontiers between the ministries. He also found that the Russian bureaucrats had little idea of the distinction between administrative and legislative action; the ministries frequently made the laws which they subsequently enforced.

This sort of confusion soon fell into order – at any rate on paper – before Speransky's logical industry. He divided the operations of government into five main sections and apportioned the ministries to each subdivision. He greatly extended the range of government activity. He held that it was not enough merely to gather taxes; the State must encourage economic activities: the State must not only punish crime when it was committed but must also take active steps to encourage virtuous behaviour in its citizens. To achieve this latter end, Speransky created a Ministry of Police which was intended to keep a fatherly eye upon individuals and to prevent them from being maltreated by other citizens or by bureaucrats. Such friendly intentions could and did degenerate into a more sinister concern with the doings of individuals. Speransky also started the complete internal reorganisation of the ministries; each was subdivided, routine procedures were adopted, each minister was provided with a council which could give advice and ensure administrative continuity. Provision was made for the regular consultation of interested groups of citizens. In brief, Speransky achieved whatever can be achieved by the introduction of regularity and efficiency; his system lasted until 1917 and would have worked very well had it not been for one fatal flaw – the fact that it was impossible to secure an adequate supply of educated and honest men to operate this beautifully constructed machine. As it was, Speransky's bureaucracy became noted for its corruption, tardiness, servility and inefficiency; it became the favourite target of literary men seeking an easy laugh.

Speransky and Alexander understood the importance of efficient educational institutions in a bureaucratic State and quite a lot was done during the reign to improve facilities. Three new universities (St Petersburg, Kazan, Kharkov) were added to those already existing (Moscow, Vilna, Dorpat) and an effort made to increase the small number of secondary schools. It was understood that a bureaucracy could be effective in Russia only if the nobles – the only educated class – could be persuaded to staff it. With this end in view a lycée which had the avowed object of educating noble boys for the bureaucracy was started under imperial patronage at Tsarskoe Selo. But the noble boys were not in general attracted towards the bureaucratic corridors; life offered prospects more alluring than quill-pushing in St Petersburg.

In the interests of bureaucratic efficiency Speransky tampered with the provisions of the Petrine Table of Ranks, a measure from which the nobles had virtually escaped during the eighteenth century. In 1809 he announced that in future those nobles who held the honorific title of Gentleman of the Chamber would either have to perform the duties appertaining to the office or lose the position which it gave them in the Table. Even more unpopular was another measure of the same year. During Catherine's reign the nobles had won the privilege of quicker promotion up the Table of Ranks than their non-noble competitors, thus negating the intentions of Peter I who had envisaged promotion based upon merit alone. Speransky now announced that any official, irrespective of origin, who wished to be promoted to the eighth grade, that of Collegiate Assessor, had to pass an examination. With his usual passion for exact definition Speransky detailed the subjects in which the candidates would be examined. They included natural, roman and civil law, economics, statistics, arithmetic, geometry and physics. This ordinance aroused consternation: no longer could the noble bureaucrat look forward to a life of undistinguished indolence, punctuated by regular promotions upon the ground of seniority alone; he might be mortified by the promotion of his juniors and social inferiors over his head because they could pass the examination and he could not. This attempt to open a career to talented persons added greatly to Speransky's unpopularity; he was accused of trying to copy Napoleon. From this time the pressure upon Alexander to get rid of him increased; he was accused of being a French agent and consequently, as the danger from France grew greater in 1812, Alexander was forced to dismiss and exile his most valuable minister.

But before his fall Speransky had completed and presented to Alexander the most ambitious of his projects – nothing less than the complete remoulding of all the institutions of state. His plan was studied by Alexander in 1809 but was never, except in one important respect, put into practice. It lay unpublished and unknown in the imperial archives until 1905. This reluctance to make known Speransky's project was the result not so much of the plan itself – which proposed to maintain intact the autocratic power – but of the preface in which Speransky freely criticised the existing organisation of Russia and explained that to leave it unchanged would lead to revolution. 'The universal discontent, the predisposition to negative interpretations of the present, are nothing but a general expression of boredom and satiation with the existing state of things', he wrote. His plan provided for a complete division of powers. The bureaucracy would confine itself to the execution of the laws, a newly modelled Senate would supervise an independent judiciary, and elected Dumas, at the provincial and the national level, would check and supervise both the executive and the judicial systems. Ministers would have to answer for their departments to the state Duma, and local officials would have to answer to the provincial Dumas. Thus Speransky, who has been accused of being nothing but a bureaucrat, obviously saw the need to provide a constant check upon the activities of the bureaucrats.

The most important of the institutions envisaged by Speransky was the Council of State. This was the one part of the plan which Alexander put into operation: the first session of the new Council of State took place in 1810. The new Council lasted until 1917. Its chief function was legislation, or rather the preparation of legislative measures for the approval of the Emperor, who was not bound to accept any of the Council's legislative recommendations and could, if he liked, accept the views of the minority of its members. The Council numbered about thirty-five; ministers were appointed *ex officio* and the other councillors were chosen by the Emperor. The Council divided into subcommittees and at first worked as Speransky had intended. But soon the failure to implement the other parts of the plan threw additional burdens upon the Council. Instead of concerning itself exclusively with legislative business, it undertook judicial work as well; the Senate was only too pleased to find a body with the authority to make decisions in the tangled thickets of Russian law.

Speransky's main purpose was to establish the rule of law in Russia. Absolutism, he saw, did not necessarily mean the situation which actually existed in Russia, a situation in which it was impossible to find out what the law was and in which the law might be changed at any time by the whim or the ignorance of an official. In other words, he wanted to give Russia a formal framework of institutions which would operate logically but only at the command of the Tsar. The latter would cease to be 'the little Father of his people': he would become the impersonal motive power of the machine. This was a noble plan and one which could have been put into effect. But Speransky was sacrificed to the nobles in 1812; in Herzen's words, 'his body was exiled to

Nizhni-Novgorod, while his spiritual children were exiled to the Archives'. Speransky, however, survived; in a more servile frame of mind he brought his immense gifts to the service of the State both during his exile and under Alexander's successor.

During his four years of intense administrative reform (1808–1812), Speransky had turned his attention to the public finances. He found them in total disorder. Alexander had inherited a large debt, he had taken no steps to reduce it and his wars had of course swollen the annual deficit. In 1803 the deficit stood at about 15 million paper roubles; by 1809 it had increased to 150 million paper roubles. Alexander did not resort to his grandmother's device for increasing revenue – the sale of state lands complete with their serf populations – for he did not want to place the state peasants into bondage. He consequently had to renew her policy of printing his way out of financial trouble; during his reign the issue of paper currency increased enormously. The result was the further depreciation of the paper rouble which fell from 72 per cent of silver in 1801 to 43 per cent in 1809. He also floated foreign loans, but the unstable condition of the Russian finances made this expedient very expensive. The Treaty of Tilsit deprived the State of the Customs revenue earned from trade with England. This was the situation which greeted Speransky.

His financial plan (presented to Alexander in 1810) was a masterpiece and would have altered the course of Russian history had it been fully adopted. Its main recommendations were as follows: the government should publicly recognise that the paper roubles were a form of state debt and should be redeemed at face value. This would create confidence and enable the government to borrow from the Russian public more easily. To achieve this end, Speransky suggested that no more paper money should be put into circulation and that the government should sell sufficient land to provide itself with the money to redeem the paper roubles. Secondly, Speransky examined the public income. The main tax was that started by Peter the Great, the poll tax. This, Speransky pointed out, was unsatisfactory because it was levied at a standard rate upon the number of 'heads' counted at the last official census, and bore no relation to the capacity of any particular person to pay. In practice it was a tax upon peasants; the landowners did not contribute. Speransky advised that it should be replaced by a graduated land tax which would be based upon the revenue obtained from private estates, and would rise and fall according to the individual's ability to pay. Finally, Speransky, who was a disciple of Adam Smith and the English *laissez-faire* economists, advised the creation of private banks which would be able to draw out and put into useful circulation the capital which at that time was lying idle. Some parts of Speransky's plan (which very much annoyed the nobles) were cautiously put into effect by Alexander. He promised to stop the issue of paper but broke his word almost immediately. In 1812 he brought in a land tax; but he allowed the landowners to assess themselves and it produced very little. It was abolished in 1820. This was a turning point in the history of nineteenth-century Russia. The enforcement of an equitable taxation system in which the nobles relinquished their freedom from taxation, could have reformed the whole social structure more effectively than the emancipation of the serfs – a project which continued to claim Alexander's attention until the end of his life.

After 1812, the government fell even more deeply into debt. The quantity of paper roubles was reduced and no more were issued after 1823, but this was only done at the cost of heavy borrowing. The total public debt increased annually in spite of an increase of revenue from the Customs after 1815 and in spite of the end of the war. The paper rouble remained depreciated; the peasants were already so impoverished that some taxes had to be remitted in 1826; officials were so badly paid that they were forced to resort to extortion and the acceptance of bribes. Such were the financial conditions transmitted by Alexander the Blessed. Russia could hardly afford to be a great power without a radical transformation of her social system. The problem was to effect such a transformation without weakening the power of the autocrat by angering the privileged nobility.

After the great crisis and the victories of 1812–14, it was impossible to take up Speransky's bureaucratic plans. The heightened conservatism of these years was well expressed by the historian Karamzin in his *Memorandum on Ancient and Modern Russia*. He deplored the abolition of the old collegiate ministries; he asserted that the autocracy would be weakened by the united action of single departmental heads, unrestrained by the collegiate principle. He called for less bureaucracy and for more political power to be given to the nobility. Alexander would not meet the conservatives by sharing political power with the nobles; but neither could he offend them by carrying on with the Speransky reforms. Besides, he too was well aware of the danger of bureaucracy to the autocratic principle. A machine acting efficiently within a legal framework might soon become the master of the Tsar, not his servant. After 1815 Alexander kept what Speransky had given him but he did not dare to make it any more powerful.

Yet this dilemma did not prevent Alexander from continuing to take a keen interest in reform. He got Novosiltsov to draw up an elaborate scheme (1819) which he subsequently failed to implement. This proposed an elected consultative body (state Duma) based upon an elaborate federal structure. The power of the autocrat was to be retained but it was to be more widely based upon the ascertained will of his most powerful subjects. In the new Russian-dominated Kingdom of Poland Alexander did more than merely consider draft constitutions: he created one, and hinted that it might later be extended to Russia. The Polish Sejm (parliament) had an elected lower house. The franchise was given not only to the landowners but also to the rich townsmen. Alexander did not curb its powers even when it became clear that Poland was not reconciled to Russian rule. Nevertheless, Alexander's rule after 1815 became less enlightened. It evolved in a spirit of rather brutal discipline which is symbolised by the name of Arakcheyev.

THE POST-WAR REACTION: ARAKCHEYEV

Arakcheyev, wrote Herzen,

> was undoubtedly one of the most loathsome figures that rose to the surface of the Russian government after Peter the Great . . . he was made up of inhuman devotion, mechanical accuracy, the exactitude of a chronometer, routine and energy, a complete lack of feeling, as much intelligence as was required to to carry out orders, and enough ambition, spite and envy to prefer power to money. Such men are a real treasure Tsars.

The hold of this pedantic drill-sergeant over the liberally educated Alexander has caused much speculation. Arakcheyev had served Paul faithfully and it may be guessed that Alexander assuaged the ache of his own conscience by employing his father's favourite. Arakcheyev was exiled for a short time at the beginning of the reign, was restored to favour in 1803 and made Inspector-General of Artillery, a post for which he was well qualified by his knowledge of ballistics. He was not a courageous soldier in spite of the prodigies which he performed on the parade ground; at Austerlitz he asked to be removed from the front line because the near view of the realities of war disturbed his sensibilities. However, this did not prevent Alexander from making him Minister of War in 1808. He used his position to intrigue against Speransky, whom he hated, and to prepare the artillery for the coming conflict, which he did very efficiently. After 1812 Alexander turned to Arakcheyev more and more until by the end of the reign he had become the chief intermediary between the Emperor and his people. The tales told about the cruelty of this arbitrary bully are many; they are no doubt exaggerated but they give an impression of the fear and horror which he inspired. He is said to have bitten off the ear of a soldier who made some error on parade; he flogged and murdered his village serfs until he discovered the murderer of his drunken mistress. Like many cruel men he was sentimental. He loved nightingales to such an extent that he ordered all the cats in his village to be hanged.

This was the man charged by Alexander to administer the scheme of military colonies, the most important post-war plan. It was necessary to maintain a large army in order to carry out Alexander's grandiose foreign policy, but money was short. Alexander consequently decided to turn his soldiers into part-time peasants and his peasants into part-time soldiers. Money would be saved and no doubt both peasant and soldier would benefit morally from the strict rules of these military encampments. An early version of the scheme was tried out in 1810, when it had been supported by Speransky and consequently opposed by Arakcheyev. But in 1816 the plan was renewed and Arakcheyev found his *métier*. The first colonies were created in the Novgorod province. All the male peasants under forty-five years of age were required to don uniform; young boys were sent away to special camps for full-time training, girls were ordered – as a matter of military discipline – to marry and to bear children at a prescribed rate. Serving soldiers and their wives were settled in the villages; agricultural operations were carried out to numbers; the housewife must have the regulation number of pots and pans and they must shine with the regulation gleam. Washing must be done on the prescribed day and the housewife who failed to obey the rule was flogged. In fact, the military settlements gave Arakcheyev the excuse to flog everybody and he availed himself of it with gusto. Alexander closed his ears to the complaints. When he visited 'his dear friend' at some military village, he found everything in splendid order. Neatness reigned throughout; a less peremptory inspection would have revealed that the clothing of the neatly uniformed peasants concealed the marks of flogging and that crops were neither sown nor gathered. In spite of the evident inefficiency of the system, Arakcheyev persisted and by the end of the reign nearly three-quarters of a million peasants had been gathered into the military colonies. Revolts were frequent and they were vigorously put down.

The brutality of Arakcheyev was matched by the spread of obscurantism in religious and educational ideas. For this Prince Golitsyn, Procurator of the Holy Synod and Minister of Education, was responsible. He encouraged conservative ideas by backing some of the manifestations of the post-war religious spirit. The Bible Society had great success. Levity was to be cured by the study of the scriptures in modern Russian. Some pious ladies, however, went further than Golitsyn desired. Such was Madame Tatarinova who encouraged her disciples to dance in circles until they fell into a trance and 'spoke with tongues'. The universities were scoured for signs of subversive ideas. Lecture notes were carefully examined and offending professors dismissed if the material revealed anything offensive to the social order or the spirit of Christianity (which were thus quaintly linked). Pushkin's first works were published during this era. He fell foul of the authorities, who forced him to leave the capital and reside in Bessarabia.

THE DECEMBRISTS

Alexander died suddenly at Taganrog in November 1825. 'Our angel is in heaven', wrote his wife with pardonable optimism. His unexpected departure had, however, left much confusion about the succession. The law of primogeniture had been restored by Paul and consequently the official heir to the throne was Alexander's younger brother Constantine, as Alexander himself had no legitimate son. But Constantine had committed the imprudence of divorcing his German wife and marrying his Roman Catholic Polish mistress, an act which deeply distressed the Orthodox hierarchy. His children by his second wife could not succeed to the Russian throne whereas the third and youngest brother Nicholas had already married a suitable German – who had turned Orthodox as these German princesses always did – and had fathered a male heir. Alexander therefore got Constantine to renounce his claim to the succession (1822) and hinted to Nicholas that he would be called upon to mount the throne. Unfortunately Alexander shrouded these transactions in secrecy in spite of the fact that they closely concerned his millions of subjects. He elaborately prepared three sealed packages which contained copies of Constantine's renunciation and of a manifesto in which Nicholas was declared heir. These packages were deposited in the Senate, the State Council and the Holy Synod with instructions that they should be opened immediately upon Alexander's death. The original version of Constantine's renunciation was placed reverently in the tabernacle of Russia's holiest church, the Uspensky Cathedral in the Kremlin. This series of events illustrates why law was so little respected in nineteenth-century Russia. If the imperial family could treat the law of succession with so little respect, why should humbler persons accord it any more?

The sudden death of Alexander found Nicholas in St Petersburg and Constantine in Warsaw, where he commanded the Polish army. Nicholas was very unpopular among the Guards; he had justifiably earned the reputation of being a harsh military martinet. Constantine, however, had not been in the capital for many years and he had earned a quite unjustified reputation for 'liberalism'. Officers and soldiers alike noted that the pay of the Polish army was higher than that of the Russian and that service in the ranks was for eight years, not twenty-five as in Russia. Nicholas recalled that it was the Guards who had made and unmade sovereigns in the eighteenth century. He was not sure whether an unpublished command of the dead Alexander carried any weight; he decided to disobey the dead Emperor's orders and to proclaim Constantine Emperor. The Guards took the oath on 27 November with much enthusiasm. Loyal citizens hastened to display pictures entitled 'Constantine I, Emperor and Autocrat'.

Constantine, however, remained consistent to his original position and prevented the Warsaw garrison from taking an oath of loyalty to him. He refused either to come to St Petersburg himself or to issue a public proclamation explaining his renunciation. By this time the high officials and clergy had read the various secret documents scattered about by Alexander but they nevertheless took the oath to Constantine. If Nicholas published the secret documents the troops and the populace would think that he had forged them in his own interests. For three weeks this tragi-comedy lasted with Michael, the fourth royal brother, speeding back and forth between St Petersburg and Warsaw. This was a humiliating time for the bastion of European legitimacy. At last, on 12 December, a final refusal either to rule, or to appear in the capital or publicly to renounce the throne, was received from Warsaw. On the same day Nicholas heard that a serious plot against the throne had been discovered. It seemed that a ridiculous dynastic muddle was about to become a revolution.

The revolutionaries of 14 December – known to history as the Decembrists – were young men of very different types. Many came from the lower ranks of the impoverished gentry but some had aristocratic names. Some were serious students and thinkers who had matured plans for the future of Russia; others were gamblers and men of pleasure, discontented or down on their luck, excited by the prospect of tumultuous action.

The most original thinker among the Decembrists was Pestel (born 1793). He was the son of a Governor-General of Siberia and had had a brilliant military career which included service at the battle of Borodino and promotion to the rank of colonel at the age of twenty-seven. He was an assiduous member of some of the many secret societies which became popular after the napoleonic wars. Many of these societies were Masonic in type, others aimed to help their members to lead a more virtuous life, and others were little more than an excuse for prolonged conversation. The existence of these societies was well known to Alexander who at first did nothing about them because he sympathised with their ethical aims. Pestel, however, differed from the ordinary society member; he spoke openly about the need to regenerate Russia and at length founded the exclusively political 'Southern Society' in 1821. The aims were, 'by decisive revolutionary means to overthrow the throne and in extreme necessity to kill all persons who might represent invincible obstacles'. But Pestel did not stop short at a declaration of means; he also wrote and distributed a constitutional project, 'Russian Justice'. In this, the earliest manifesto of the Russian revolutionary movement, Pestel advocated a republican form of government which, he said, would be dominated for many years after the revolution by its leaders. Even after this long initial period, the power assigned to the elected representatives of the people was small; they were merely to supervise the executive, vested in five persons (State Council) who would collectively exercise supreme power. Pestel also worked out a plan for the reorganisation of land tenure. The peasants were to be emancipated and given a share of land. The nobles – whom Pestel detested – were to be expropriated but the smaller landowners were to be compensated. All land was to become state property. Pestel strongly criticised any federal constitution: he envisaged a Russia one and indivisible and looked forward to the Russification of the subject nationalities. He believed that a *laissez-faire* type of economic structure should be adopted but otherwise personal freedom would be restricted, if necessary by a secret police. Pestel's plan bore little resemblance to those constitutions which were later fashionable among the Russian liberals. He despised the English constitution because, he claimed, it was merely a device for perpetuating the domination of the aristocracy. He sought for the establishment of a more efficient absolutism, a plan which offered good prospects of success in Russia.

Pestel's book was not well received by his fellow conspirators. The rich young Guardsmen of liberal tendencies suspected that Pestel intended to act the part of a Russian Napoleon. Nikita Muravyov, one of the leading lights of the 'Northern Society' centred in St Petersburg, provided a reasoned outline of a constitution less absolutist than Pestel's. Muravyov proposed a constitutional monarchy with full safeguards for personal liberties. Russia was to be divided into thirteen federal states, each of which was to be represented by three members in the upper house of a bicameral national assembly. The lower house (about 450 members) was to be elected by universal manhood suffrage and to be the sole source of legislation. The Emperor was to continue to wield supreme executive power, but subject to the consent of the upper house in matters of foreign policy. Only those who could satisfy a high property tax would be entitled to seek national or local office. The peasants were to be freed without land and would thus remain virtually the property of the landowner. This scheme owed much to the constitution of the United States but it was adapted to the interests of the Russian landowners.

So long as Nikita Muravyov remained the dominant figure among the conspirators in St Petersburg there was little hope of a concerted movement. Pestel tried to persuade him to accept his plans in 1824, but Muravyov thought them 'unfeasible and impossible, barbarian and morally repellent'. But early in 1825 Muravyov had to pay close attention to family matters and his absence from the meetings of the 'Northern Society' enabled a number of men who were more sympathetic to Pestel's view to infiltrate into it. They were not serious political thinkers like Pestel and Muravyov but rather restless Byronic figures of a kind referred to by Pushkin (who was himself on the outskirts of the Decembrist movement):

'Twas all mere idle chatter
'Twixt 'Château-Laffitte' and 'Veuve Clicquot'
Friendly disputes, epigrams . . .
Penetrating none too deep,
This science of sedition
Was just the fruit of boredom,
Of idleness –
The pranks of naughty grown-up boys!

Such a man was Yakubovich, a celebrated campaigner in the Caucasus, a lady-killer and a notorious dueller. He had once fought in a four-handed duel in which he had carefully shot the dramatist Griboyedov through the wrist because he knew that he liked to play the piano. For this he had been exiled to the Caucasus and continued to nurse a resentment against the Emperor. Kakhovsky was another adventurer who attached himself to Ryleyev, the poet who became the centre of the St Petersburg circle during Muravyov's absence. Kakhovsky was a gambler and desperately poor. He was quite willing to murder Alexander, Nicholas or any other member of the royal family, but his motives were low. A short time before the revolt he wrote to Ryleyev as follows: 'Aside from boredom and dissatisfaction, I haven't even the means to satisfy my hunger. I haven't eaten a thing since Tuesday!' Alexander Bestuzhev, another of this group, was an army officer with literary inclinations. Some passages from his diary give an impression of his calibre:

> I am back in St Petersburg! Spent the evening at Sofia Ostafievna's [who kept a well-known 'House']. Dropped in at Komarovsky's where we played silly games. From there to a magnificent ball at the Verghins. Danced a lot but got bored presently and left. Had a lively talk with the Englishman, Foche, about Byron. Have decided to learn to speak English. Stayed at Akulov's until midnight. Danced, but without a single heartflutter for W. was not there . . . .'

So these bored, Byronic, raffish Guardsmen drifted into the plot.

There had been a vague plan to murder Alexander in 1826. But the Emperor's sudden death, followed by the confusion about the succession, encouraged the hotter heads to strike at once. Their indecision ceased when they discovered that the names of all those involved in the Decembrist societies were known to Nicholas. Consequently there was nothing to lose by immediate action. The chosen day was 14 December, the day upon which the troops of the St Petersburg garrison were due to take an oath of loyalty to Nicholas. It was hoped that many of the soldiers would refuse to do so because they had already taken an oath to Constantine. Those who mutinied could then be used to seize control of the capital.

The conspirators' expectations as to the number of soldiers who would refuse to take the oath to Nicholas were disappointed. In all some three thousand mutineers gathered in the Senate square around the equestrian statue of Peter I. The majority of the troops, as well as the Senate and the Council of State, quietly took the oath and surrounded their disobedient comrades. The civilian onlookers, and especially the workmen who had a fine view of the whole affair from the scaffolding of the unfinished St Isaac's Cathedral, showed some sympathy with the rebels. But no attempt was made to enlist their aid, for the Northern leaders were frightened of provoking a general social revolution. The behaviour of many of the leading conspirators was inexplicable. Prince Trubetskoy, who had been appointed 'Dictator', appeared briefly upon the square, complained of a headache, retired to a neighbouring office and at length sought refuge in the Austrian Embassy. Yakubovich the daring adventurer, a terrifying figure swathed in bandages, approached Nicholas to assure him of his loyalty, then went home where he sat for the rest of the day nursing a loaded pistol. Only Kakhovsky displayed any revolutionary ardour; he shot General Miloradovich, an elderly and distinguished soldier who had come across to the rebels to persuade them to lay down their arms. By noon on the short winter day the rebels had lost the initiative and the rival forces stood facing each other in the square. A large and menacing mob was gathering in the surrounding streets; they openly encouraged the rebel officers to hold on until dark. Nicholas was at last persuaded to order cannon fire. The first few shots broke the rebel ranks and cracked the ice on the Neva, thus making retreat impossible in that direction. The final spark of the attempted revolution was extinguished on 3 January 1826 when a handful of rebel troops, commanded by Sergei Muravyov-Apostol, surrendered to a large government army near Kiev. Pestel had been arrested on 13 December; his name had been found in some papers left by Alexander at Taganrog.

Nicholas was deeply shaken by the rebellion. He appointed an investigating committee and interviewed many of the ringleaders himself. Some he bullied without effect but others, including Trubetskoy, broke down completely, confessed everything and asked for pardon. Several of the leaders gave detailed explanations of the reasons which had impelled them. Kakhovsky wrote: 'Liberty, that torch of intellect and warmth of life, was always and everywhere the attribute of peoples emerged from primitive ignorance. We are unable to live like our ancestors, like barbarians or slaves.' Many of the conspirators wrote of the inspiration which they had been given by visiting the West after the war of 1812–14. Bestuzhev wrote: 'The army, from generals to privates, upon its return, did nothing but discuss how good it is in foreign lands. A comparison with their own country naturally brought up the question, Why should it not be so in our own land? . . . Did we free Europe in order to be ourselves placed in chains?'

Speransky was brought back to high office in order to sift through the mass of evidence unearthed by the commission. There was no trial; Speransky merely exercised his bureaucratic gifts to place each of the 121 accused into a predetermined category. Five of the leaders were condemned to death (in spite of the fact that the death penalty had been abolished by Catherine II) and more than one hundred were deported to Siberia. The hanging of the five was bungled. The ropes were too long and Muravyov-Apostol, Ryleyev and Kakhovsky survived the first drop. Kakhovsky swore and Muravyov-Apostol said: 'Poor Russia, she cannot even hang decently.' The fate of the Decembrist exiles, many of them accompanied by their wives, provoked compassion and their fortitude inspired admiration. Many preserved their self-respect through study, participation in the Siberian administration and farming; some retained undimmed their spirit of opposition. Lunin, for example, said that although he had only one tooth left, it was directed against Nicholas. This earned him the Emperor's especial disfavour; Lunin was removed to the silver mines at Akatui on the Chinese border where 'the atmosphere was so heavy that no bird could live within a radius of a hundred and fifty miles.' The survivors, twenty-nine in all, were amnestied by Alexander II in 1856 but several decided to remain in Siberia.

When allowances have been made for the fiasco of 14 December and for the romantic excrescences, the Decembrist movement will be found to contain much of importance. For the first time a group of the governing class had tried not just to remove one Emperor in order to make room for another – that had happened often during the eighteenth century – but to apply critical reason to the political and social problems of Russia. Pestel's contribution is of particular importance for it showed how a revolutionary party could mould the institutions of Russia into a pattern in accordance with Russian traditions. An absolute State did not necessarily have to be run by a medieval autocracy. Another generation of revolutionary thinkers quickly replaced the scattered Decembrists; their plans differed greatly from those of Pestel and Nikita Muravyov, but they praised the Decembrists for daring to be the first to assault the autocracy in the name of an idea.

THE TSAR NICHOLAS I (1825–55)

Under Nicholas I the St Petersburg autocracy received an impress which it was not to shake off before 1917. He completed the system which had already been foreshadowed by Alexander's later policy. He realised, correctly enough, that the Decembrist revolt had been merely the Russian version of a general European malaise. Consequently he threw all his energy into the maintenance of the 1815 treaties which, so he believed, helped to protect Russia from the contagion of revolutionary ideas. Under Nicholas, the Russian government waged a sort of ideological warfare against the European opponents of legitimacy and autocracy. But in practice the power of Nicholas did not extend far beyond the marching range of the Russian army. Much to his sorrow, he was unable to suppress the hostile and vocal thinkers and writers of Western Europe. But within Russia his power to crush and to mould was limited only by the energy and intelligence with which he could act. Of energy Nicholas had an unlimited supply; if his intelligence had matched it he would have turned Russia into a graveyard in which the only voices to be heard were chanting the praises of the 'Most High' Tsar. Nicholas was the first ruler of Russia to attempt to prescribe exactly what his subjects were to think. His educational system, his police, his bureaucracy, his censorship, were constantly goaded into action against the students, professors and writers who during the thirties and forties were thrown up by the greatest intellectual ferment in the history of Russia. The reward which Nicholas earned was the whole-hearted hatred of his victims who, in spite of the activity of the agents of repression, turned Nicholas into a loathsome but comic despot even during his own lifetime. Nicholas forfeited the respect of the most intelligent of his subjects and no subsequent Tsar ever regained it. This was the price exacted for thirty years of police work in Western Europe. Nicholas himself believed that he was only carrying on the Petrine tradition of Russian autocracy, but in fact the difference between the two rulers was profound. Peter had used his absolute power to force his subjects to accept the most progressive ideas of Western Europe; Nicholas used his to prevent any of the advanced ideas of the West from entering his country. The consequence was that by the middle of the century Russia, which claimed to be the bulwark of a divinely inspired world-order, had fallen into the ranks of the second-class industrial powers. For the rest of the century she lacked the wealth to support her pretensions.

The character of Nicholas was well suited to the rôle he played. He missed the wide, if superficial, education which Catherine had given Alexander. He was brought up on the parade-ground, he loved the minutiae of military matters. The only subject which he studied with passion was military engineering. He loved to watch the way in which a body of soldiers moved as one man; he hoped to instil the same sort of discipline into his subjects. He had no ability to distinguish between the trivial and the essential. He worked assiduously throughout his reign at the mountains of paper thrown out by the bureaucratic machine. He loved detailed reports and he loved to regulate the most personal details of the lives of his subjects. He personally interviewed all the leading Decembrists and he continued to take a keen interest in the fate of all political exiles. When a certain lady secured the annulment of her marriage Nicholas wrote on the file: 'the young lady shall be considered a virgin'. When some peasants claimed that they had discovered a cellarfull of treasure the 'Most High' took a close interest in the affair and in the end condemned the peasants to confinement in a lunatic asylum. But Nicholas was not always at his desk. He liked to make sudden and unexpected journeys of inspection. Those who failed to meet his requirements were swiftly punished. A doctor who had mismanaged a hospital was dispatched to the guardhouse; an officer who failed to meet him was reprimanded. But Russia was so vast and the treatment of individuals was so hasty that Nicholas's vigour did little to cure slackness and abuse. Nicholas also descended upon the schools in the capital:

> Today Nicholas visited our First High School and expressed dissatisfaction. . . . He entered the fifth class where teacher Turchaninov was teaching history. During the lesson one of the pupils . . . listened attentively to the teacher, but leaning on his elbow. This was taken to be a breach of discipline. . . . The curator was ordered to fire teacher Turchaninov.

The Emperor's narrow zeal was fortified by a happy family life and a rigid belief in God. Nicholas was an excellent husband and father who earned the approval of Queen Victoria. The succession to the throne was assured and court life displayed none of the distressing irregularities of the previous reigns. As for God, Nicholas was not only pious in his outward life; he seems to have genuinely believed that he had a special relationship with Providence. 'I am a sentry at outpost', he said, 'on guard to see all and observe all. I must stay there until I am relieved.' He would have agreed with the peasant who asked Ivan Aksakov: 'Is it true that our Tsar has a special paper from the Good God and that from time to time he goes up to heaven to see him?'

The paternalistic rule of 'the Russian Louis XIV' could not be carried on through the existing machinery of state. Nicholas wanted an instrument through which he could directly impose his will. This he found in the Imperial Chancery, an office which he enlarged until it became the centre of the bureaucratic machine but which complicated the smooth functioning of government because it reproduced many of the functions which were already being carried out elsewhere. The 'second section' of the Chancery was created by Nicholas in 1826; it was charged with the task, already attempted by previous rulers, of codifying the laws of Russia. Speransky, who had survived his years of exile, became the effective chairman of the new section. Speransky's original plan was to prepare first a collection of all the laws made since 1649, second a digest in which this great mass of laws would be rationalised, and third a new code based upon the digest. Nicholas disliked the third plan because it would permit Speransky to subject the Russian laws to a rational analysis, but he gave permission for the first two steps to be taken. Working with his customary swiftness, Speransky finished the first part of his work in 1830. The laws of Russia were printed in forty-five volumes, exclusive of appendices. The digest (fifteen volumes) was finished in 1833 and came into effect in 1835. This should have been one of the most important events of the reign; but for several reasons it failed to make a significant contribution to the much needed reform of Russian legal practice. First, it was discovered that Speransky had added many concepts to the body of Russian law which he had not found in any of the historical collections but which he had taken from the codes of Western European countries. His tendency had been that of the professional bureaucrat – to simplify. But the result was that the judges were uncertain about the correct line to take and this confusion was the more serious because Speransky had concentrated his efforts to modernise upon such subjects as wills, contracts, and property rights – subjects of vital importance if

Russia was to be able to develop a capitalist industrial economy. Secondly, no provision was made for the Digest to be brought up to date by the application of judge-made rulings. The Emperor and the bureaucracy remained the sole source of new law and no machinery existed to ensure a rational development of new law. In fact, fresh legislation continued to pour out in a completely haphazard fashion and this tended to undermine any stability and confidence which the Digest might have created. Thirdly, the publication of the Digest was not made the opportunity to reform the system by which the law was administered. Judges were still no more than civil servants whose chief aim was promotion and who secured this aim by ensuring that the State and its ministers never suffered in their courts. In fact, the Digest was a measure typical of the reign; it imposed a superficial orderliness and clarity, rather like an infantry drill manual, but immediately below the well ordered surface impression, the citizen found the old arbitrariness and confusion. These shortcomings must not be blamed upon Speransky, for Nicholas maimed his original scheme just as Alexander had maimed his great scheme of constitutional reform.

The 'third section' of the Imperial Chancery (also founded in 1826) enjoyed the greatest notoriety; it has come to symbolise the reign of Nicholas. It was through this department that Nicholas exercised his godlike powers to control and dominate the private affairs of all his subjects. Throughout his reign it was controlled by his closest associates – first Benckendorff and then Orlov. This section was chiefly known for the fact that it controlled the secret police; but in the eyes of its founder its true function was not merely to ferret out political subversion but to act as the eyes and ears of the Tsar in all the corners of Russia. It is said that when Nicholas appointed Benckendorff he presented him with a white handkerchief 'to dry the tears of the widows and orphans'. In other words, Nicholas expected this *élite* to check and supervise the working of the bureaucratic and judicial machinery, to punish the guilty official and the corrupt judge and to turn into a living reality the theory that the Tsar was the Father of his people. In their sky-blue uniforms and with their impeccable manners, these guardsmen were to be the visible manifestation of the will of the 'Most High'. The Third Section certainly succeeded in inspiring universal terror and hatred but it is very doubtful whether it improved the quality of the administration. It proved a boon to those who had private grudges because they could inform against their neighbours, but its police work was so inefficient that the discovery of one of the largest revolutionary circles of the reign – the Petrashevsky group – was the work of the ordinary police, which remained under the control of the Ministry of the Interior. Nevertheless a visit to the Third Section might result in sudden imprisonment or exile; the terror inspired has been described by many writers. Krylov, for example, relates his fear of sitting on the armchair provided for he was convinced that it was poised over a trap-door opening into a dungeon. Herzen paid several visits to headquarters in St Petersburg. There he was interviewed by a general with the most elaborately polite manners, who informed him that it was the decision of the 'Most High' that he should depart at once for Siberia. Herzen's offence was that he had discussed a rumour that some sentry had shot a passer-by. On another occasion Herzen handled his own dossier: 'On the margin was written in big letters in pencil: "too soon". The pencil marks were glazed over with varnish, and below was written in ink: "*Too soon*", *written by the hand of his Imperial Majesty. – Count A. Benckendorff.*' In these offices Herzen witnessed the misery and the glory of Russia under Nicholas. The misery was represented by the petitioners, some of whom flung themselves upon their knees at the feet of the various functionaries when the latter deigned to emerge from their offices. The glory was represented by the following scene witnessed by Herzen as he sat outside Benckendorff's office:

> Next there walked into the room a general, polished up and highly decorated, tightly laced and stiffly erect, in white breeches, with a scarf across his breast. I have never seen a finer general. If ever there is an exhibition of generals in London as there now is a baby exhibition in Cincinnati, I should advise his being sent from Petersburg. . . . He was attended by the most elegant cornet in the world . . . a fair-haired youth with incredibly long legs, a tiny face like a squirrel's and that simple-hearted expression which often persists in mamma's darlings who have never studied anything. . . . This eglantine in uniform stood at a respectful distance from the model general.

The fifth section of the Imperial Chancery was directed for much of the reign by another general, Kiselyov, a brave man and a fine administrator. Its task was the immense one of preparing for the abolition of serfdom, an ambition which Nicholas inherited from his brother. In addition to all these new sections already described, Nicholas also appointed numerous secret committees to enquire into a wide range of problems. These committees were protected by imperial edict from interference by any of the regular administrative offices, even the Council of State. Their activities further disrupted the machine which had already been thrown out of gear by the new sections of the Chancery. In addition, Nicholas made much use of special emissaries, who were usually generals. Posting incessantly about the Empire, these persons added to the characteristic blend of fear and confusion.

As well as providing his country with a bureaucratic strait-jacket, Nicholas also gave it an official ideology. This was later known as 'Official Nationality' and was summarised under the words Autocracy, Orthodoxy and Nationality. The first two words gave little trouble to the writers and officials who were given the task of explaining the true doctrine to the Russian people. 'Autocracy' meant that the will of the Tsar was to be exalted above all else, even the rule of law. Gogol explained that law is harsh and unbrotherly; it must be mitigated by the supreme grace of the most high. Pogodin asserted that the Russian peasants would not become human beings unless they were forced into it by the will of the autocrat. The force of human will and not the light of human reason must be the basis of the State. The official apologists of autocracy praised with enthusiasm the work of Peter the Great. Kankrin proclaimed: 'If we consider the matter thoroughly, then, in justice we must be called not Russians but Petrovians. . . . Everything we owe to the Romanov family. . . . Russia should be called Petrovia. . . .' Only autocracy, it was argued, could overcome the force of original sin; by rejecting this form of government the Western constitutional states had condemned themselves to a way which left out of account the basic fact of human nature. The chauvinism of this first principle was reinforced by the way in which the official spokesmen interpreted the idea of 'Orthodoxy'. Membership of the Orthodox Church was considered to be a *sine qua non* of respectability. Religious toleration was no part of Nicholas's programme and during his reign the Church, under the energetic leadership of a former cavalry general, Count Protasov, persecuted the non-Orthodox groups. Chief among these were the Old Believers, whose religious scruples forbade them to pray for the Tsar or to do military service. Such communities were transported to Siberia or their children taken from them and made the wards of the State. Jews were also encouraged to give up their national dress and were forbidden to maintain their own schools. The Polish revolt of 1830 gave Nicholas a good opportunity to destroy the power of the Catholic Church; but in this matter he proceeded with caution. It is true that during the thirties the Russian government closed about two hundred monasteries, forbade mixed marriages and, in 1839, forced the Uniat Church back into the Orthodox fold. But strong action against the Catholics led to a breach with the Papacy and that disrupted the conservative alliance with Austria. Similarly, attempts to force the conversion of the Protestants in the Baltic provinces aroused the anger of one of the most important classes in the State – the Baltic German nobility, without whom the maintenance of the bureaucracy was impossible. In other words, Orthodoxy was aggressive in all cases in which powerful retaliation was impossible.

But the third principle of the official trilogy was fraught with danger and led to some disagreements among the government apologists. For during Nicholas's reign the idea of 'Nationality' was being debated not only by the official writers but also by the Slavophils; the ideas of the latter had, for Nicholas, an unwelcome effect upon the views of the former. What Nicholas wanted was a dynastic interpretation of nationalism: that is to say, the assertion that the subjects of the Russian Tsar (who were in fact of many different races) were one people simply because they were ruled by one autocrat. But as the reign progressed it became increasingly difficult to get this interpretation accepted. Romantic nationalism was as powerful in Russia as in Germany. The Romantics started to make claims for the Russian people as opposed to the Russian autocracy. Such a distinction had been possible since 1812 and it was to prove to be one of the roots of the Russian revolutionary movement. The Romantics demanded a domestic policy of Russification and a foreign policy of supporting the ambitions of brother Slavs who were under foreign domination. Such a claim was directed against Austria and was consequently directly opposed to the Tsar's foreign policy. With the outbreak of the Crimean War, the Romantics had it all their own way. The formerly obedient spokesman of Official Nationality, Pogodin, wrote in rapturous vein: 'As Russians we must capture Constantinople. ... As Slavs we must liberate millions of our older kinsmen'. Official Nationality had started to give way to Panslavism.

One of the most celebrated of the spokesmen of official Russia was the Minister of Education, Count Uvarov; he laboured to direct his department according to the ideological precepts. Uvarov's problem was to produce a sufficient number of trained minds for the bureaucracy while preventing the infiltration of subversive ideas. In part, the problem was solved by increasing the number of specialist courses and institutes. A large number of such places were founded or expanded during the reign; it was hoped that the students of veterinary science, technical drawing and forestry would not be inclined towards original thought. But some general education there had to be. Uvarov provided it by expanding the size of the existing universities, by giving a wider choice of curriculum, including the natural sciences, and by founding one new university. But he also kept a close watch over professors and schoolmasters, textbooks and students. From 1834 full-time inspectors kept watch over the out-of-class activities of students; they were to encourage morality as well as the avoidance of dangerous ideas. In many universities and above all in Moscow they found plenty to do, for the student bodies, far from being satisfied with the doctrines of official Russia, were intoxicated by philosophies with implications dangerous to the régime. In spite of Uvarov's best efforts – which included the attempt to restrict university entrance to the sons of noblemen – the students carried on and transmitted the Decembrist tradition. After the events of 1848 Nicholas dismissed Uvarov, on the grounds that he was too liberal, and subjected higher education in Russia to an even fiercer repression. The size of the student body was cut down to 300 per university, constitutional law and philosophy were eliminated from the curricula and logic was to be taught only by professors of theology. Students and teachers were forbidden to study abroad and the historian Granovsky was not allowed to lecture about the Reformation because it was feared that his audience might be perverted by learning about a succcessful revolution. It was officially declared that the aim of education was to prepare 'loyal sons for the orthodox church, loyal subjects for the Tsar and good and useful citizens for the fatherland'. This policy seemed to have succeeded when, in 1848, Russia remained free from a serious revolutionary outbreak. But in the long run Nicholas and Uvarov were helping to prepare a native revolutionary movement which displayed more obstinacy than that of 1848.

But perhaps the most serious legacy of the reign lay in Russia's failure to keep pace with the Industrial Revolution. This was especially marked in the iron industry. At the beginning of the century Russia was the leading producer of pig-iron in the world; but by 1860 she had fallen to the eighth place, outstripped even by her autocratic neighbours, Prussia and Austria. This was because the Urals iron industry failed to take advantage of the technological changes which enabled Great Britain to increase her iron production twenty-four times during the period 1800–60. The Urals iron industry was worked by serf labour and much of it was owned by the nobles. They lacked the capital to modernise and feared that modernisation would spell the end of their privileged position. The threat of competition from cheap British iron was met by the tariff of 1822 which placed a duty of 600 per cent upon imported pig-iron, and behind this barrier the nobles continued to enjoy their economic advantages and the Russian iron industry continued to stagnate. The same fate overtook the woollen cloth industry and for the same reasons. But the cotton industry provides a striking example of what could be achieved even in the Russia of Nicholas by a manufacture which remained largely outside the control of the nobles. This was because cotton manufacture was a comparatively new industry and because it required a large quantity of capital. At first the Russian cotton factories worked with imported English yarn, the demand for which trebled between 1812 and 1840. But a sudden transformation came upon the industry when in 1842 the British government allowed the export of cotton machinery. In a decade English machines and English capital made immense progress in Russia. Raw cotton was now in heavy demand; the quantity imported increased by more than thirty-five times in the period 1812–51. By the end of Nicholas's reign, the cotton industry was starting to look for new markets in Asia and for a domestic supply of raw cotton. The striking difference between the sluggish growth of the noble-dominated iron industry and the capitalistically organised cotton industry became a powerful argument in favour of the emancipation of the serfs. The same difference shows that behind the seemingly all-powerful façade of autocracy, the nobility retained the most important of their economic privileges.

'The main failing of the reign of Nicholas consisted in the fact that it was all a mistake.' This was the verdict of a contemporary who was far from being a revolutionary. The effect of the Crimean War (1853–6) upon Russia may surprise an English reader, who is more likely to remember it as little more than a series of glorious blunders. But for more than a generation Russians had been accustomed to believe that their country had defended European civilisation against the perils of the revolution. Its political institutions may not have made for a joyful life for the individual but, so it was believed, they did create strength and effective power. But now that system which had defeated Napoleon was unable to defend the soil of Russia against a weak and ill-led Anglo-French expeditionary force. Yet the force of Russian nationalism was not aroused. Fashionable crowds watched without anger while the British fleet bombarded Kronstadt, and the attitude of the peasantry was revealed by a sharp increase in the number of agrarian disturbances. Tolstoy described the mood of the defenders of Sevastopol as one of dour courage rather than exalted patriotism. For it was well understood that the allied armies had come to destroy a system which had ceased to have much support but which was too powerful to be overthrown by the divided forces of the Russian revolutionaries. The fall of Sevastopol seemed a small price to pay to get rid of the hollow, pretentious, petty and inefficient disciplinary code which Nicholas had mistaken for government. Defeat cast doubt upon the security of the foundation of the edifice of imperial bureaucracy set up by Alexander and Nicholas. That foundation was Russia's 'peculiar institution' – serfdom. For half a century the enlightened had doubted whether this institution should be retained. The Crimean defeat brought vociferous criticism into the open. When Russia defeated Europe, Russia despised European institutions: but when Europe defeated Russia, European ways were imitated.

3

The Peasants: Serfdom, Emancipation, Discontent and Revolution: 1800-1917

CONDITION OF THE SERFS AT THE BEGINNING OF THE NINETEENTH CENTURY

By the beginning of the nineteenth century the structure of serfdom was complete. It was the work of the Romanov dynasty; its profounder results were active until 1917. The condition of the peasants was a tragedy for themselves, an inspiration for the conscience-stricken intelligentsia and an increasingly difficult problem to successive governments. The peasantry were Russia. They paid nearly all the taxes, they provided the food, they were the hordes of domestic servants, they died in the wars, they starved frequently and suffered always. In this great sea of village folk the buildings and civilisation of Moscow and St Petersburg reflected perhaps in the glitter of some local manor house, seemed an irrelevant intrusion. The rulers of Russia might be planning war or peace, their attention might be fixed upon Paris, Berlin, London or Manchester, but their living was being earned for them by the ploughman on the steppes and the peasant craftsman in the forests. It is not surprising that one of the great themes of nineteenth-century Russian literature is the superficiality of town life in an essentially peasant land. Tolstoy is the great master of this contrast and both in his novels and his own life constantly stressed the need for a social conscience which would place the peasant and the countryside at the centre of interest. It was, however, a part of the tragedy of nineteenth-century Russia that the condition of the peasantry could only be improved by large-scale urbanisation. It may be that those who, like Tolstoy, admired and loved the villager, did much harm by trying to preserve a picturesque way of life which the developments of the century made increasingly productive of poverty.

Throughout Russia the land was cultivated on the open-field system. In the forest areas of the centre and the north, the peasants' wooden houses were grouped together in broad clearings and the village fields were usually conveniently near. In the open plains of the south and the south-west, the villages of clay and wattle houses straggled along the watercourses and the village fields might be many miles from the peasant's home. In both areas the arable fields were divided into strips and each peasant household worked many such strips in different parts of each field. The system was inefficient; much time was consumed by travel between house and strip and the whole tempo of agricultural operations had to be adjusted to the capacity of the slowest and most ineffective farmers; improvement was impossible on the unenclosed strip. The inefficiency of the open field was increased by the generally prevalent custom of periodical repartition. It was held that only the house and the garden plot belonged to the individual peasant household. The arable strips were the property of the village community which consequently had the right to ensure that they were held by those families which needed them most. Thus, at the time for division (which was not exactly fixed by peasant custom) a family which had decreased in size could expect to lose some strips to other families which had grown larger. It more often happened, however, that all the families had become more numerous. Then the village would have to create new holdings by reducing the number of strips in some of the old holdings. This process of equalisation was much encouraged by the government during the early nineteenth century because it prevented the emergence of a class of landless labourers. But it made little headway in White Russia and the Ukraine where the peasant strips were held in perpetuity.

Throughout Russia, the peasants concentrated upon the cultivation of grain. Rye was the staple food but wheat and barley were grown increasingly as the export market expanded during the century. The climate and soil of the centre and the north made it impossible for the peasants to satisfy even modest needs by agriculture alone. The gap was filled partly by a widespread and intensive peasant handcraft industry and partly by emigration to the new lands of the south and south-west. It was only during the nineteenth century that these new lands filled up; overpopulation was to be one of the major factors in the peasant problem. During the centuries when serfdom was growing, the exact contrary had been the case; in fact the main reason for the growth of serfdom was to prevent the scarce labourers from drifting away towards the frontiers. South of the forest line, the soil was well suited to grain growing. But although this black earth was rich it was so wastefully farmed by hordes of peasants that its yield per acre remained substantially below that of the much poorer land of North America. None of the changes of the nineteenth century had much effect upon the productiveness of peasant agriculture. Even in 1917 the primitive *sokha*, a plough which scratched the surface of the soil rather than turning it over, was still in general use.

In 1800 there were about 34 million peasants in a total population of 36 million. Of these about 19½ million were subject to private landlords and the rest were state serfs. In general terms the government, and especially Catherine II, had entered into an unwritten contract with the nobility: in return for giving up their claim to political power the nobles had been allowed to enjoy complete power over their serfs. A nobleman might own a dozen or a hundred thousand serfs. It made no difference to his power over them. He was a little autocrat within the great autocracy. The State used him as its agent for peasant affairs, its local 'gratuitous chief of police'. The landlord could make any change he liked in the agricultural arrangements of the village; he could, for example, take for his own use the valuable meadow and forest rights. He could seize the peasant's movable goods, order him to marry or not to marry, sell him away from the land without his family, take his sons and daughters into domestic service, send him to the privately owned mines in the Urals, command him to become a musician or actor. The aged and the sick could be turned out of the village; the recalcitrant could be transported to Siberia, sent into the army for twenty-five years, imprisoned or knouted – that was the customary punishment. The murder of one's serfs was prohibited by the law; but since the peasants were forbidden to complain to the officials about their lords, murder frequently occurred and went unpunished. In the forest regions the landlords forced the payment of rent (*obrok*) at an arbitrary rate which took into account the earnings of peasant craftsmen. In the steppe lands, however, where the produce of the soil was marketable, the landlords compelled their serfs to do *barshchina* or forced labour. Both Catherine II and Paul I attempted to limit the amount of *barshchina* which could be legally exacted: in 1797, for example, three days was declared the maximum. But in practice the landlords got as much free labour as they liked. Since this was usually demanded at harvest time, the peasants' own crops suffered. The peasant had to provide not only his labour but also the necessary tools and animals. The owner of the soil was thus spared the need to make capital investment. During the eighteenth century the real value of both *obrok* and *barshchina* increased greatly. This meant that the landlords were exploiting their serfs more and more. They could live better by merely applying the laws more strictly. Unlike English landlords of the same period, they did not have to apply capital, knowledge and skill if they wanted to make more money. Serfdom made life too easy for them; it destroyed their capacity to adapt to changing circumstances. Not all landlords benefited equally from serfdom, for there was a great gulf between the richest and the poorest. At the beginning of the nineteenth century about 45 per cent of the total number of serfs belonged to about 2500 individuals. At one end of the scale was Prince Sheremetyev with 60,000 serfs: at the other were thousands of rural gentry with less than ten 'souls'.

The position of the state peasants was not as bad as that of the serfs in private ownership. It was unlikely that they would be removed for domestic service; and since much of the state land was in the north and north-east of European Russia and in Siberia, in inhospitable countryside far removed from any town, local conditions made it possible for the peasants to be comparatively free of external authority. The state peasants enjoyed a more stable rent and were not subject to the many uncertainties attached to private ownership. Like private serfs they were bound by a passport system designed to keep them fixed to the village. In spite of their rather happier position, the state peasants were far from contented. The memory of freedom was closer. They had been reduced to servile status only in the eighteenth century as a result of the financial needs of the Petrine State. Peter's most successful tax was the poll tax (first levied in 1718). This was levied on all non-noble males and had the result of dividing the whole population into two great classes – those who did not pay (accounted 'free') and those who did pay (accounted 'serf'). The taxes paid by the state peasants were as essential to the government as the rent and service of the serfs to the landlords. Towards the end of the eighteenth century, the proportion of serfs increased in relation to the state peasants. This was because Catherine II, always short of money, found it convenient to reward soldiers and courtiers by giving them tracts of land complete with peasant cultivators.

In spite of being the victims of the arbitrary system described above, the peasants had retained not only the memory but even some of the institutions of the days before serfdom. The villagers were used to acting together to further their common interests. The open-field system demanded co-operation; the days for sowing, harvesting and mowing had to be agreed among the individual farmers. Arrangements had to be made about meeting the communal obligations to the State and the landlord – poll tax, rent and labour service. The villagers had to operate the partition of the common fields, a task which required much organisation. Each household had to be given its fair share of the good and the bad land and room found for new families. These matters were regulated by the village assembly and its elected officers; only in the serf-villages was election regularly undertaken in nineteenth-century Russia. The assembly consisted of the heads of each household, each of whom enjoyed almost despotic power over his own family. The peasant family was frequently very large because the father was able to force his adult sons to stay under the paternal roof even after they were married. The peasant woman was not highly regarded; the daughter had no share in the family property and the wife was regarded as a household slave. Wife-beating was a common occurrence which later in the nineteenth century the law tried to stop. 'Whom, then, is one to beat?' is the answer of the peasant accused of this crime. The peasant household tended to reproduce on a small scale the general ordering of Russian society – unquestioned obedience to a despotic head. In the villages survived the customs and morals of pre-Petrine Russia, a society which was largely unknown to the educated classes but which had, to a surprising extent, conserved its own primitive traditions.

The memory of earlier freedom caused frequent peasant revolts. The Russian peasant was noted for the gentleness of his manners, but a particularly brutal landlord or unusually bad harvests could drive the villagers into violence. Frequently this took the form of sporadic thieving from the landlord's forests or meadows, but sometimes landlords were murdered and their houses burned. The greatest of such revolts had been that led by Pugachov, who in 1773–4 had terrorised the whole Volga region, captured and burned the large town of Kazan, and driven thousands of landlords in terror into Moscow. Pugachov had called himself Peter III and explained to the serfs that their miseries were caused by the usurping government of Catherine II. During the early nineteenth century discontented peasants often supposed that their condition would be alleviated if only the true Tsar knew the facts about their lives. Hatred was aimed not so much at the throne as at the landlords and the local bureaucrats. The following lines from a peasant song express this feeling:

> He has destroyed us,
> The evil *barin*, the seigneur.
> It is he, the evil one, who chose
> The young men,
> The young men
> To be soldiers,
> And us, the beautiful girls,
> To be servants;
> The young married women
> To suckle his children,
> And our fathers and our mothers
> For labour.

The landlords had good reason for remembering Pugachov. Hardly a year passed without a military expedition to one or more of the provinces and the summary execution of recalcitrant peasants. Disturbances became more frequent as the nineteenth century progressed and became the more serious when the army was absent on campaign. Although their economic position was obviously injured by the prevailing uncertainty in the countryside, the landlords were almost to a man opposed to any change in the system of serfdom. But the government could not remain indifferent to an evil which threatened the security of the State; and from the beginning of the reign of Alexander I steps were taken which, in the end, were to lead to emancipation.

EARLY ATTEMPTS AT REFORM

Alexander I and Nicholas I both wanted to abolish serfdom but they were unable to do so. They feared the opposition of the gentry, they suspected that all social bonds might be weakened, they did not know what to put in its place; above all, they suspected that they would have to give the gentry political rights in return for the loss of power over their servile peasants. In spite of all these difficulties and fears both monarchs attempted reform. Measured by statistics they had little to show for a half-century of effort; but they did ensure that the gentry never settled down into a complacent enjoyment of their privileges, and they helped to form public opinion against serfdom, an opinion which was invaluable when at length, after the Crimean War, Alexander II abolished it. In thus acting against serfdom the autocrats were using their power in the tradition of Peter the Great – that is, as a revolutionary force. Another Peter would not have hesitated for as long as Alexander and Nicholas; their scruples show to what extent the autocracy had lost its revolutionary vigour.

Alexander's most important measure was the Law of Free Agriculturalists (1803).. This enabled landlords to free their serfs, either as individuals or by whole villages, on condition that the peasants were given some land and not reduced to mere labourers. This was thought to be the prelude to a general forced emancipation – and it is possible that Alexander saw it as such. But so loud was the outcry that the Emperor went no further along that line. Only a handful of serfs was liberated under this law but a very important precedent had been set: emancipation must be with land, Russia would not tolerate the English type of landless labourer. In many other measures Alexander tried to mitigate the evils of serfdom. Public auction and advertisement were forbidden, serf-owners could no longer send their serfs to penal servitude in Siberia and an attempt was made to prevent the wealthy from purchasing serf-substitutes for military service. Alexander also tried an experiment in emancipation in the Baltic provinces. There, in the period 1816–19, the serfs were freed but without any land allotment. The land became the outright property of the nobles who were able, following the English model, to establish a new middle class of tenant farmers on their estates. The experiment did not satisfy the peasants and the Baltic provinces remained a centre of discontent.

Nicholas, too, was much concerned with the serf problem. Its humanitarian aspect did not trouble him, but he hated the disorder caused by peasant revolts (which were frequent in the thirties) and the loss to the State of poll tax and rent. In 1842 Nicholas is reported to have said: 'Serfdom is an evil; no one can have any doubts about this. . . . At the moment, even to think of the emancipation of the serfs would be to aim a blow at public discipline. But a way must be found for the gradual passing over to another order of things.' It is altogether characteristic of Nicholas that this 'way' was a bureaucratic one. A fifth section of the Imperial Chancery was created especially to deal with the problems of state peasants. It was entrusted to the 'liberal' Kiselyov, an energetic general of the sort much employed during this reign. It was hoped that the improved condition of the state peasants would encourage the private owners to improve the lot of their serfs. Kiselyov removed the state peasants – henceforward declared to be 'free inhabitants' – from the control of the existing administrative system, and placed them directly under the Ministry of State Properties. This seemed in itself a very large change since it concerned about one-third of the total population of Russia, but its significance should not be overrated. Such a change was typical of the bureaucratic mentality of the period. It was thought that to create a new ministry was in itself a major achievement. In fact, all that happened was that a few officials and files moved out of the Ministry of the Interior into the new ministry. Kiselyov's new system gave a good deal of self-government to the state peasants: it also offered them much technical advice; it enabled some of them to emigrate to less populated regions; it built some hospitals and schools and encouraged better farming with prizes and model farms. But none of this worked very well. Perhaps the peasants recalled the experiments of Arakcheyev: Kiselyov merely offered a slightly more civilised version of the military colonies. Kiselyov's bureaucrats were just as dishonest, cruel and unjust as the ones formerly employed by the Ministry of the Interior. The peasants continued to revolt, and the bayonet and the whip continued to be the motive force in Russian agriculture.

Nicholas and his advisers had little effect in the reforms

directed towards the private serfs. The Polish landlords were punished for the part which they had played in the revolt of 1830 by being forced to accept 'inventories' in which the relations between serf and landlord were distinctly stipulated; but this system was not introduced into the Russian provinces of the Empire. The prominent measure of the reign (1842), introduced after lengthy deliberation and with infinite precaution, was little more than a repetition of the law of 1803. Once again the government insisted that voluntary emancipation must be accompanied by a land-grant to the peasants; once again, the law created much complaint among the landowners and had virtually no result. Other minor measures of the forties – one, for example, gave serfs the right to purchase land (if they could) – were forgotten under the impact of the 1848 revolutions. The stability of Russia compared with her neighbours provided a strong argument for retaining serfdom.

EMANCIPATION 1861

The resistance of the landowners to change was certainly not the result of their general prosperity under the serf-system: it was rather a reflection of their uncertainty as to what would take its place. The large-scale farmer in south Russia had very different interests from the petty serf-owner of the forest region who lived upon the manufacture of his twenty serfs. It is impossible to give a generally valid answer to the important question: was serfdom seen to be an unrewarding method of organising agriculture by 1861? Did the landlords hope to gain financially by emancipation? The answers given to these questions would no doubt differ according to the location of the estate and the number of serfs owned.

In south Russia, near the Black Sea ports, the foreign market was already stimulating grain production in the forties and fifties. Here large gangs of serfs could be profitably employed in farming and carting operations. This cheap labour could take the place of capital investment. The world market for Russian grain – like that for American cotton – gave a new reason for the retention of a monstrously unjust social system. But fortunately for Russia, the economic advantages of serfdom were not as sharply regionalised as the economic advantages of slavery in the United

States. Some landowners made a large sacrifice in 1861. However, there is much evidence that the landowners in general were not doing well in the years before 1861. Before that year, two-thirds of the private serfs had been mortgaged to state institutions, the number of small estates had declined sharply and there had been a small decline in the total number of private serfs. There is no evidence of significant investment in agriculture. The rents extracted from the peasants had increased greatly since the beginning of the century. All this indicates that the landlords (especially of those areas in which the peasant *obrok* was of great importance) might not be averse to a change which would introduce a money economy into the countryside, especially if the peasants were simultaneously deprived of the use of much land which could be subsequently let to them for a money rent. In the steppes, on the other hand, the landlords wanted both free (cheap) labour and the rich land. The laws of 1861 attempted to harmonise these discordant interests with those of the State. In the process the peasant was almost forgotten and the crucial question of agricultural efficiency neglected.

The emancipation of the peasants in 1861 was preceded by five years of intense, embittered and fundamental discussion. This reform touched every interest, every group and every region of the Empire. By it, all the parts would be brought into a new relation with each other and with the whole society. It was a social reform with profound political implications. If the peasants were to be liberated, what about their masters? Would they receive political liberty? Could capitalist development take place in a State which could alter property relations in this sweeping way? Might not private investment in industry be subjected to equally high-handed treatment? Who would govern the emancipated serfs, the gentry or the bureaucracy? Should Western models be followed or should the peasants be treated (as the Slavophils desired) as the guardians of a specifically Russian and wholly precious tradition? Should the emancipated peasantry be given full rights as citizens or left as a legally inferior caste? Should they be given land and if so how much, and how much should they be made to pay for it? These and many other questions formed the subject of the great debate on emancipation which occupied the period 1856–61. To the anxiety of wondering whether the right destiny had been chosen for Russia was added the private fear that emancipation would mean financial ruin. The reaction of the peasants themselves could not be exactly gauged. They might rebel if they got too little and they might get out of hand if they got too much. In spite of the long debate, in spite of the effort and goodwill put into it, the steps eventually taken in 1861 were the wrong ones. The Romanovs had a further fifty-six years to rule. During that time the peasantry became a progressively more revolutionary force until in the end they brought down not only the dynasty but the whole social order. Before 1861 they had been troublesome but manageable: after 1861 they turned into a destructive force.

'It is better to abolish bondage from above than to wait for the time when it will begin to abolish itself spontaneously from below.' With these words, addressed in public to a section of the nobility, Alexander II opened the great debate in 1856. He spoke as if he were another Peter the Great: he spoke as if the peasants were just about to rise as a mass. Neither implication was true. Alexander was a rather weak and indecisive ruler. He had to be kept up to the mark on emancipation by members of the family who were much stronger than he was – by his brother Constantine and his aunt, the Grand Duchess Elena Pavlovna. He had a conscience about serfdom; he was horrified by the weakness of Russia revealed by the Crimean War (1853–6); he wanted to build railways and knew that it would not be possible to do so in a serf society; he was an obedient son of Nicholas and wanted to complete what his father had undertaken. He knew that there was no immediate danger of a peasant revolt, for the peasant army had fought bravely in the Crimea and continued to suppress obediently the revolts which were nothing new in Russia. He used the strong words quoted above because he wanted to get the co-operation of the serf-owning nobles in the great task which he had undertaken. In itself this desire shows that he was no Peter the Great. It might have been better for Russia if he had been.

After a brief pause the Tsar's initiative had the result which he desired: it provoked a dialogue between the liberal bureaucrats and the more progressive (or self-seeking) serf-owners. Together, with the help of a press which was for a time liberated from censorship, they forced the conservatives in both camps to accept a change. They carried along with them the many timid spirits who desired some reform but who feared to give the peasants too much land. They also opposed the more radical spirits (like Herzen and Chernyshevsky) who thought that the peasants were not being given nearly enough. Among the liberal bureaucrats were Lanskoy, Minister of the Interior, Rostovtsev (whose views changed in a liberal direction during the debate) and N. A. Milyutin (a nephew of Kiselyov). Among the reforming landowners who played an important part in the editorial commissions which discussed the details of reform were the Slavophils Samarin and Cherkassy. Notably absent from the committees were any representatives of the peasants themselves. By 1861 the influence of the liberal bureaucrats was already on the wane. Rostovtsev died in 1860: Lanskoy resigned in 1861. The publication and the application of the laws took place when reaction had already replaced the more progressive spirit of the beginning of the reign.

From about the end of 1858 the main outline of the projected reform became clear. The peasants would not be liberated without land; they would have more than just a house and garden – exactly how much they were to get would depend upon the regional conditions; they would not be given full rights as citizens; they would run their own village affairs according to traditional law and custom; they would have to pay the landlord for such land as they received but they would not have to compensate him for the loss of labour services; they would not be encouraged to break out of the communal form of landholding into a more individualistic (and efficient) mode of agriculture. Such were the terms granted in 1861. Personal serfdom was brought to an end; the individual ceased to be the chattel of his owner. Abritrators (usually local landowners) were appointed to supervise the complicated local arrangements. Each peasant village had to reach a separate agreement with the landowner. The arbitrators were given guidance but they, in the end, were responsible for seeing that both sides had a fair deal, that the peasants got enough land and that the landowners got the right price. The house-serf was free to leave his master and (in theory) had to be paid for his work. The house-serfs were not provided with land and the lack of other employment ensured that the landowner continued to enjoy a large supply of cheap labour. The law made elaborate provision, region by region, for the grant to the peasant cultivator of some part of the village land. Usually the allotment received by the peasant was smaller than that which he had cultivated before the reform; on average the peasants lost about one-fifth of their pre-1861 holdings. It was up to the landlord to select which parcels he would keep and frequently these were the meadows and forests in which the peasants had pastured their draught animals. The peasants now had to pay rent (or labour service) for land which they had previously had for nothing.

The emancipation law kept the peasant tied to the soil. The former authority of the landlord was to be exercised by the *obshchina* or peasant commune. This body (which might or might not correspond to a single village) was empowered to prevent a peasant from selling his allotment, from leasing it, and from moving away from the village permanently. The commune was also collectively responsible for the payment of taxes. Within the commune, authority was vested in the heads of the households who still enjoyed almost despotic power over their families. It was, for example, impossible for a son to leave the village without the permission of the head of the household and the passport which could be granted only by the commune elders. The emancipation did not intend to create a free, mobile labour force. The peasant remained a class apart, bound by laws which did not apply to the rest of society.

But the most onerous arrangement, so far as the peasant was concerned, was that concerning the payment of redemption to the landlord. Before 1861 the peasants had thought that the land was their own; 'We are yours but the land is ours'. They found, after 1861, that they had to pay, over a period of forty-nine years, the sum which the State had already advanced to the landlord to 'compensate' him for the land which had been alloted to the peasants. To make matters worse, these redemption dues (for which the commune was collectively responsible) were calculated on a basis which gave the landlord far more than the land was worth. In other words, contrary to the original intentions, the redemption dues compensated the landlord for the loss of his serfs as well as the loss of his land. This was particularly marked in the central and northern provinces where peasant rent rather than peasant labour was important. In the Polish provinces, however, the redemption cost corresponded more closely to the land value; for in this area the government trusted the peasant more than the landlord. But even where the valuation was realistic, the peasant found the greatest difficulty in paying. Dismayed by the unexpected burden placed upon them, many former serfs were unwilling to enter into redemption agreements with their masters. Negotiations dragged on in many cases up to 1881, when a new law made such agreements compulsory. Many peasants (about 6 per cent of the total) took advantage of a clause inserted at the last minute into the 1861 law, which enabled a landlord to discharge his obligation by giving a peasant free of charge an allotment one-quarter the size of the minimum laid down in the emancipation law. This was much done in the black-earth provinces; it created a class of cultivator bound to live below starvation level.

The steps initiated in any village by the emancipation can be summarised as follows: the minimum and maximum size of the peasant allotments had been defined, area by area, in the emancipation law. The peasants had to agree with the landlord as to the exact extent of what he was going to give them, assisted if necessary by the arbitrators. A charter was then drawn up to enshrine the agreement and to record the sum to be paid by the peasants. After this, the peasants paid 20 per cent of the purchase price and 80 per cent was advanced by the State in the form of bonds. The peasants then had forty-nine years in which to pay their debt to the State. A proportion was collected annually from the commune. If the peasants did not want to accept the terms offered by the landlord, the latter had the right to insist on the sale but in this case he lost 20 per cent of the price. The peasants had no means of forcing a reluctant landlord to sell. Until the redemption arrangements were completed the peasants remained under 'temporary obligation'. This was a condition which suited many landlords so well that they refused to come to an agreement until forced to do so in 1881. The reaction of the peasants to the law of 1861 was generally one of incredulity. Many believed that the real terms of the law had been concealed from them. There were many risings: in the suppression of one of them (1861) 102 peasants were killed.

THE RESULTS OF EMANCIPATION

After emancipation the peasant might be free but the landscape did not change. The open fields remained, primitive agricultural implements were still used, the houses still clustered in the villages. It was hoped that agricultural efficiency would be increased by that class which had obviously benefited by the emancipation – the landowners, now in possession of the enormous capital provided by redemption. The fate both of the peasantry and of the dynasty depended upon the success of capitalist farming.

Reliance upon the manor house proved, however, to be the major miscalculation of 1861. At least two-fifths of the redemption payments disappeared into paying the landlords' private debts, and most of the rest was spent as income. The mortgage of estates continued at a great rate, especially after the establishment of the Nobles' Land Bank in 1885. By 1904 about one-third of the private land was mortgaged to this institution alone; much more was in the hands of private creditors. The sale of private estates was also rapid; between 1877 and 1905 the quantity of land owned by the nobility decreased by 37 per cent. An increasing proportion of what remained was leased out to peasants. Although private agriculture was more efficient than that of the serfs, by comparison with foreign yields the manor-house lands were pitifully inadequate; they produced less per acre than any other European country, and less than the dry, empty grasslands of North America. In 1900 the yield on private land in Russia was about 12 bushels of wheat per acre, in Germany 27 and in the United Kingdom 35. Such figures indicate that capitalistic farming had not taken root in Russia. In spite of all that the government could do, the landowning nobility was in obvious and catastrophic decline by 1900.

The fate of the peasantry in the post-emancipation era was closely linked with that of the nobility. A growing number of families, tied to too little land, had to find too much money to pay a few highly inefficient cultivators. By 1900, in spite of some spectacular famines, the peasant population had increased from 50 millions (1860) to 79 millions. During the same period, the average size of peasant allotments had decreased by about 20 per cent. Little relief was had before 1905 from colonisation or emigration. The crop yield of peasant land remained very low;

animals were more difficult to keep after 1861 and their absence decreased the supply of manure and made deep cultivation impossible. The net income of peasant allotments was absurdly low. At the end of the century, the average in the province of Vladimir was about £2 per allotment per annum. The maximum was about £6 in the province of Novgorod.

It is surprising that the peasantry remained fairly quiet until the end of the century, but obviously the explosion would have come much sooner had there not been some safety-valves. Perhaps the most important was the ability of the peasantry to lease and buy the large quantity of private land which was unloaded by the nobles. The leases were short and economically damaging; the yield per acre of the ground leased hardly covered the rent and certainly did not pay a wage to the peasant renter. Peasant purchasers, both individuals and co-operative groups, bought huge amounts of land especially after the establishment of the Peasants' Land Bank in 1883. The peasant demand for land pushed up the prices charged to far beyond its economic value. But the peasant was not concerned with such abstract ideas; he only knew that each extra acre meant so much less hunger.

Wage labour, either in the towns or in the grain lands of south Russia, was an increasingly important source of extra income. A common sight on the muddy spring roads was a horde of men and women, often destitute even at the beginning of their journey, hurrying south for the first ploughing. The wages paid were very low, but the lucky wanderer might return to the village in the early winter with a few roubles saved – a few coins which might make the difference between life and death for some member of the family. As for town labour, this was, in the end, the only solution to the peasant question. But before 1905, Russian towns grew too slowly to make a substantial difference. Urban population doubled between 1860 and 1900, but its total in 1900 was only 14 million.

Successive governments could hardly remain unaware of the deepening crisis in the Russian countryside. But changes, when they came, were timid and superficial; they tended towards making the laws of 1861 function better. More radical action was taken only when the peasant risings of 1902–5 combined with the shock of defeat in the war against Japan. This strengthened the conviction that only by violence from below could change in Russia be effected.

As early as 1872, the Valuyev commission, which consisted for the most part of landowners, reported some distressing symptoms. One nobleman wrote: 'Anyone who looks at it from the outside might well think that the district had been ravaged by the enemy, so pitiful has it become.' But it was easy enough for such noblemen to ascribe the lot of the peasant to drunkenness, laziness and ignorance. Other evidence was not so easy to ignore. The government found increasing difficulty in collecting the redemption dues. By 1875, the arrears amounted to 22 per cent of the annual assessment; by 1900 this had increased to 119 per cent. In other words, the State was owed more in uncollected arrears than it proposed to raise in one year. This position was reached in spite of some attempts to reduce the load of taxation upon the peasantry. Some portions of the redemption dues had been reduced or even cancelled, and in 1886 the poll tax had been abolished. Nevertheless no radical effort was made to lighten the load of indirect taxation which fell mostly upon the peasantry. Most of the state revenue came from taxes levied upon vodka, tobacco, matches and paraffin. The introduction of a graduated income tax would have done much to lighten this burden.

Again, it was recognised that the peasants must be helped to obtain more land if starvation and revolution were to be avoided. To this end, a Peasants' Land Bank was created and started operations in 1883. Bunge, the Minister of Finance, proposed that it should give generous loans at low rates of interest. But his plans were foiled by landlord interests in the Council of State which insisted that the rate charged should be 7½ per cent and that no more than 75 per cent of the purchase price should be advanced. The meagreness of the Bank's contribution towards solving the problems of the peasantry can be shown by the fact that by 1892 it had purchased only 34 per cent of all the land bought by peasants in that period. In other words, it had failed to drive the moneylender and his extortionate interest rates out of the village. After 1892 the Bank became very much more active, thanks to the energy and vision of Witte. He gave it a new statute in 1895 which reduced the interest rate to 5½ per cent, extended the length of time for which the mortgage could run, and increased to 90 per cent of value the sum which could be advanced. The Bank was also empowered to buy land as it came on to the market and, by judicious purchases, was able to build up self-contained holdings for later mortgage to individual peasants. From 1893 to 1905 the Bank's share in peasant purchases rose to 75 per cent of the whole. Witte's enlightened management of the Bank was a small but (as will be seen) significant step towards the agrarian policy which was put into effect after 1905. He used it to build up a class of peasant proprietors which, he thought, would be more efficient and more stable than the communally organised peasantry envisaged by the 1861 statues. Witte's handling of the Bank rested upon the assumption that the commune – so dear to the reformers of 1861 and to the populists – was not the most desirable organisation of Russian agriculture.

The obvious distress of the peasantry also turned the government's attention to colonisation. There was plenty of room in Siberia and there were good strategic reasons for trying to fill it up. Conditions for agriculture were, however, very difficult; not until the nineties was there any railway system to the east of the Urals, and the landlords were suspicious of any scheme which would deprive them of cheap labour. For a long time after emancipation emigration remained spontaneous; it was neither forbidden nor encouraged by the State. But in 1889 a cautious beginning was made to state-assisted colonisation. This made it easier for intending colonists to escape from their communal obligations in European Russia, and gave some financial assistance to emigrants. The building of the Trans-Siberian Railway encouraged the State to fill up the area through which the line passed and gave some assurance that the peasants would not escape the attention of the central government. By the turn of the century emigrants were crossing the Urals at the rate of 100,000 per year, compared with about 10,000 per year before the law of 1889. But although this was an improvement, it made little contribution to the problem of land-hunger. Even when the emigration was running at its highest figure, the natural increase of population in the European provinces was about fourteen times greater than the number of emigrants. Only forced emigration could contribute significantly.

The difficulty experienced in collecting redemption dues

caused the government at first to strengthen the power of the nobility over the village. In 1889 a new law provided for the appointment by provincial governors of local 'land captains' chosen from among the nobility. These 'land captains' were to supervise the activities of the village communes and to punish those elders who failed to obey government orders, especially in matters concerning tax payments. They also had the task of seeing that the commune remained intact. During the eighties the highly conservative government of Alexander III had decided, contrary to the wishes of those who had framed the emancipation laws of 1861, that the peasant should not be encouraged to hold his land in free ownership even when the redemption duties had been paid. Various laws, therefore, made it very difficult or impossible for a peasant household to withdraw from the commune, or to sell, lease or mortgage allotment land. In this way the 'emancipated' peasantry was firmly tied to the land and forced to remain a class apart from the rest of the community.

This policy was adopted at a time when it was becoming clearer that a process of economic differentiation was gathering momentum among the peasantry. Within the villages, so it seemed, some peasants were becoming richer while others were being turned into landless labourers. The rate at which this was happening was difficult to assess, and for the historian the whole matter is complicated by ideological conflict. The Marxist writers, beginning with Lenin's *The Development of Capitalism in Russia* (1899), have been at pains to show that the movement towards the 'proletarianisation' of the peasantry had advanced far by 1905. For the Marxists it was essential to demonstrate this in order to show that the true interest of the peasant lay in identification with the town proletariat. But however much the doctrinal bias of the Marxists may be suspected, it is impossible to deny that some degree of differentiation had occurred before 1905. The official statistics show only about two million full-time wage workers in agriculture in 1900; no count was taken of the millions of seasonal labourers nor of the large number working instead of paying rent. Again, a study of the figures about peasant renting and buying of land shows that only a small proportion of the total number of peasant households could afford this means of evading land-hunger. But perhaps most convincing of all is the evidence about the owning of horses; without them, in most of the provinces of Russia, arable farming was virtually impossible. A survey of 1893–6 showed that nearly one-third of all the peasant households lacked even one horse. It must therefore be assumed that many millions of peasants had not the means to cultivate their own land and must have lived, at least in part, upon working for somebody else. This implied that the commune now chained together not persons of roughly similar economic status, as had been the case in 1861, but rich and poor peasants. The latter could hope for nothing from the commune and so looked with increasingly eager attention at the landlord's estate. Only from that quarter could land-hunger be appeased.

The post-emancipation governments at first adopted a conservative attitude towards the commune. The Slavophils (whose views were accepted by the establishment during the eighties) taught that the repartitional system went far back into the Russian past and that its continued existence was a guarantee against the outbreak of class struggles in Russia. Of more immediate importance was the fact that without the commune it would be difficult to collect the taxes and redemption dues. In 1893 a law attempted to stir the communes into life; the 'land captains' were empowered to force a village into making a repartition where this had not been done during the previous twelve years. Apparently the law was not very effective; even by 1905 a large proportion of repartitional villages had not divided up the land since 1861. The reason for this inertia of the commune was the unwillingness of the richer peasants to give land to the poor. By the end of the century, however, Witte had grasped the fact that the commune was dying. In laws of 1899 and 1903 he took the first steps to loosen its grasp – steps which were confidently followed by Stolypin after 1905.

THE PEASANTS AND THE 1905 REVOLUTION

But Witte's discovery that the peasantry was not as conservative as had been thought, came too late. From 1902 to 1906 occurred the most widespread and the most violent peasant revolts since the time of Pugachov. It is highly unlikely that this *jacquerie* had any political motive and it would have been more easily put down had it not been for the war against Japan and the revolution of 1905. The revolutionary parties had been active in the villages but their influence had been little greater than that of the populists in the seventies. One peasant, questioned about the circulation of the 'little books' of the socialists, replied: 'No rumours came to me about any little books. I think that if we lived better, the little books would not be important, no matter what was written in them. What's terrible is not the little books, but this: that there isn't anything to eat.' The village remained remarkably insulated against the weapon of the intellectual – ideas; it was probably more influenced by the revolutionary surge in the cities. The peasant risings were concentrated in the southern and western provinces – in just those areas in which land-hunger and class differentiation had been most obvious. However, even the Marxist writers do not claim that the peasants showed much political consciousness. For the most part they behaved like the villagers of Kholzovki in the province of Kursk. One night in February 1905 they took their carts along to the forests of the local landowner, cut some wood, brushed aside the estate guards, and returned home, no doubt to warm themselves with the landlord's logs. There was little destruction of manor houses and few occasions upon which the peasants seized and cultivated the private land. For the most part, it seemed, cold and hunger had passed the limit of endurance and the simplest steps were taken to achieve relief. In some areas there was organised refusal to pay taxes, rent and redemption dues. But generally the peasants did not follow the advice given in the Socialist Revolutionary manifesto of March 1905: 'Drive the landlords and the rich villagers out of their warm, profitable places; beat, cut, choke the lackeys of the Tsarist government . . . the land ought to belong to the whole people.' Instead the peasants seem to have adopted the idea that, while they were entitled to help themselves to the land of the State and the Church, private landlords were to be compensated for what they lost. Such demands were framed by the All-Russian Peasants' Union – a body which met near Moscow in the summer of 1905 and which probably expressed the hopes of the revolutionary parties as much as the actual desires of the peasants.

The government tried to restore calm in the countryside with the usual punitive expeditions. But it is plain that Witte had already lost faith in the old ordering of agriculture. The Conference on the Needs of the Agricultural Industry under his chairmanship (1902) had collected many volumes of evidence to show that the open-field system was wasteful and inefficient. If the open-field system was to go, the commune could go too; and if the commune were to be abolished, the redemption dues might as well be cancelled. In fact, the *jacquerie* of 1902–6 served Witte's purpose well for it helped him to prove that the commune was far from being a stabilising influence. By November 1905 he had persuaded Nicholas II that the peasants could be tranquillised by the abolition of redemption dues; this was consequently enacted and the way cleared for 'the wager on the strong'. It seemed (but only for a short time) that the *jacquerie* had produced political results advantageous to the peasantry. Time was to show that in November 1905 Witte had merely carried out the classical manoeuvre of dividing the revolutionary forces.

In retrospect, the forty years since emancipation had witnessed a deepening of the crisis in the Russian countryside. Official policy was only partially responsible for this. It could not account for the population increase which was the main cause of land-hunger and consequently of discontent. But for much else the bureaucrats and the landowners were responsible. Both these classes had to choose between a prosperous peasantry or a docile peasantry. They chose the latter. For the bureaucrats this solved the problem of tax collection. Russia followed a great-power policy. This luxury was much more expensive at the time of the Japanese War than it had been during the Crimean War. The peasant, nominally free but in fact still tied to the land, provided the ambitious State with the money for aggressive policies. If the peasant were prosperous he would produce more money for the State, but he might be less docile and be strong enough to resist payment. For the landowners, both self-interest and ideology encouraged the retention of a docile peasantry. They had little capital to invest in agriculture; consequently they wanted cheap labour. They benefited from peasant industry; consequently they resisted the idea of labour mobility, which would tend to attract the peasant to the towns. As Slavophils they had ideological and emotional reasons for admiring the peculiar state of affairs in Russia. The Russian village community was unique. It saved Russia from the evils of capitalism. The Russian peasant might be poor but (so it was alleged) he was happy to live in the picturesque communities in which primitive socialism satisfied his deepest needs. Both the bureaucrats and the landowners were wrong. By 1905 the peasants were neither docile nor prosperous. They were better educated than they had been in 1861. Their grievances were focused and expressed by educated revolutionaries. They had closer links with the towns. They travelled more, they were less superstitious, less fatalistic. They were disappointed by the results of the long-awaited emancipation. Since that had evidently not worked, what was to come next? They witnessed the flight of the landlords from the countryside, and as the landlords disappeared, the peasants grew used to helping themselves. More and more land came under peasant control and was subjected to inefficient peasant agricultural methods. The peasants distrusted authority, for they knew from experience that authority meant trouble. The landlord flogged, the bureaucrat cheated and extorted and the soldier shot. Witte was among the first to grasp the meaning of this rising tide of anarchy. His solution was to create a new class of rural capitalists, who would remain loyal to the social order because it was in their economic interest to do so. Lenin, a mind equally acute, observed the same symptoms but drew quite different conclusions. The peasantry had become in his view a revolutionary class. He saw that he could use them to destroy the bonds of society and thus give him the chance to seize political power.

THE FOREIGN GRAIN TRADE

Before describing the profound changes in agriculture promoted by Stolypin after 1905, it is necessary to examine Russia's foreign grain trade. This was yet another burden laid upon the peasantry by a remorseless State. Russia under the later Romanovs presented the same strange spectacle as visitors to Ireland witnessed during the potato famine in 1846; while the people were starving, a brisk export trade in food was being carried on at the ports. The cargo ships of the world carried away the grain which was cultivated in Russia and without which hunger was general and starvation not uncommon.

The growth of Russia's grain exports may be gathered from the following table:

Year	*Value of grain exports in million roubles*
1800	12
1825	17
1850	36
1861	56
1880	282
1900	299
1905	435
1913	655

Expressed as percentages of the total value of exports, grain accounted for about 10 per cent at the beginning of the nineteenth century, 15 per cent in 1835, 30 per cent in 1850, 50 per cent from 1870 to 1900 and 55 per cent in 1913. During most of the century, Russia was the greatest grain-exporting country in the world. This position was lost to the U.S.A. from about 1880 to 1900; but by 1914 Russia had regained it. During the mid-century period, Britain was the biggest buyer of Russian grain – the repeal of the Corn Laws in 1846 had a marked effect upon production in Russia – but by 1914 Germany had become the largest customer. Most of the trade went through the Black Sea ports (90 per cent in 1914) and in consequence Russian statesmen became increasingly sensitive to the problem of the Straits. When they were closed by Turkey in 1914, Russia's foreign trade dwindled to nothing in spite of efforts to open new export routes through the northern ports. The growth of the grain trade had put Russia into the position of depending for her economic existence upon the control of the Bosporus, a strip of water a few thousand yards wide.

Some of the exported grain was the produce of large estates; many such existed, especially in the south, and were organised for the export market. But the great bulk of the exported grain came from peasant farms. Primitive Russian agriculture had to compete in the world markets with the mechanised and fertilised grain lands of the old and new worlds. The following table gives some idea of the disadvantages of Russia as a grain-exporting nation:

	Wheat yield (*in* poods *per* desyatin) *1910*
Russia	48
Great Britain	142
Holland	152
India	48
U.S.A.	64
Canada	72

It will be seen that even the empty lands of North America had a higher yield than the overcrowded arable area of Russia. India and Russia had the same level of agricultural efficiency.

Under anything like free market conditions, the Russian peasant would not have had any grain for disposal on the foreign market. But he had to pay his taxes and his redemption dues; therefore he had to have cash; and in order to get cash he had to sell some part of his crop even though he went hungry himself. By increasing the tax burden on the peasantry the State could increase the quantity of grain available for export. In addition the railway system was arranged so that large areas of Russia were brought within the range of the export market, and preferential tariffs for long hauls to the ports or the frontiers encouraged peasants and dealers to seek export. The effect upon peasant husbandry was bad, for the farmer was compelled to concentrate upon the production of wheat even where neither the soil nor the climate favoured its growth, to put every available acre of land under the plough, and to do without cattle to an extent which deprived the land of natural manure. 'We shall eat a little less but we shall export', said Vyshnegradsky, a Minister of Finance; what he meant was that the peasants would eat a little less. But the obvious symptoms of agricultural distress convinced Witte that the export trade in grain – which he wanted to expand in order to pay for industrialisation – would have to be based upon a more efficient agriculture. To achieve this the commune and the open-field system would have to be abolished. It would be necessary to organise the peasant for efficient production, not in accordance with some antiquated notions about the desirability of a certain type of social stratification. Whether the peasant would benefit from this more utilitarian concept was to be seen;

for the historian it is unfortunate that the new order in the countryside failed to last sufficiently long to make possible a confident verdict.

THE STOLYPIN REFORMS

The peasant disturbances which had heralded and coincided with the revolution of 1905 continued into the next year, but by then the government had recovered its nerve and punitive expeditions discouraged further outbreaks. This rapid recovery was remembered by the peasants in 1917 and caused them to proceed cautiously. In November 1906 Stolypin published the law which (with later amendments in 1910 and 1911) was to make profounder changes in the Russian countryside than even the emancipation of 1861. As in 1861, recent defeat had made reform necessary. But in 1906 the question was much more urgent. The peasants were more unreliable and the future of the State more immediately threatened.

Stolypin's new line had been prepared by Witte; the experience of the 1905 revolution placed the majority of the conservative landowning class behind it too. The landowners had for a time been faced by the prospect of all their land being seized from them by force; what they really wanted was to be allowed to sell it to the peasants, preferably at a price well above its market value. The delegates to the First Congress of Representatives of the Nobles' Society, which met in May 1906, blamed the revolutionary ardour of the peasants upon the commune, 'the nursery of socialist bacilli'. The peasants must be taught to respect private property, and how could they when the system of the repartitional commune did not permit them to have any? Therefore, argued the frightened nobles (who for half a century had been in ecstasies about the purely Russian beauties of the land commune), 'the strengthening of property rights among the peasants . . . will increase their attachment to that which is their own and their respect for that which belongs to others'. In other words, the peasant must be taught the joys of landowning so that he will refrain from stealing the private estates. Nicholas II approved of the arguments of the nobles. He and the court circle repressed their dislike for much of what Stolypin did because they reflected that this was, after all, the only way to preserve the landowning class, the traditional prop of the monarchy. The fact that the landowning class was (in every sense of the word) bankrupt appeared not to worry Nicholas. A more astute man would have wondered whether the landowners were worth preserving. It is important to notice that Stolypin's reforms, although they were radical, received full support from the conservatives. In this way they differed from the reforms of 1861, but they resembled the latter in that they were imposed from above in order to prevent revolution from below.

In outline Stolypin's plan was as follows:

a. In the repartitional communes (about three-quarters of the whole) any householder could demand to have his strips in full hereditary ownership. In the hereditary communes the strips were declared to be from henceforward under the full ownership of the head of the household.

b. Full hereditary ownership was established in all the repartitional communes in which no repartition had occurred since 1861.

c. Peasant land which thus came under full ownership could be bought, sold or mortgaged; but it could only be bought by peasants and no peasant was to own more than six allotments.

d. The title to ownership gained under *a* or *b* above was vested in the person of the head of the peasant household. The joint ownership of the other members of his family was extinguished.

e. In the hereditary communes, the consolidation of the scattered strips of each household could be secured by a simple majority vote.

f. In the repartitional communes, any single householder could demand consolidation of his strips and his share in the common meadows and waste; the commune had to fulfil the demand if the local Land Commission thought it desirable.

These measures enabled and encouraged individual peasants first to seek full private ownership and then to consolidate their scattered holdings into an enclosed farm. They constituted a direct attack on the commune and an open invitation to the richer peasants – known as *kulaks* – to break free from the commune and set up individual farms. Those who were unable to do this would either emigrate to the towns or become landless labourers, working for their richer peasant neighbours. Stolypin's laws would complete the process of class differentiation in the countryside.

The whole energy of the State was thrown into the promulgation of the new laws. The land commissions were instructed always to act on behalf of the peasant who was seeking separation. Government grants were available for building the isolated farm houses, fences and barns which the new system made necessary. The activities of the Peasants' Bank were increased greatly in scale and changed entirely in scope. Before 1906 the Bank had advanced loans to communes; after 1906 the Bank did its business almost entirely with individuals. Most of the land handled by the Bank was that sold by the nobility; the Bank sold 3¼ million estates in the period 1906–17, at prices which were well above the economic value of the land. Indeed, a rapidly growing mortgage debt before 1914 suggests that the peasant was unable to meet the high prices charged by the Bank. But it satisfied the conservatives by giving the landowners the chance to sell their land and to make a profitable exit from the stage of Russian history, which they had occupied for so long. The sale of the 'Cherry Orchard' no doubt brought a tear to a well-fed theatre audience; the historian cannot afford such sentimentality.

By 1917, in spite of a slackening of effort during the war years, Stolypin's reforms had been widely effective. At least one-half of the peasant land was held in full hereditary tenure and on about 15 per cent of the land enclosure of some sort had taken place. The *khutor*, or fully independent farm, was still a rarity but the *otrub* (where the peasant's house remained in the village but his strips were enclosed) was becoming fairly common in some provinces. There can hardly have been a commune which failed to feel the threat of impending change; without the war and the revolution Stolypin's reforms would soon have affected the whole of European Russia.

The status of the peasant was altered during the pre-war years; the gap between the ex-serf and the free classes of society was not abolished but it was narrowed. The authority of the noble 'land captains' was diminished and corporal punishment for peasants abolished. Inevitably, however, much flogging was still inflicted by administrative order. The passport system was revised so that it became much easier for the peasant to go to the cities. By 1914 the urban population of the Russian Empire was about 26 million out of a total of 175 million; it had been about 14 million in the census of 1897. In the same period the proportion of urban to rural population had increased from 12·8 per cent to 14·6 per cent. Another safety valve which must have tended to increase the sense of freedom in the villages was emigration. This was strongly encouraged by Stolypin's government. Literature describing the pleasures of a farming life in Siberia was distributed in the villages; financial inducements were offered to would-be emigrants. Since most of the land in Siberia belonged to the State, there was no difficulty about providing allotments of a suitable size and at low rents. Under these stimuli, emigration reached three-quarters of a million in 1908 but fell to less than a half of that by 1914. In all, the population of Asiatic Russia increased by 56 per cent from 1897 to 1916 and reached 21·1 millions in the latter year. Impressive as this achievement seems, however, it must be remembered that these millions transferred to Asia or the cities hardly absorbed the natural increase of the European peasant population. The effect of slightly improved medical care was to worsen the population problem; the birth rate increased and the death rate diminished during the pre-revolutionary decade. 'The wager on the strong' did not effect much improvement in the lot of the poorer peasants.

Another aspect of Stolypin's policy was the encouragement given to peasant co-operatives. To an increasing extent they took the place formerly occupied by the now suspect village commune, and their success illustrates the profound need in the countryside for communal effort. These co-operatives, or *artels*, provided credit, purchased machinery, fertiliser and seed, and in some cases provided the means for marketing. Even among the bands of migrant labourers it was not uncommon to find an *artel* which housed and fed the members. This capacity for spontaneous common action was to be shown later during the 1917 revolutions when the peasants, as well as the urban workers, were to create soviets through which they managed both their economic and political affairs. A class which had endured centuries of oppression retained the capacity for democratic self-government. The peasants had good reason to distrust any authority that came from outside the village. When given the chance to express their will through their actions, they did not choose one form of government rather than another: they chose to have no government at all.

THE PEASANT REVOLUTION OF 1917

It is possible that the Stolypin reforms could have created a conservative peasantry if they had been given enough time. But the war which broke out in 1914 produced an upheaval in the countryside which eclipsed anything done by Stolypin: in fact, it was the greatest upheaval since the Mongol invasions. A vast conscript army (fifteen million men were called up) stripped the villages of the young men. The cultivated area shrank in size because only women and old men were left to work the land. Hordes of refugees from the western provinces disorganised life in the rear areas. The export trade in grain was suddenly ended: the failure of the Dardanelles campaign (1915) meant that the ports would remain closed until Germany had been defeated. But there was much more food in the village than there had been before 1914; the Russian peasant fed himself rather than feeding the city dwellers in other countries. The distillation of vodka was forbidden in 1914, and this increased still further the supply of grain. But agricultural efficiency declined. It was impossible to get mineral fertilisers from abroad. Machinery could not be repaired or replaced. Horses, requisitioned for the front, disappeared from the countryside. Gangs of women harnessed themselves to the plough, just as for centuries they had dragged the barges up the Volga.

The towns had to be fed. The government wanted no conflict with the peasants at this point of Russia's history, and paid them well for their grain. There had never been so much money in the village but there was nothing to buy with it; the production of consumer goods for civilian use had ceased, and there was rapid inflation. The result was that the peasants became more and more self-sufficient. They would not grow more food than they needed themselves. They would eat well and let the others do what they could. By the end of 1916 there were food shortages in the towns, and the government resorted to forcible grain collection. The peasants disgorged at the point of the bayonet what they would not sell on the market. The trouble was that a peasant held the bayonet too. Wartime conditions intensified the natural anarchism of the Russian peasant. The slender bonds between town and country snapped and were not to be firmly restored until the Stalinist revolution.

The peasants did not make the first revolution of 1917. They watched cautiously: they remembered how quickly the State had recovered in 1906. They observed that the new Provisional Government promised a radical land reform to follow the meeting of a democratically elected Constituent Assembly. They noted that this assembly was not to meet until after the defeat of Germany. Meanwhile the new government looked very much like the old one. The peasant was exhorted to produce grain and to sell it. He was forbidden to seize the landlord's land or to pasture his beasts on the landlord's meadow. Could this new government be trusted to give the peasants the land settlement which they wanted? Evidently not, even though the new government was closely in contact with the political party (the S.R.s) which claimed to represent the interests of the peasants. It soon became apparent, however, that the revolution had made one important change in the countryside: the Tsarist terror had been removed. The soldiers who came down to requisition grain no longer shot and flogged. They fraternised with the peasants and ignored the commissars of the Provisional Government. The peasants took the hint. Now there was nothing to stop them from doing what they wanted. From April 1917, rather quietly, without violence but with irresistible force, they took over the countryside. They formed village committees; they refused to pay rent; they prevented the use of paid labour; they cut down the landlords' forests; they refused to sell grain; they occupied church and state lands. All this was well-organised anarchy. The peasants acted together for their own good, obeying their own village committees. If they looked towards the politicians at all, it was merely to distinguish that group which would give them a legal title to what they had already done.

On this point the Provisional Government spoke with a confused voice. Some of its members, like the S.R. Minister of Agriculture, Chernov, said that the peasants could have all the land after the meeting of the Constituent Assembly. But who could be certain what would happen when the assembly met? It seemed likely that Chernov's more conservative colleagues in the Provisional Government would try to protect the rights of the landlords. They would surely insist upon compensation being paid. If the decision was left to any government, the peasants would be cheated; centuries of experience had taught them that hard truth. Meanwhile, only one political party told the peasants what they wanted to hear: seize the land now. That was the Bolshevik party. Lenin alone actually wanted anarchy in the countryside; his party alone could survive it. His party alone wanted an immediate end to the war. Even the S.R.s were tied to the continuation of the war and consequently to at least a postponement of the peasant demands. The peasants had no idea that a great gulf lay between them and the Bolsheviks. The peasants thought that when they had taken the land they would live happily ever after. Lenin saw the land reform as a temporary, tactical device. When his party was strong enough, it would take the land back under state control.

The nature of peasant demands is well illustrated by the following resolution of the Congress of Soviets of Peasant Deputies (June 1917):

> The right of private property in land is abolished forever. Land can be neither sold nor bought nor pledged nor alienated in any way. All land is taken over without compensation as the property of the whole people and passes over to the use of those who work on it. The right of using land is enjoyed by all citizens (without distinction of sex) of the Russian state who desire to cultivate it with their own labour, with the help of their family or in a co-operative group and only so long as they are able to cultivate it. Hired labour is not permitted.

These words clearly establish what was in the minds of the peasants. Only those who directly used the land could dwell on it. Neither landlord nor bureaucrat should be allowed to set foot in the village. The whole rural apparatus of the Petrine State should be eliminated.

After the harvest had been gathered, the revolutionary movement in the countryside gathered momentum again. Now it was more violent in tone. The landlords' houses were thoughtlessly burned, whereas previously the village committees had been apt to make careful inventories of the material possessions found in the manor houses. Few landlords were killed, even in the autumn. The Russian peasant was gentle even in the moment of victory. The purpose of house-burning was to ensure that the landlord and his family never returned. The countryside was being purged. The new violence of the autumn was also caused by a fear that the days of freedom were limited. Surely somebody would send down some punitive expeditions into the country soon: they always had in the past. In addition, the villages had received back many of their young men. The peasant soldiers at the front had heard what was happening at home. The desire to share in the agrarian revolution was the most powerful reason for the collapse of front-line discipline in the summer of 1917. It was not the words of the revolutionary propagandists which took Russia out of the war. It was the wish of the soldiers to be present at home to witness and assist at the most astonishing event which any peasant could imagine: the end of the landlord–bureaucrat tyranny. Thrones and empires could topple, the Allies and the Germans could fight it out as they liked. What really mattered to the peasant soldier in 1917 was the fate of a few fields, woods and meadows in his home village. The Russian revolution was largely the product of the triumph of these local emotions.

This was the situation when the Bolsheviks seized power in November 1917. They did so at a moment when most of Russia was looking the other way. That is the secret of their success. They wrote a lot of nonsense afterwards about a 'proletarian' revolution. In fact, the essence of 1917 was a peasant revolution. This primitive, backward mass had provided the social breakdown without which the Bolsheviks could not possibly succeed. The meek had inherited the earth – or so it seemed. The peasant had turned out to be the true autocrat of Russia. But his victory was not to last long. After a decade an even more powerful State was to harry him into collectivisation.

VILLAGE LIFE IN THE EARLY TWENTIETH CENTURY

The following passages give an idea of conditions in a couple of black-earth villages in the neighbourhood of Voronezh at the beginning of the twentieth century. The villages were close to a main road and to a large town and so enjoyed unusual opportunities to benefit from non-agricultural activities. In 1861 the villagers had chosen to accept the 'pauper holdings' of one quarter of the norm. For twenty years they had not ploughed their allotments at all and had subsisted upon land rented from the local landlord. But in 1884, rents had suddenly risen by 150 per cent as a consequence of the overcrowding of the black-earth area and its exploitation for the grain trade. The peasants were forced to plough the marginal land, most of which was quite unsuited to grain production. Even so the allotment lands did not yield sufficient grain even to feed the villages, let alone provide a surplus to pay the rent and taxes.

> The obvious deficit in grain was made up partly by renting land. The standard rent was rather less than 2 roubles an acre (say 4*s*), or a slightly smaller amount in cash, plus about 7½ loads of manure delivered on the landlord's land. These payments in manure had considerable importance in an area in which animals were few. . . . Manure was so scarce that none of it was used for fuel. . . .
>
> The deficit in straw – a very important commodity where it is the main source of fuel, bedding for humans and animals, and roofing – was partly made up by buying the stubble from the landlord's estate, at the rate of about 20 kopeks an acre. Wood is necessary to the baking of bread, and the people had not enough wood for this daily need. But there was a large 'Economic Forest' belonging to the landlord; and the villagers, heckled by the investigator on this subject, could only tell him with sheepish grins: 'The baking must be done: so we go to the *Barin* in the forest: he has got wood rotting'. In other words, they stole wood.
>
> There were seventeen households only, out of 159, whose grain lasted from the end of one harvest to the time for sowing the next. All the rest had to buy grain, and thirty-nine households did not cultivate their allotments at all.
>
> . . . most of the livelihood was derived from wage-earning occupations. Nearly a third of the men were so occupied in the neighbourhood and about one-sixth of them outside it. A much smaller number of women were thus employed: but in both sexes the numbers tended to rise. The large-scale agriculture of the landlord gave occupation both to men and women: but stone-quarrying, a dangerous and unhealthy

occupation, with no sort of provision for occupational disease or accident, was the main local industry apart from agriculture. Grain- and wood-carrying was common, and some had recently taken to dealing in these commodities on their own account in a small way.

An agricultural labourer by the year received 60 roubles (say £6) for a male, 36 roubles (say £3 12*s*) for a female, plus food. Day labourers who found their own food got from 20 to 40 kopeks (say from 5*d* to 10*d*) for males, from 15 to 30 kopeks (females): according to the season: the summer day, at harvest time, being eighteen hours, with three hours off for food.

The village also took in orphan children from a *zemstvo* orphanage in Voronezh. But the mortality was very high – not only because of the diet which lacked meat entirely and was very deficient in fats, but also because of the wide prevalence of syphilis among the foster parents. There was a hospital on the other side of the Don and another 8 miles away to which syphilitic patients could be taken but which lacked a sufficient number of free beds.

Malaria was common in both villages. Along with this fact we read that each village has a pond formed by damming a small stream on its way down to the Don. It is in these ponds that the cattle drink. The dam is formed of dung. . . . Dead horses and cows are skinned, and sometimes carried out and left to the dogs, but sometimes buried in the house-yard. . . .

In one of the two villages there were *no* privies: in the other there were eight. Excrements were left in the outer passages of the house . . . and *eaten by swine, dogs and fowls*. . . . There were only two beds in the whole of one village. . . . The stove, benches, shelves and the floor were the sleeping-places. In the vast majority of the *izbas*, the cattle wintered inside.

Our doctor has the true scientific passion for statistics. The ordinary red and the Eastern black cockroach were observed in 90 per cent of the dwellings, but bugs in only 15·5 per cent. This fact gives occasion for an interesting observation. *The bug is a natural aristocrat*. He does not like the miserable bedding of the poor man. The black beetle is most democratic. But there is a point at which he too draws the line because there is too little food to be got. The peasants have a saying about the most extreme poverty: 'There are no black beetles here'.

The investigator thought that these were typical black-earth villages, perhaps rather fortunate in having non-agricultural sources of income. Voronezh was among the areas with a high incidence of peasant disturbance both in 1905 and in 1917.

4
Official Russia in the Reign of Alexander II: 1855-81

THE TSAR LIBERATOR

Alexander II was born in 1818. His mother was one of the many German women who gave the later Romanovs a distinctly Teutonic make-up. From her Alexander inherited a lasting respect for things Prussian. His father, Nicholas I, was a man against whom the heir to the throne might have been expected to rebel. In fact nothing of the sort happened in the realm of political ideas; it is true, however, that the young Alexander early established a claim to arrange his own emotional life. His early amorous adventures had been conventional enough – a ballerina had been involved. But in 1839 his parents were horrified to learn that he had fallen in love with Princess Mary of Darmstadt, a young lady of fifteen. It was not her age which alarmed the Russian court; it was the fact that she had been born to her mother some fourteen years after that lady's open and public breach with her husband. The father was known and he was of shockingly low birth. Alexander threatened to live abroad if his father's consent was not given. Nicholas capitulated in 1840 and Princess Mary married Alexander in 1841. Their union was certainly fruitful but it is sad (and characteristic of Alexander) that having caused so much trouble in gaining his bride he should, in later years, have been frequently and flagrantly unfaithful to her. The most celebrated of his mistresses was Catherine Dolgorukaya. He had several children by her and exercised the supreme autocratic power to legitimise them in 1874. Four years later he installed his second family in the Winter Palace; the Tsarina, so passionately wooed in 1839, could hear her rival's children playing while she lay dying. In 1880 he married

Catherine and proposed to have her crowned, a prospect which horrified the court and the Tsarevich. The romantic irregularities of Alexander's life alienated the court and educated circles and had much to do with the complete indifference with which his assassination was greeted.

Alexander's formal education was held to have ended in 1837. He showed little aptitude for intellectual work. In the manner of his romantic age he wept easily and frequently, and the sight of suffering, of which there was so much in Russia that not even a Tsarevich could escape it, caused him pain. 'Well-intentioned but weak as water', was Granville's terse summary. He loyally carried out his father's wishes and at no time acted as the centre of a court opposition faction. He dutifully accepted his father's military enthusiasms, attended numerous parades in carefully chosen uniforms and developed a conventional passion for hunting.

Alexander was not of the stuff of which state-builders are made. He had neither the energy, nor the will, nor the patience to recast the State. It may therefore appear puzzling that he is described as 'liberator', 'moderniser', and the reforms of his reign compared with those of Peter the Great. In fact, inasmuch as these labels imply that he was an innovator, they are misleading. Alexander merely put into effect changes which had been frequently discussed since Catherine's reign; even Nicholas had contemplated the emancipation of the serfs and had taken some important steps to that end. Nothing Alexander did altered, or was intended to alter, the fundamental political fact of a God-created autocracy. On the contrary, all that he did was conservative in intention. By rearranging the social relations of the classes he planned to restore the full power of the autocracy. To suppose that Alexander meant to give Russia a constitution and was only prevented by either the selfish conservatism of the landowners or the foolish violence of the revolutionaries, is a myth.

Alexander, then, did not aim to reverse his father's policy nor did he propose to copy Western constitutionalism. But he was faced by a serious crisis when he came to the throne in 1855. The State which had defied Napoleon, and which had stood firm when other autocracies toppled in 1848, was about to see a part of its territory conquered – and conquered by a small, inefficient expedition operating far from its bases. In British history the

Crimean War is a series of gloriously inexplicable blunders immortalised in execrable verses. But in Russian history it was a most alarming disaster. What could have gone wrong? What should be done to put matters right? Alexander was the last Tsar to receive the backing of educated opinion in an attempt to solve these problems.

Perhaps the greatest difficulty facing Alexander was not that of the peasantry but that of the relationship between the throne and the landowners. This was the governing class and upon its co-operation the throne depended. How could it be kept loyal after it had been deprived of the benefits of serfdom?

In the view of some members of the class the Tsar's course was simple: the landowners were to be compensated for the abolition of serfdom by being given a share in political power. This view was widespread among the 1400 landowners who took part in the various regional and other committees which discussed the terms of emancipation. Nothing, however, was accomplished. This was partly because the landowners were themselves divided. The conservative majority wanted an elected assembly because they saw in it a means to prevent the Tsar from emancipating the serfs on any terms other than those highly favourable to themselves; the liberal minority were uncertain whether they wanted a consultative or legislative assembly and were frightened of playing into the hands of their conservative opponents. Alexander remembered the French assembly of notables of 1789 and set himself firmly against any such proposals. The liberal point of view was most forcefully expressed by the nobility of the province of Tver. In 1859 and again in 1862 they petitioned the Tsar to grant an elected assembly. On the first occasion they were rebuked and on the second their leaders, among whom was Alexei Bakunin, brother of the anarchist, were imprisoned in the Peter and Paul fortress. This refusal to permit the most powerful class in the land to exercise its due influence at the centre of power has been seen as a turning-point in the history of Russia. The landowners continued to be a ruling class which did not rule. They failed even to make use of whatever economic opportunities emancipation had given them. The great estates remained very inefficient; the debts of the landowners increased while the extent of the land which they owned decreased. The most active and intelligent turned against a system which gave them position without power, and joined the ranks of the revolutionaries and the critics. By rendering the class politically innocuous Alexander made it socially subversive.

LOCAL GOVERNMENT REFORMS

The most that Alexander was able to give the landowners was a place in local government. The position formerly occupied by the serf-owner might have been filled after 1861 by the bureaucracy, but Alexander wisely decided that to bring the bureaucrat directly into contact with the peasant would be imprudent. The landowners could continue to be used as a screen between the central institutions and the people. The problem was to give the landowners sufficient power to enable them to be useful without giving them so much that they would be in a position to help themselves to more.

The result of these calculations was the Zemstvo statute of January 1864. This created elected bodies at both the provincial (*gubernia*) and the district (*uyezd*) level. Following the Prussian model, the electors were divided into three classes, nobility, townsmen and peasantry. In order to prevent the peasantry from overwhelming the *zemstvos* by weight of numbers, they were elected indirectly and their representatives were fewer in number than those of the nobility. The district *zemstvos* elected the provincial members. Full *zemstvo* meetings took place once each year and were confined to generalities. The business was transacted by an elected board of three members who remained in office for three years. The president of the provincial board had to have his appointment confirmed by the Minister of the Interior; the Governor of the Province had to confirm the presidents of the district boards. These arrangements disappointed the liberal bureaucrats who had helped to make them. They had hoped for a juster representation of the peasantry and more opportunity for local self-expression. They were also disappointed by the way in which the powers of the *zemstvos* were defined. They were restricted to 'local economic needs': public health, maintenance of prisons, hospitals and lunatic asylums, improvement of agriculture, prevention of famine, relief of the poor, road maintenance and (to a limited and grudging extent) primary education. These tasks needed money. The central government unwillingly allowed the *zemstvos* to raise a local rate which was levied chiefly on land. Even then, a large proportion of what was raised could not be spent on the tasks listed above but had to be used for jobs which should have been done by the central government, such as raising and transporting army drafts and paying pensions to the families of those killed in action. The *zemstvos* could appoint and pay people to carry out professional functions – doctors, nurses and teachers – but they had no executive powers and had to rely upon the co-operation of the local police.

In spite of what might be regarded as crippling difficulties, the *zemstvos* achieved much in those areas in which they were established. They alone brought some of the less controversial benefits of civilisation to the peasantry. But it should be stressed that the main benefits of the *zemstvos* were not felt until after the great famine and pestilence of 1891. During the reign of Alexander II they suffered under too many disadvantages; not only were they entangled by the bureaucracy but they were dominated by the landowners. The latter ensured, for example, that peasant land was taxed much more heavily than the noble estates. To the peasantry, the *zemstvo* appeared to be an institution which perpetuated, under another form, the authority of the serf-owner.

Similar principles were applied to the 509 towns and cities by the statute of 1870. The urban populations were divided into three classes, according to the amount which they paid in trade taxes. Each class elected one-third of the city *duma* (*zemstvo*). This system gave the rich a majority in the *duma*, but such a property qualification was common throughout Western Europe at the time. A more serious weakness was that the franchise excluded the professional classes, who paid no trade tax. The *dumas* nevertheless did much good work, especially in the field of education. About 1880, Moscow and St Petersburg were more rationally governed than London.

Many of the liberal bureaucrats and landowners were bitterly disappointed by the meagreness of these local government reforms. They had expected a 'crowning of the structure' by the creation of a national assembly. They did not mind how narrow the franchise was at first; they would even have been willing to accept a purely consultative body as a beginning. The important thing was to get the elective principle accepted on a national scale.

Alexander II was far less inclined to accept such a scheme than his uncle, Alexander I. To many Western writers this has seemed to be a turning point in Russian history. It is argued that if Alexander II had taken the plunge, it would have been possible for Russia to develop along Western lines. Bourgeois liberties would have broadened down from precedent to precedent and Russia and the world would have been spared the horror of communist totalitarianism following Tsarist autocracy. Such a view is rather provincial. It is based upon the assumption that it would be desirable for all mankind to be ruled as the United States and Britain are ruled. After about 1865 the demands for a constitution faded away. Many of its early supporters changed their minds. They wondered whether the Western model should be followed. They feared that a restricted franchise would merely be an excuse for the social domination of one class – the conservative landowners. They noted that the Western constitutions in fact encouraged such class domination. They wanted to wait until all Russians were in a position to enjoy constitutional rights. Above all, they thought that the best means of rapid social change was by way of a powerful autocracy. That was the Russian way. They were disappointed by the timidity of Alexander II, but they were not shocked by his refusal to adopt Western forms.

One of the most damaging critics of Alexander's attempt to liberalise Russia was Leo Tolstoy. He clearly revealed his position in *Anna Karenina* (1873–6). It is a gloomy book, filled with scepticism and doubt, quite unlike *War and Peace* published a decade earlier. Levin, its hero, evidently a portrait of Tolstoy himself, is agitated by doubts about the political and social future of his country. He completely rejects Western models. Democracy, respect for law, freedom of the press, universal education, capitalism – these things all strike him as blind alleys which will merely lead Russia whither the West had already gone. Levin's solution to contemporary problems is anarchical. It is to seek private happiness in the fulfilment of such tasks as lie immediately to hand. Tolstoy paints a brilliant portrait of Anna's husband, Alexei Alexandrovich Karenin. He is a highly-placed official, obviously closely concerned with the liberal reforms. He is an excellent man – honest, intelligent, industrious. Yet Tolstoy makes us see him through Anna's eyes. She admires him but she cannot love him. Neither she, nor Tolstoy, nor Russia could endure what he represented.

LEGAL REFORMS

The emancipation of the serfs inevitably led to some reorganisation of the legal system. But the changes which were introduced by the law of 1864 were far greater than this cause alone required. Alexander was probably driven further than he intended to go by the enthusiasm of the liberal bureaucracy who, in this field, were not restricted by the prejudices of the conservative nobility. The hand of the liberals was nowhere more clearly seen than in the legal reforms of 1864. Zarudny, the official chiefly concerned, prepared himself for the work by several years of study in the West. In the end he decided upon a completely new departure; Russia was to be made to accept legal institutions which had taken centuries to evolve in Britain and France and which fitted very different social conditions. It was a repetition of the heroic impatience of Peter the Great; the results were a bitter disappointment to the liberals and provided useful material for the attacks of the Slavophils.

The law of 1864 provided for three types of court. The *volost* courts were retained for peasants only. They administered customary law; the judges were elected by the communes from among the peasants. Sessions would take place in the judge's hut and would frequently end in the reconciliation of the parties. These village judges were usually illiterate but could decide aptly. Marital brutality was a common complaint. On one occasion a woman refused to live any longer with her husband who had beaten her. The peasant judge refused to pronounce for either party; instead he sentenced both to a brief imprisonment, knowing that there was only one room in the village which could serve as a jail. These courts heard only the petty cases; they were popular because they gave quick and cheap decisions. But it should be noticed that they preserved the old class system and made nonsense of the preamble of the 1864 law which asserted the Western idea of the equality of all before the law.

The second type of court was copied closely from the English model; the Justice of the Peace court. The *zemstvos* were empowered to elect (and pay) justices who were to concern themselves with judicial matters only. The J.P. could hear cases between peasant and peasant if one of the parties so desired; he had to hear cases in which one or both of the parties was not a peasant. His jurisdiction was hardly wider than that of the *volost* judge. In civil cases he could take cases in which no more than 500 roubles worth of property was involved; in criminal cases he could fine up to 300 roubles or imprison for one year. The bench never achieved the prestige of its English model. The working J.P.s had frequently failed at other occupations before taking up this ill-paid job; the great landowners did not lend their authority to the office. The reformers had perhaps forgotten that in England the standing of the J.P. was the result of his administrative function; he represented the royal government in the shires.

The third type of court was based upon the French plan. The centre of the system was the Chamber of Justice. There were ten of these in Russia and, to emphasise the fact that they were quite separate from the bureaucratic machine, they stood in areas distinct from the administrative divisions. Attached to each Chamber were several regional courts which passed on appeals to the Chamber of Justice. It was possible to appeal from the Chamber to the supreme court, which was the legal department of the Senate. Judges were appointed by the Minister of Justice and could not be removed. They were not appointed from the Bar but usually from the staff of the Ministry. As bureaucrats, they were amenable to the usual pressures – lack of promotion or appointment to some particularly unattractive spot. Perhaps the most striking result of the 1864 reform was the creation of a Bar. Trial by jury was introduced in criminal cases and the accused was given the right to employ an advocate. The profession rapidly attracted many brilliant men who would elsewhere have aired their talents in parliaments. Even the great political trials at the end of Alexander II's reign were usually reported in the press. The speeches for the defence were intended to be an indictment of the government just as much as a legal argument. Members of the Bar were the freer in their utterances because they knew that no promotion awaited them; alone in Russia they lived by popular esteem.

These reforms were prefaced by declarations culled from the fashionable jurists, insisting that the subjects of the Tsar would henceforth enjoy all the personal rights so valued in the West.

But there was nothing to stop the autocrat from taking away the rights which he had given, and nothing was done to abolish the right of administrative arrest. In fact, the police, and in particular the Third Section, continued to operate an extra-judicial legal system. They continued to arrest on suspicion and to banish without trial. If they did think it necessary to bring the suspect to trial it was possible to do so before a military tribunal or a special court in which the usual safeguards of the law were not applicable. Such courts proliferated during the great wave of terrorist activities at the end of the reign. Another important respect in which the law of 1864 failed to live up to its principles was its failure to protect the citizen from the bureaucrat. The state servant could not be prosecuted in the public court for abuse of his power except with the consent of his superior. These discrepancies between declared intention and practical fact brought the law and the lawgiver into contempt. The reform of 1864 had much to do with the discontent of the seventies.

The punishments known to the law were, in theory, milder than those in use in other European countries. Capital punishment had been abolished in the eighteenth century and whipping was made illegal in 1863. But political offenders were not admitted to these merciful provisions. The Decembrists had been hanged and so were the terrorists of Alexander II's reign. More usual, however, was exile to Siberia – a punishment less feared than imprisonment in European Russia. Every year, summer-caravans of thousands of prisoners – criminals, peasants expelled from the communes and 'politicals' – made their way along the Volga, across the Urals to the collecting station at Tiumen. Thence the most unfortunate were sent to the mines at Nerchinsk; this was virtually a sentence of death. The maximum length of forced labour was twenty years, after which the prisoners had to settle in Siberia, although the more determined frequently managed to escape. The anarchist Bakunin, for example, got himself employed by the East Siberian administration, boarded a ship at Vladivostok, and whiled away the voyage to the United States by making friends with and borrowing some money from an English clergyman. He at length arrived in London and one evening burst into Herzen's lodgings in Paddington just as the latter was about sit down to a meal of oysters with his wife and his mistress. The astonished Herzen was forced to describe the fate of the revolutionary movement in the West to his hungry friend (whom he supposed to be in Siberia) before he could get anything to eat. Few were as lucky as Kropotkin who escaped from prison in St Petersburg. The pains of exile were often alleviated by humane guards and warders. Wives usually followed their husbands and were sometimes allowed to live in the jails with them. The courage of the aristocratic ladies who followed their Decembrist husbands was remembered throughout the century. It should also be noticed that many Russians shared the attitude towards exile in Siberia expressed by Dostoyevsky in *Cime and Punishment*. This was that such punishment was really merciful because it permitted and encouraged spiritual regeneration. The crime figures for Siberia hardly provide statistical evidence in favour of Dostoyevsky's assertion; but then such regeneration as he wrote of would not be expected to make its mark upon a mere compilation of figures.

EDUCATIONAL REFORMS

In the field of education, Alexander's bureaucrats also laboured hard. Here one of the chief problems was to render the student body docile. The Russian legislators must have wished that they could have inculcated the social conformism of English students; perhaps they should have introduced compulsory team-games. But even this form of corporate activity would have seemed dangerous to a government which in 1867 prohibited student concerts and dramatic productions.

The beginning of the reign saw a rapid removal of some of the restrictions imposed by Nicholas. Study abroad was permitted, uniforms were abolished, chairs of philosophy were restored and the universities were allowed to elect their own rectors. But the ungrateful students merely wanted more and in 1861 there were disturbances in nearly all the universities, so serious that they had to be closed. The appointment of a retired admiral (Putyatin) to the Ministry of Education did not help; in December 1861 Alexander turned to the liberal Golovnin who set about the preparation of a new university charter. He made a thorough study of foreign models and his labours (revised by a suspicious committee) were embodied in the statute of 1863. This gave the universities as much independence as any institution could hope to have in

Russia. The professors were given the right to lecture in whatever way they chose and the elected university councils had some control over the appointment of university staff. But since the government retained an ultimate power of veto in academic appointments, it can be seen that Golovnin had not altogether triumphed. Women were excluded from the universities on the grounds that they were likely to impede the academic progress of the young men. All forms of student organisation were forbidden. This simply led to the formation of large numbers of secret organisations, some revolutionary, others merely recreational. Even this unpromising law was revised after 1866. In that year the student Karakozov fired at the Tsar; he fired in the middle of the day, from a large crowd. It appeared that the youth was a member of a student revolutionary circle. The reactionaries made all they could of the attempt and ten days later Golovnin was dismissed and Count D. A. Tolstoy replaced him as Minister of Education.

Tolstoy was already Procurator of the Holy Synod. He retained this job and it was consequently thought that education had fallen under the dead hand of the Church. Tolstoy did not achieve the repeal of the statute of 1863, although he tried hard, but he did ensure that it would not have the effects which Golovnin had expected. He restored police supervision of students and harried the more independent professors. Nevertheless distinguished minds continued to inhabit the Russian universities and the number of students increased. In spite of Tolstoy's best efforts, the student body remained recalcitrant and proved to be one of the main sources of revolutionary opinion. Tolstoy was not opposed to the higher education of women. Since their rebuff in 1863 many Russian girls had gone abroad to study; there was a large colony of them in Zürich. There they imbibed dangerous opinions and Tolstoy judged that it was better to lure them home. He made no objection when courses were opened to them at St Petersburg University in 1869. Other universities did likewise and by 1881 there were 2000 women in places of higher education, most of them medical students.

It was in the field of secondary education that Tolstoy had the greatest influence. By the 1864 statute, Golovnin had reorganised this sector. This divided the gymnasia into two classes, classical and modern, both with seven-year courses. But only the pupils at the classical gymnasia could qualify for the university; the others had to be content with going on to one of the higher technical institutes. In theory, these schools were open to all but since most of them charged fees the student body was restricted to the wealthy. Golovnin gave the secondary system its classical bias; in so doing he was merely following the pattern of Western universities of the time. Tolstoy strengthened this bias. He believed that subversive influences could be eliminated by concentrating upon classics, mathematics and drawing. He also made efforts to preserve the class nature of the system by permitting only the sons of the upper classes to enter the gymnasia. In 1871 he altered the timetables in the classical schools. Latin received 15 more hours per week, Greek 4 hours and mathematics 4 hours. History was considered dangerous because it encouraged teachers to indulge in criticism and generalisation; consequently it lost 2 hours. Even then, the history teachers were instructed to concentrate upon Russian victories and to pass lightly over the less glorious aspects of national history. Text books had to receive the approval of the ministry. The teaching of literature was also held to be subversive and Tolstoy replaced much of it with instruction in Church Slavonic. In case any student should fail to flourish upon this wholesome diet, he revived an inspectorate which was to visit student lodgings and ascertain whether student morals were in good order. It was found at first that Russia lacked a sufficient number of classical masters to carry out the minister's programme. But luckily a fresh supply was found in Eastern Europe; soon Czech and other scholars were at work and Tolstoy could view with pleasure the increasing number of Russian youths who were able to write faultless Greek verses and, so it seemed, were inoculated against ideas for life. In fact, Tolstoy's system bred revolutionaries among both the students and the masters. Not only was any individual with a glimmering of intelligence revolted by the intellectual fare offered but the system ensured a large number of failures. The student who was expelled had hardly any choice but that of joining the revolutionaries.

In 1872 Tolstoy reformed the modern schools. Here he deducted 10 hours from the study of the natural sciences and added 21 to 'drawing, etching and handwriting'. It is to be feared that these peaceful pursuits did not render the young men docile.

Tolstoy's régime helped to imprint deeply upon the mind of the Russian intelligentsia the idea that scientific studies alone were dignified, fruitful and free; if they did not liberate, why should the government be so set upon repressing them?

Primary education made little progress during the reign, in spite of the expression of pious hopes and the publication of numerous statutes. This was partly because of the obstructive attitude of the Church which (as in Britain and France) wanted to preserve a fine bloom of ignorance among the masses. But there was also the indifference of the local communities which had to pay for the schools, and an acute lack of teachers. Tolstoy naturally brought the primary schools more tightly under the control of conservative forces (1874). He also encouraged the creation of more training schools for teachers. By 1881 only 9 per cent of the children of primary school age were receiving education. In the same year there were about 125,000 pupils in the secondary schools, including nearly 47,000 girls. The universities had nearly 10,000 students (3600 in 1855). One new university – Odessa – had been founded during the reign by the great surgeon, Pirogov. Something had been achieved but at an enormous cost in disruption and discontent.

ARMY REFORMS

Of all the institutions which required reform during the reign of Alexander II, the army was of outstanding importance. It was the disaster of the Crimean War which had launched the Tsar and the bureaucracy upon the way of change; it was hardly useful to alter society if the army was still unable to defend the frontiers of Russia. Such a reform should have been fairly easy to secure since it did not arouse the controversy and confusion which, as we have seen, grew from the other efforts to modernise. All Russians were agreed upon the necessity for an efficient army, and an efficient army was compatible with the retention of autocracy. In this matter the Tsar could reform without undermining his own position. During the reign further impetus to reform was given both by the emergence of a powerful German army and by the growing tide of Panslav feeling which seemed certain to provoke a Balkan war.

D. A. Milyutin was appointed Minister of War in 1861 and

retained the office until the end of the reign. He had time to be both thorough and consistent. His first efforts were directed towards changing the immense bureaucracy of the central command. He actually reduced the number of bureaucrats in the Ministry of War – a remarkable achievement in itself – and set up fifteen military districts. Each was equipped with a commander and he was provided with a council whose advice he generally had to take. But the effectiveness of this measure of decentralisation must have been much reduced by the fact that the heads of the various military subdivisions – artillery, engineers, etc. – were in all matters affecting their speciality subject not to the local commander but to the Ministry of War. However, the system worked well in Central Asia where it permitted a determined local commander to take the initiative.

In other directions, too, Milyutin reformed the upper ranks of the army. The status of the General Staff was raised and in 1865 the post of Chief of Staff was created. This officer was given charge of a department of the War Ministry but was not necessarily a member of the supreme War Council. The eighteen members of that body were appointed directly by the Emperor; the only *ex officio* member was the Minister of War himself. Under Milyutin's inspiration the Council produced a new military code in 1869. This comprised fifteen volumes and should have covered every possible contingency. Much attention was given to the supply and medical departments but without much success: both broke down during the war of 1877–8.

Like other army reformers Milyutin pondered long on the matter of filling the upper ranks with suitable officers. He did not, like his British contemporary Cardwell, have to tackle the abuse of the purchase of commissions, but he did have to revitalise the old Cadet Corps which, under Nicholas, had specialised in the training of gentlemen and courtiers without demanding too much modern military knowledge. Count Peter Kropotkin entered the most exclusive of these schools – the Imperial Corps of Pages – in 1857 just as it was beginning to feel the effect of the post-war reforming spirit. There were then 150 pupils, the sons of court nobility, aged from twelve upwards. Those who passed the final examinations could select any regiment of the Guard, irrespective of the number of vacancies in that regiment; the top sixteen pupils were nominated *pages de chambre* and had to perform honourable but hardly military functions in the royal palaces. This of course made them familiar with the Tsar's entourage and was the gateway to a profitable and easy career. In 1857 the chief authority in the Corps was wielded by Colonel Girardot, a man of the time of Nicholas. He allowed the eldest pages to run the Corps as they liked:

> . . . one of their favourite games had been to assemble the 'greenhorns' at night in a room, in their nightshirts, and to make them run round, like horses in a circus, while the *pages de chambre*, armed with thick india-rubber whips, standing some in the centre and others on the outside, pitilessly whipped the boys. As a rule the 'circus' ended in an Oriental fashion, in an abominable way. The moral conceptions which prevailed at that time and the foul talk which went on in the school concerning what occurred at night after a circus, were such that the least said about them the better. The Colonel knew all this. He had a perfectly organised system of espionage and nothing escaped his knowledge. But so long as he was not known to know it, all was right.

The striking resemblance to a Victorian British public school is strengthened when we read: 'The *pages de chambre* severely punished any of the other boys whom they caught smoking but they themselves sat continually at the fireside chattering and enjoying cigarettes'. Kropotkin goes on, however, to describe how the atmosphere at the Corps changed:

> . . . the Colonel's influence was rapidly vanishing. . . . For twenty years Girardot had realised his ideal, which was to have the boys nicely combed, curled, and girlish-looking, and to send to the court pages as refined as courtiers of Louis XIV. Whether they learned or not he cared little; his favourites were those whose clothes-basket was best filled with all sorts of nail-brushes and scent-bottles, whose private uniform . . . was of the best make and who knew how to make the most elegant *salut oblique*.

Kropotkin subsequently describes how, under the new régime, he received an excellent education at the school – an education so wide and so humane that he decided to leave the army to become a scientist, and finally joined the ranks of the revolutionaries. Milyutin was as powerless to abolish this venerable institution as Cardwell to abolish Eton, but he was able to ensure that those cadets who intended to pursue a military career had to pass through two special classes which concentrated upon matters military rather than upon the accomplishments of a courtier.

In other spheres of army education Milyutin was much more successful. He was able to organise schools for boys from the age of twelve upwards in which the pupils received an excellent general education before specialising in military subjects. So powerful was Milyutin's influence that he was able to protect the broad curricula of his schools against the dead hand of Tolstoy. The military pupils were learning history and science when their contemporaries in the schools of the Minister of Education were concentrating on the classics. By the end of the reign, there were fourteen such military schools with 4200 pupils. Milyutin also enlarged the specialist schools and academies. He gave particular attention to artillery and engineering. By the end of the reign the railway specialists had clearly reached a very high level of competence; it was the military engineers who constructed the very difficult railway through Turkestan.

In the matter of weapons it must be confessed that Milyutin did not achieve much, but this may be attributed to the general backwardness of Russian industry. The Crimean War had been fought with a muzzle-loading rifle. During the sixties Milyutin merely had this weapon converted into a breech-loader by the addition of the Krenk mechanism. The majority of the troops were thus armed during the Turkish War of 1877–8. Its rate of fire was slow because the spent cartridge generally had to be pushed out with a ramrod. In 1877, only the Guard was armed with the new Russian breech-loading rifle, the Berdan. These were at first manufactured in Birmingham; later the machinery was sent over from England and set up in the Russian ordnance factories. The field artillery was all made of bronze until 1877. This was an inefficient metal for guns; range and accuracy suffered. On the eve of the Turkish War, 1500 steel guns were ordered from Krupps, but they did not arrive in time. It was discovered that in some respect the Russian army was not so well equipped as the Turkish.

But Milyutin's chief difficulties as a reformer were a direct result of the emancipation of the serfs. After 1861 the army could no longer be regarded as a penal settlement for difficult or disobedient peasants. Milyutin tried to humanise it slightly by reducing the length of service from twenty-five to fifteen years and by abolishing flogging, but he was still unable to get recruits in sufficient numbers or of sufficient quality. During the sixties the army numbered about 800,000, which made it larger than any other in Europe, but the reserves were far smaller than those of the other military powers. Milyutin wished to introduce conscription to get over this weakness. This proposal was blocked by the conservatives who increasingly dominated Alexander II. The Franco-Prussian War (1870) helped Milyutin. Prussia won it with a very large army of high quality mostly composed of reservists. The day of the peasant army was over; if Russia wished to remain a great power she must have a national army. Milyutin at last got his way in 1874. The conscription law of that year began with the following statement of principle: 'The defence of the throne and the country forms the sacred duty of every Russian subject. This regulation calls upon the entire male population to participate in military service, without exemption by purchase or by providing substitutes'. Service was fixed at fifteen years, of which six were to be spent with the colours. University students who had completed their course had to spend only six months with the colours. Secondary education also entitled a man to exemption. Even a man who had finished the primary school course only did four years active service. Nevertheless the reform of 1874 created the only democratic institution in nineteenth-century Russia. It was noticed that the spirit of the army changed rapidly. A link between the rich and the peasantry had been formed; at last the intelligentsia could 'go to the people' legally. The draft of 1874 showed only 11·8 per cent literacy rate among the recruits and Milyutin hastened to set up schools for the elementary education of the rank and file. It seemed that the army might become a real school for the nation. It could have provided a focus for the one force which united the majority of Russians in the seventies – nationalism.

The experience of the Turkish War was bitterly disillusioning for Milyutin. It is true that there were few complaints about the spirit of the troops in the field: but the rear organisation and the supply system were as chaotic as they had been during the

Crimean War. There were no maps; there were no spades; there were no boots for the troops who crossed the Balkans in the winter of 1877–8; the hospital service was cruelly ineffective; the contractors made huge profits out of defective supplies. That the Turks could defend Plevna at all was regarded as a criticism of the régime. The gloomiest conclusions were drawn about the ability of the Russian army to fight against a Western power. Although Alexander was not in personal command of the Balkan army, he was blamed for everything; his reputation never recovered from the fiasco of the early months of the war. It was noted against him that he even failed to prosecute the firm of Greger, Horwitz and Kohan, the main army contractors, who had made exorbitant profits out of bad supplies. Yet again, war had shaken the autocracy to its foundations.

ECONOMIC DEVELOPMENT

At the beginning of Alexander's reign, Russia was relatively much poorer than in 1815. The other great powers had, to varying degrees, enjoyed the benefits of the Industrial Revolution while Russia had stagnated. Alexander was aware of the problem; he could hardly fail to notice the effects of the Crimean War which had driven the government into extensive use of the printing press in order to meet current expenses. But Alexander did not understand either the causes or the consequences of Russia's economic backwardness. He failed to see that Russia could no longer support the expense of being a great power; he failed to understand that her economic ills could not be cured by a dose of liberal economic doctrine.

Reutern, another Westernising liberal bureaucrat, was Minister of Finance from 1862 to 1878. His ambitions were limited but he achieved some success. In 1863 he published the first budget in Russian history. This was no doubt a real advance and one which would have met with the approval of Mr Gladstone, but it seemed pointless to inform the public about the details of an account over which they had no control whatever. Reutern did not reform the tax system. Indirect taxation accounted for 69 per cent of revenue in 1880; this was raised largely from the tax on vodka. The poll tax was the chief direct tax and that was so assessed that the peasantry paid most of it.

It appeared that the autocracy was too weak to tax the rich. In spite of all his efforts, and an increase in revenue from 432 million roubles in 1863 to 793 million in 1880, Reutern managed to balance the budget only five times. The printing press had to be used again, especially during the Turkish War. Reutern had tried hard to secure convertibility for the paper rouble; in 1862, it stood at eighty-seven kopeks and he might have achieved his objective had he not bungled a conversion operation on the London market. By the end of the reign the value of the paper rouble had fallen to sixty-three kopeks. Perhaps Reutern was most successful in the promotion of private banking. This attracted a surprising quantity of capital, most of which was invested in Russian industry. He did not encourage the flow of foreign capital. It was left to Witte to see that only such a flow could lift Russia from the economic doldrums.

Reutern accepted the liberal dogma about tariffs. They were regularly reduced from 1867 to 1868. This policy certainly encouraged imports which increased from about 200 million roubles in 1861 to about 520 million in 1880. The fiscal result was no doubt excellent, but there was also an adverse balance of trade during most of the sixties and seventies. It ran at the rate of nearly 100 million roubles a year during the period 1871–5. In 1877 the free-trade tendency had to be reversed, and in 1880 a tariff was at last imposed upon imported iron and pig-iron.

The most serious result of the liberal tariff policy was upon the development of Russian heavy industry. It is likely that Alexander and his advisers were uncertain as to whether they really wanted the growth of large industrial towns and populations. They can be forgiven for not understanding the crucial importance of heavy industry in the first stages of industrial development. At the beginning of the reign, Russia produced about 0·2 million tons of iron and steel. For a short time this level of production actually declined because of the disruption caused by the emancipation to the serf-worked Urals mines. By 1880 production approached 0·6 million tons; but this was not nearly enough to meet domestic needs. Iron and steel production was for the most part inefficient. Much of the industry was still organised along *kustar* lines: that is, small-scale domestic production in the villages. Most of the pig-iron was still smelted with charcoal; many of the blast-furnaces continued to operate with the inefficient 'cold-blast'; and the

Russian iron-masters were slow to introduce the open-hearth furnace. The only really efficient plants were built and operated by the Welshman Thomas Hughes in the Donets basin. He brought with him a complete crew of Welsh artisans who settled in the town pleasantly named Yuzovka. Hughes fired his first furnace in 1872 and so initiated a great industrial revolution in south Russia, but his plant did not contribute significantly to total output before the end of Alexander's reign. Engineering also suffered from extreme technical backwardness – a defect which the educational system did nothing to cure. Much ingenuity was shown by Russian inventors like Chernyshev and Yablochokov, but there was no market for their skill. Most of the machinery for the cotton factories had to be imported from Britain; it was said that even the nuts and bolts required for railway construction had to be imported.

The cotton industry benefited from the liberal tariff policy and progressed rapidly during the reign. The value of the goods produced doubled during the seventies, and by the end of the reign the industry showed a marked concentration of labour in large factories. There was no labour legislation, wages were low and conditions very bad. Many of the workers were peasants who had scarcely broken the link with the village. Unfortunately, Russia's most buoyant industry made little contribution to solving her most pressing economic problems. The cotton industry produced almost entirely for the domestic market and its increased demand for raw material merely aggravated the adverse trade situation. The need for a domestic supply of raw cotton encouraged the advance into Central Asia.

Railway development was keenly encouraged by Alexander and his advisers. The figures reveal a striking advance in this field. At the beginning of the reign Russia had about 700 miles of railway; in 1881 there was about 14,000 miles. In keeping with the liberal sentiments of the ruling circle, it was decided to hand over the task of railway construction to capitalist enterprise. At first foreign capital alone was allowed to enter this presumably lucrative field; but lack of foreign enthusiasm led the government to permit Russian capitalists to build railways from about 1870 onwards. In both cases the companies received state backing and state guarantee of profits. These guarantees were used up to the hilt. The lines failed to show a general profit (although some individuals made large fortunes) and by the end of the reign the companies owed the State more than 1000 million roubles. Russia seemed to have got the worst from both private enterprise and state intervention. The lines were not built with military needs alone in view although it is possible that the use of the broad gauge would make it more difficult for an invader to use the system. It was clearly stated in 1857 that the main purpose of the railways was to facilitate grain movement from the interior to the ports. This aim was certainly achieved. The railways carried increasing quantities of wheat and barley to the Black Sea ports and thus encouraged the emergence of a specialised agricultural economy in south Russia. It is to be wondered whether the development of Russian waterways would not have served the same purpose and saved some part of the enormous imports required for railway building.

In the end, these imports had to be paid for by the export of grain. In 1861 the grain exports were worth about 56 million roubles, or one-fifth of the total value of exports; in 1881, they were worth about 300 million roubles, or about three-fifths of the total value of exports. This grain export had to be achieved in spite of legislation which prevented Russian agriculture from becoming more efficient through investment or reorganisation. In other words, Alexander and his advisers expected a semi-feudal agriculture to provide the foreign purchasing power with which a modern industrial system was to be built. The task was impossible from the beginning: it was made doubly so by the fall in world grain prices which started to influence Russian trade from the mid-seventies. By 1890 Russian grain was selling abroad at about half the price per ton paid in 1870.

During Alexander's reign the population of Russia increased from sixty-two to seventy-nine millions. The urban population grew from 11·3 per cent to 12·9 per cent of the total. Marxist historians assert that Russia, during his reign, took the first decisive steps towards becoming a capitalist society. Those who are not doctrinally aligned may doubt the validity of this assertion. The weakness of the domestic market, the reliance upon the export of agricultural products, the technical backwardness, all point to the continued existence of a pre-industrial economy. Neither Alexander nor his advisers grasped the seriousness of the situation. They continued to behave as if Russia were a great power although, economically, she was little more advanced than India. Although it may be possible to describe Alexander as a 'liberator' in other fields of activity, in economic matters he failed to liberate the potential strength of Russia. It was left to Witte, in the nineties, to act with the decisive ruthlessness which alone could lift Russia out of backwardness. Then the timid liberalism of Alexander's reign was firmly set aside, and the half-hearted attitude towards industrialisation rejected.

THE POLISH REVOLT (1863) AND THE CONSERVATIVE REACTION

Although many of the reforms described above were timid or half-hearted, Russia by 1880 was very different from Russia in 1860. Whether the difference constituted 'progress' depended upon the viewpoint of the observer. For the liberal bureaucrats who had been active at the beginning of the reign, a promising start had been nipped in the bud. For the radical sons of the liberal fathers, Russia was still an intolerable prison-house, only to be cleansed by fire and the sword. But for conservative opinion, Alexander had made some bad mistakes at the beginning of his reign. He had backed the wrong horse. He had encouraged aspirations towards liberty when he ought to have followed the tradition of paternalistic discipline so successfully revived by his father. Russia was to be bullied and whipped out of backwardness – if, indeed, Russia was backward at all. Alexander was not strong enough to resist the reactionary current. Events seemed to show that the conservatives were right. Everything went wrong. Russian society was in a turmoil. Secret societies of young people created disturbances which made the Decembrist rising look like a petty street brawl; the Tsar's life was in constant danger; educated public opinion was indifferent to his fate. He would not go back on his reforms but he would not go forward either. He turned more and more to the traditional guardians of conservatism.

The most influential was Shuvalov. He became head of the Third Section in 1866 and retained this post until taking up the London embassy in 1874. He was a brilliant, charming man who exercised a powerful influence upon Alexander. D. A. Tolstoy was another powerful conservative figure. He became Minister of

Education in 1866 and his work in that field has already been examined. The third important reactionary was the journalist Katkov. He started the reign as a liberal, a supporter of serf emancipation, but his views were changed by the Polish rebellion of 1863. Thereafter, he became a rabid nationalist. In his papers he skilfully expressed distaste for reform combined with a glorification of Russia and of Russia's duty to impose herself upon the external world. He thought that imperial conquest could unite the Russians behind throne and altar. Panslavism was the name given to this conservative nationalism. It was very different from the cautious patriotism advocated during the reign of Nicholas I. Unfortunately it was influential. During the remaining years of the old régime, Katkov's spiritual progeny continued to act upon a virulent nationalism probably nastier than anything else in Europe at the time. Katkov's disciples had the doubtful honour of fathering the modern version of antisemitism.

The Polish revolt of 1863 gave the conservatives the chance they needed to frighten Alexander II away from further reform. They could argue that this event had been sparked off by the reforming movement in Russia. There was a certain amount of truth in the assertion. The Polish landed gentry had not forgotten the revolution of 1830–1. After its suppression, they had lost the constitution granted by Alexander I. The Sejm was abolished and the Polish army disbanded. The Kingdom ceased to have an independent existence although the Polish landowners continued to enjoy considerable powers. One of their chief grievances was that Lithuania was ruled separately from Poland, as a Russian administrative division. They considered that even if Poland were not free, it should at least be united. Most educated Poles were, of course, nationalistic. That meant that they were all anti-Russian though some were less violent than others. This was the curse of Polish history, a sort of disease which it has taken much bloodletting to cure. At the beginning of Alexander's reign, the Polish nobles were nervous about what sort of peasant emancipation would be forced upon them by the Russian government. Most of them, like their Russian counterparts, did not want any reform at all. They feared that the Russian government would give the Polish peasants a much better deal than it gave the Russian peasants. A contented Polish peasantry would provide a satisfactory basis for Russian rule.

Alexander approached the Polish problem in a conciliatory spirit. He appointed liberals to important positions – his brother Constantine as viceroy and Wielopolski to head a reform commission. The latter was a specimen of that rare type, a Polish nobleman willing to put Polish prosperity before Polish nationalism. He published plans for reform of the peasant question on lines acceptable to many of the landowners: that is, the peasants were to become tenants, not freeholders. He announced his intention of setting up institutions for local selfgovernment. But to no avail. Even those who had previously agreed that such plans would be sensible did not dare to do so when it was known that they received the backing of the Russian government. Such a backing was certain to kill any proposal in Poland. Students demonstrated; obscure but emotive battles fought long ago against forgotten foes at strange places were remembered; the whole sad story of Polish history was paraded. The Roman Catholic Church entered the struggle: in the course of some street battle sanctuary had been violated. Wielopolski, in despair, tried to exile some of the leaders. In 1862 he announced that conscription would be applied in Poland. The purpose of this was to get most of the inflammable young men out of the way, but it was possibly a mistake. The ardent patriots fled to the woods in order to escape conscription. Arms were obtained; the moral support of the comfortably situated middle classes of France and Britain secured; and the futile rebellion started in January 1863. It took a year to suppress. It was poorly led, and was handicapped by violent quarrels between the leaders who spent much more time fighting each other than fighting the Russians. Exiles sent reports of imminent French intervention but they were all lies. The peasants looked on, hardly less enthusiastic about the Russian troops than about their own masters. There was the usual individual heroism, so romantic to read about and so useless in fact. The Russian reaction was severe, especially in Lithuania where M. N. Muravyov distinguished himself by efficient brutality. But in Poland itself, milder councils prevailed. N. Milyutin carried out a land reform rather favourable to the peasants, but it did not have the desired effect: it did not make the Polish peasants happy to be Russian citizens. Nationalism remained the strongest force. All that remained now was for the Russian government to try to turn the

Poles into Russians. They failed, and eventually Poland won its independence. The story of that twenty-year freedom from Russian domination (1919–39) is, if anything, even more melancholy than the story of Poland under Russia.

But more significant than the Polish revolt in driving Alexander into the conservative camp was the rapid development of a revolutionary movement in Russia itself. Its roots went back to the previous reign. It would probably have declared itself whatever sort of régime Alexander II had decided to adopt, but its vigorous existence strengthened the conservative argument for repression. Alexander could ignore the various student disorders of the early sixties; he could not ignore the attempt made on his life by Karakozov in 1866. This was a declaration of war. For the remainder of the reign the government was at war with the revolutionaries. In the end they killed the Tsar but they failed to provoke a revolution. This was a menacing development, much more dangerous – and much more unexpected – than Polish nationalism. The Romanov autocracy had been threatened by many enemies; by foreign powers, by the nationalism of the subject peoples, by the class ambitions of the noblemen, by peasant revolts. Never had it been threatened by young students armed with little more than ideas. This was the most original development of nineteenth-century Russia. Its effects were eventually to involve the whole world. Karakozov's pistol shot marked the beginning of a new era in Russian history. The Russian underground declared its presence. Until 1917, it would not allow itself to be forgotten. In 1917, it triumphed.

5
Unofficial Russia in the Nineteenth Century: 1825-81

THE SLAVOPHILS

The complete annihilation of the Decembrist movement, the comparative stability of Russian society under the despotism of Nicholas I, the silence imposed by the censorship and the Third Section, could not conceal the existence of a deep current of revolt and doubt beneath the comparatively placid surface. This current is of particular interest because, after passing through many unruly and turbulent channels, it was to flow through the disciplined banks of Marxist ideology.

The most active and influential critics of the thirties and forties were the Slavophils. They owed little to the Decembrists and they were horrified by the radicalism of the younger men who followed them. Before the end of the century their teachings, in a modified and impoverished form, had been transmuted into a base, violent and conservative nationalism. Nevertheless, in the beginning Slavophilism was a genuine and fundamental criticism of the régime of Nicholas I; and such was the richness (and confusion) of their ideas, the breadth and relevance of their questions, the twist that they gave to the traditional problems of Russian society, that the Slavophils had the peculiar honour of being not only the progenitors of Russian nationalism but also the grandparents of the Russian revolutionary movement. By 1848 the arguments which they had provoked had clarified the issues; their opponents went a different way but carried with them if not the ideology at any rate the attitudes of the Slavophils. Slavophilism was a seedbed of nineteenth-century political and social thought in Russia.

The principal Slavophil thinkers and writers were Khomyakov

(1804–60), the brothers Ivan and Peter Kireyevsky, the brothers Ivan and Konstantin Aksakov, and Samarin (1819–76). They were all of ancient and prosperous gentry families surrounded by numerous serfs and relatives. They lived comfortable lives, well nourished, secure, respected and even liked. The Slavophil families were interlinked by marriage; several lived as neighbours in the same suburb of Moscow. They were well educated, gifted linguists, travelled much, wrote voluminously and played little part in official life. Their interests were in theology, history, folklore, philology and indeed in anything which was or which was alleged to be specifically Russian. Two of them even died of the same disease (which was not confined to Russia) – cholera. Their insistence upon wearing clothing which they claimed to be the ancient garb of the Russian folk excited the contempt of their younger contemporaries. They sported beards of patriarchal length and volume; they sang the praises of cabbage soup (so long as it was made from Russia cabbages) and spent a lot of time on their estates totally absorbed in the virtues of (Russian) dogs and horses. Whenever they travelled abroad they wrote at length to one another, complaining about whatever country they happened to be in. They particularly disliked France and Germany but had quite a soft spot for England. Khomyakov thought that the Angles were of Slavic origin and rationalised his affection for the more ancient parts of Oxford upon the grounds of this affinity.

The Slavophils were nationalists who owed much to the romantic and idealistic German philosophers of the previous generation. Their nationalism, however, was quite different from the brand sported at St Petersburg by the official apologists of Nicholas I's government. For the Slavophils, St Petersburg was a blight and Peter the Great had been the greatest disaster in Russian history. By creating a 'German bureaucracy' he had diverted Russia from her true destiny. He had torn Russia away from her cultural roots and, by making her a rational and utilitarian State, he had deprived her of her special and individual personality and had made of her nothing but a pale imitation of a Western State. Peering fondly back into the pre-Petrine past, the Slavophils saw the image of a society which they wished to recreate in the future. This aim was a revolutionary one, for the Russia which they wished to destroy, the Russia of Peter the Great, was also that of Nicholas I. The censors did not fail to grasp the implications. Rich, respectable and 'conservative' as they were, the Slavophils were bullied by the government; Samarin even enjoyed a brief stay in the Peter and Paul fortress.

The feature of 'old Russia' which most impressed the Slavophils was the Orthodox Church. They freely admitted that, in their own time, their Church had fallen upon evil days. They were strongly opposed to the Holy Synod (created by Peter the Great) on the grounds that it made the Church too servile towards the State. In the Church, they claimed to find the perfect Christian society, a society which combined unity with individual freedom, a society in which the authority of the hierarchy was balanced by the collectively expressed will of the people, a society which was superior to the State but which had never tried to express this superiority in political form, a society which alone had preserved the spirit of Jesus Christ and the primitive Church. Contrasted with the *sobornost* or mystical Christian unity of the Orthodox Church was the divisiveness of Roman Catholicism. The Catholics had created the original schism by altering the creed; they had replaced Jesus Christ by the Pope; they had adopted the principles of legalism, authoritarianism and rationalism. They had killed faith and in its place the Westerners worshipped materialism and progress.

Closely connected with the Christian community of the Orthodox Church was the communal organisation of the Russian people. This the Slavophils did not defend upon merely economic grounds. Their claim was far higher; they alleged that it was only in the commune that the full personality of man could be realised. A peasant commune was not just a group of farmers tilling the soil together; it was in Samarin's view,

> a union of the people who have renounced their egoism . . . [it] represents a moral choir and just as in a choir a voice is not lost but follows the general pattern and is heard in the harmony of all voices: so in the commune, the individual is not lost but renounces his exclusiveness in favour of the general accord – and there arises the noble phenomenon of harmonious, joint existence of rational beings . . .'.

The remnants of this happy institution still existed in nineteenth-century Russia, but it was about to be obliterated by the twin evils of serfdom and bureaucracy. It was essential that it should be revived before Russia was subjected to the influence of Western capitalism. This prospect horrified the Slavophils. They were not opposed to technological advances – many of them were in fact improving landlords – but they hated the idea of a landless proletariat divorced from any hope of landowning. They consequently advocated the emancipation of the serfs but only if this were accompanied by the retention of the peasant commune. Slavophil opinion was harnessed by Alexander II behind the emancipation statute of 1861. This was enthusiastically received by the Slavophils; Ivan Aksakov alleged that it was 'a product of fundamental Slavic principles . . . answering to the problems of Russian history . . . we left behind us entire Western Europe with its higher culture and civilisation.' From 1861 onwards the Slavophils were more definitely aligned with conservatism. They persisted in believing, contrary to all the evidence, that the reforms of Alexander II had created the sort of society which they wanted to see.

The 'populist' trend in Slavophilism was closely linked with a distrust of the State which bordered upon anarchism. The State was artificial, external, alien; the people must be sheltered from its corrupting influence behind the wall of the commune. 'The State is in no way a preceptor. . . . Its entire virtue must consist of its negative character, so that the less it exists as a State the better. . . .' The chief job of the State was defence against external enemies. This low view of the function of the State was combined with an acceptance of the Tsarist autocracy. This was partly because they believed that it was better to be ruled by one man than by impersonal institutions; partly because they distrusted the whole Western machinery of election and parliamentary government; partly because they argued that since the power of compulsion inevitably corrupts men, the fewer who exercised this power the fewer would be corrupted. They thought the Tsar a sort of martyr to the uglier necessities of human society, 'a sacrificial symbol of self-renouncement'. The naïvety of Slavophil political thought is revealed by their hope that it would be possible to combine autocracy with freedom of the press. K. Aksakov concluded a memorandum addressed to Alexander II with the optimistic words: 'To the government, the right of action and consequently of law; to the people, the right of opinion and consequently of speech'. This irresponsible attitude towards political power was the most damaging legacy left by the Slavophils to the revolutionary populists. For at least a generation, populist thinkers continued to assert that the cause of the people could be furthered without paying any attention to the organisation of government; in fact, they alleged, it was dangerous to seize political power because its corrupting nature would inevitably divert the revolutionaries from their fundamental economic task.

The Slavophils did not stop at advocating a formula for restoring Russia to its traditional destiny: they also believed that Russia was fated to renovate the decadent society of Western Europe. In the Slavophil vision, the Russian people had a messianic duty to rescue Western man from the evil effects of rationalism, capitalism, egoism and spiritual pride. The Slavophils were convinced that the material progress of the West, far from being a sign of a superior civilisation, was in fact merely the result of one-sided specialisation. They considered that this development not only created an inhuman society in which competition and individualism had replaced 'togetherness', but also made violent and destructive revolution inevitable. The events of 1848 seemed to prove the Slavophil point. The messianic theme in Slavophilism was easily corrupted. Originally, the Slavophils had been content just to indicate and criticise the evils of Western bourgeois society; their successors, the Panslavs, thought more in terms of conquest than of the peaceful transmission of ideas. The Slavophil attitude towards the Jews – which was hostile but tolerant – was also given a sinister twist by their successors. No Slavophil would have approved of the pogroms, but it was impossible to deny the relation between the Slavophil doctrine of the absolute value of the Russian people, and the violent treatment of those who were not Russians. But the remote and ugly consequences of Slavophilism must be balanced against the fact that some of their ideas and prejudices helped to form the mind of the profoundest thinker of the period – Dostoyevsky. He is not simply to be labelled as a Slavophil: his views were insufficiently optimistic and 'populist'. He hardly concerned himself with economic and social problems but, through a genius which was entirely his own, he did reflect the Slavophil obsession with moral and religious questions, their belief in autocracy and some of their messianic fervour. His scope was far wider and deeper than that of the Slavophils, his view of man simultaneously more intense and more universal. He glimpsed the truth about the revolutionary course towards which Russian society was moving and he deplored it. But if he went further than any of the rather cautious Slavophil writers, it was always along a path which was much closer to them than to their opponents. In the end, both Dostoyevsky and the Slavophils were pitted against a revolutionary socialism which rejected the idea of God and replaced it with the prospect of creating Heaven upon earth. Dostoyevsky showed, with a clarity far beyond the pedestrian capacities of the Slavophils, that such a project would inevitably lead to the diminution of the human image and the creation of the ant-heap society. It may be strange to find Dostoyevsky in the conservative camp, but he considered that the actual alternative open to nineteenth-century Russia was likely to be even more damaging to human personality. Events may have shown that his prognostication was not without accuracy.

BELINSKY AND THE FORTIES

The great achievement of the Slavophils was that by expressing so forcefully their interpretation of Russia's destiny, they provoked a discussion and a reaction against their views. This debate took place during the forties, partly in the Moscow drawing rooms, partly in the columns of the literary reviews. Those who indulged in the polemics had to learn how to conceal their meaning from the censor; much had to be gleaned from the guarded phrase, often in a book review or theatre notice. In spite of the careful attention of the censors and the Third Section, the debate went on for a decade. It was quite unlike the frivolous discussions of eighteenth-century noblemen. True, the 'knights of the circles' were deeply influenced by foreign books. Hegel, in particular, was eagerly read and passionately debated. But the combatants were led away from mere arid academic speculation, in the manner of the German universities, because the game which they were playing had dangerous results and because they realised that their debate had the most urgent practical implications. They were concerned not with ideas alone but with ideas in relation to the destiny of the Russian people. From these enthusiastic exchanges there emerged an amazingly rich residue. The Russian realistic novel, the populist movement, the revolutionary intelligentsia – a ferment of ideas which was to influence not only Russia but all Europe – were the products of a few gifted young men who were unknown to the public even in Russia and were carefully noted only by the gentlemen of the Third Section. Of all these young men the most influential was Belinsky.

Vissarion Belinsky (1811–48) was the son of a country doctor. He was, like many members of the intelligentsia, a *raznochinets* – that is to say he belonged to none of the ranks into which Russian society was officially divided. He learned little from his stay at Moscow University and plunged straight into journalism, first in Moscow and later in St Petersburg. He suffered throughout his life from consumption, a disease which killed him young but which did not prevent him from playing an energetic and turbulent part in the debates of the Moscow 'circles'.

Belinsky's influence was felt and admitted by the whole generation of Russian writers which followed him. In his scattered and hastily-composed writings – which chiefly took the form of book reviews – he indicated the directions which Russian literature would take in its greatest age. On the one hand, he foreshadowed the nihilistic scepticism of Tolstoy; on the other the concern with the suffering individual which was at the heart of Dostoyevsky's vision. After youthful and violent love-affairs with the idealistic systems of Schelling and Hegel, Belinsky rejected all systems of thought, all attempts to explain and to rationalise, all universal philosophical structures of the sort so beloved of the German thinkers. In so doing he was certainly not led by the motives which inspired the Slavophils. Belinsky did not speak for Russia, for the beautiful Slav soul, for the alleged harmonies of the Russian tradition, for the holy mission of Russia to less fortunate lands: rather, he spoke for man, individual man, whose mysterious powers are not explained by reference to any system based upon Nature or History, who stands alone and apart from whatever 'iron laws' may from time to time be asserted to exist, whose freedom is both confined and enlarged by the fact of suffering. Man, freedom and suffering became the centre of Belinsky's thought. He became an atheist; but his confrontation of the problems raised by the Trinity of man, suffering and freedom gave his ideas a depth and range far beyond the parochial orthodoxy of the Slavophils. The themes he touched upon were later elaborated by both Tolstoy and Dostoyevsky. By the former, when he sneered at the fashionable historical theories of his day, at the preposterous belief that history is made by 'great men' or by the inevitable unrolling of non-human forces. By the latter when he emphasised the irrational tendencies of the individual, the close relation between freedom and suffering and the fact that human history is composed of individuals seeking an individual salvation.

Belinsky occupied a sphere of thought and feeling utterly different from that so comfortably inhabited by the Slavophils. He was not primarily interested in social and political questions; but it is hardly surprising that, where he touched upon them, his ideas should be in sharp opposition to those of the Slavophils. It is customary to label Belinsky a 'Westerner', and to argue that in his social and political thinking he advocated the return of Russia to a line of development already undertaken by the Western nations. It is true that he was not frightened of capitalism, that he had little use for the future development of the peasant commune, that he admired Peter the Great and that he was much interested in the writings of the French Socialists. But the distinction between Slavophil and Westerner is, in Belinsky's case, a misleading oversimplification. He understood that there was as much to hate in the West as there was in Russia. Capitalism could give rise to personal relations every bit as detestable as those which existed in Russia between serf and landowner. He had no faith in institutions as such, Western or Russian. 'A liberated Russian people would not go to Parliament', he wrote, 'but would hurry to the pub to have a drink, to smash the windows and hang the gentry.' But although he had no revolutionary programme, although it would be impossible to say what sort of social organisation he wanted to follow a revolution, there is no doubt that he thought that a revolution was a necessary precursor of the sort of revived human relationships which he wanted to see. He did not suppose that such a revolution would occur of its own accord. 'I am beginning to love mankind in the manner of Marat. To make the smallest part of it happy, I think I would exterminate the rest with fire and sword . . . people are so stupid that you have to drag them to happiness. And, in any case, what is the blood of thousands in comparison with the humiliation and sufferings of millions?' It is this belief in the saving power of revolution – almost any revolution – which is his most significant contribution to the political debate of his decade. Revolution would liberate the spirit of man from the accumulated burdens of centuries. This being so, it was almost blasphemous to predetermine what liberated man would do when the revolution was accomplished. To describe such an idea as 'Western' is absurd. Queen Victoria would have disliked it as much as Nicholas I did. Even Marx (who approved of revolution) could not accept the anarchism latent in Belinsky's political thought. But such anarchism was the inevitable result of placing man, freedom and suffering at the centre of his thought.

Perhaps Belinsky's most significant contribution to the radical tradition of nineteenth-century Russia lay in what he said about the guilt and consequent responsibility of the writer. In a society in which millions suffered and few enjoyed the material benefits which the masses provided, the writer was certainly in the ranks of the 'guilty'. But he could escape from his own guilt, he could even help to transform the unjust society in which he lived, simply by pursuing his calling in a spirit of responsibility. For Belinsky, this did not mean that the 'committed' writer must produce nothing but 'social' novels in which the 'realistically' depicted sufferings of the poor would be based upon the findings of sociological research; still less did he mean that creative writing should be turned into propaganda. The writer merely had to tell the truth. Any sort of truth would have a liberating effect. This meant that writers could not live in coteries, writing for each other and worrying about questions of form and doctrine. They must express the truth about human relationships as they in fact are and as they ought to be. Only thus can the writer break down the isolation which his personal 'guilt' imposes upon him, and help to break down the individual and class isolation which was the curse of contemporary Russia. The perception of these truths would almost certainly make the writer a rebel; without rebellion it would be hardly possible for him to break through the crust of the accepted, the stale, the false and the customary. Without such rebellion, the writer became a mere sycophant paid by the existing authorities to make the régime more attractive. Belinsky made his views upon the responsibility of the writer well known just before his death. In 1847 Gogol – whom Belinsky had previously celebrated as the foremost writer of the forties – published his *Selected Passages from a Correspondence with Friends.* Its conservative, cringing and sycophantic tone disgusted even the Slavophils. Gogol argued that whatever is, is good. This included serfdom, the Orthodox Church, cholera, poverty, the Table of Ranks and the Third Section. Belinsky's reply is couched in a language even more vituperative and bitter than his normal style, for Gogol had committed the ultimate sin of the writer: he had conformed.

> Preacher of the knout, apostle of ignorance, defender of darkness and oppression, eulogist of Tartar morals, what are you doing? . . . the Orthodox Church . . . has always been a prop for the knout and a fawner upon despotism. But Christ: why do you drag him into all this? He was the first to bring man freedom . . . and to seal and avouch his truth by suffering . . . .

Belinsky wondered whether Gogol had grasped the importance of the writer in a despotically governed State. 'Only in literature is there still life and movement, despite the Tartar censorship. That is why the vocation of a writer stands so high in our esteem. . . .' 'Were it not for the fact that you speak of yourself as a writer, who would have thought that this far-fetched, dishevelled farrago of words and phrases is a product of the author of *The Government Inspector* and *Dead Souls*.' Belinsky found a willing audience for his doctrine of the writer; the revolutionary circles contained many who believed that their duty was to serve their fellow men 'even while rebelling against them'.

HERZEN AND THE FIFTIES

Herzen was the illegitimate son of a wealthy and eccentric Muscovite. But in spite of the advantages of money and of an assured (if rather Bohemian) home life he was a critic, an outsider, 'un homme revolté' from his adolescence until his death. The details of his life, his intellectual development, his comments and opinions, should be better known than those of any other nineteenth-century revolutionary writer, for Herzen is the author of the most revealing and informative autobiography in the Russian language. This masterpiece describes not only the development of a political movement but also the triumphs, pains, pleasures and tragedies of an incomparably sensitive and intelligent man.

Before his departure for the West in 1847 – an event which marks a turning-point not only in his personal life but also in the development of populist thought – Herzen followed a path which was not unlike that of Belinsky. He formed an adolescent and passionate attachment to the memory of the Decembrists. He was present at a service held in Moscow to celebrate the Tsar's victory over the rebels; he relates that he 'swore to avenge the murdered', and with his lifelong friend Ogaryov (1813–77) he vowed to devote his entire life to this purpose. In the thirties he threw himself into the study of the French socialists and the German idealists. The circle to which he belonged attracted the attention of the Third Section; in 1834 he was arrested and exiled on the charge of 'Opposition to the spirit of the Government and revolutionary opinions imbued with the pernicious doctrines of Saint-Simon'. His exile kept him away from Moscow until 1840. It taught him an unquenchable hatred of official Russia, gave him an opportunity to explore and to reject the exalted and romantic mysticism then not uncommon in provincial Russia, and encouraged him to indulge in a romantic elopement and subsequent marriage with his cousin Natalie. She proved to be an unsuitable companion and in the end became the occasion whereby Herzen received the cruellest blow of his personal life. She deserted him for the handsome but rather ridiculous German poet Herwegh and stirred up an imbroglio which, in its ramifications, came to involve nearly all the numerous groups of Russian exiles in Western Europe.

Upon his return to Moscow, Herzen threw himself into the ideological debates provoked by the Slavophils. Like Belinsky, he is usually tagged as a Westerner; and, like Belinsky, he does not fit easily into the Slavophil–Westerner framework. It is true that he rejected both the simple faith and the pervasive religiosity of the Slavophils; indeed some of the most sensational verbal combats of the Moscow drawing-rooms were between Herzen and the Slavophil champion Khomyakov. A man as rebellious as Herzen could not tolerate the quiet acceptance of the world as it was, or the narrow nationalism of Slavophil thought. Like Belinsky, he had arrived at ideas of human freedom and dignity which made contemporary Russia far more intolerable to him than it was for the Slavophils. But he was attracted by the Slavophil teaching about the value of the peasant commune. He too was fortified by the three large tomes on this subject published by the indefatigable Prussian Haxthausen. But whereas the Slavophils saw the commune as the guarantor of order and stability, Herzen saw in it a much more dynamic principle. He believed that through the commune socialist principles, especially those preached by Proudhon, could be grafted on to the body politic of Russia. Herzen's lingering sympathy with this element of Slavophil teaching made his relations with the Westerners very difficult. He found it impossible to sympathise with the liberal historians and philosophers (like Granovsky and Botkin), who looked forward to an era of economic progress and the growth of a middle-class society. Herzen reacted violently against capitalism, seeing in it merely a fresher and more vulgar tyranny than that of the old régime. In this aristocrat there lingered a profound distaste for the jostling, competitive, utilitarian world of the bourgeois. He also disliked Granovsky's attempt to enjoy the comforts of belief in the immortality of the soul while rejecting the apparatus of Orthodox Christianity. By 1847, Herzen found his position in Moscow impossible simply because he could not agree with either party in the ideological struggle. It seemed to him intolerable that Old Russia should be dismantled merely to make room for a version of Victorian Manchester. To escape from his confusion he went abroad.

His first experience of the West was during the bloody, prolonged and apparently futile revolutions of 1848, of which he was an eye-witness, first in Italy and then in Paris. These months of high hope followed by complete despair (a despair which was heightened by the tragedies of his personal life) had a profound result upon Herzen. They brought him out of the world of the Moscow drawing-rooms and compelled him to take up a stand on the barricades. The tendencies which he had expressed in the forties became hard convictions. The spectacle of the French bourgeoisie (who claimed to be revolutionaries) hunting down and murdering the working classes of Paris in the 'June days' of 1848 revolted him. In *From the Other Shore* (1848–9) he launched a violent, bitter and impassioned attack upon the whole bourgeois world of the West. He wrote of 'the cross-eyed cretin

Louis Bonaparte', the 'walking corpse Guizot', the 'syphilitic Cavaignac'. He found in the figure of Figaro the symbol of the bourgeois world.

> In Beaumarchais' time, Figaro was an outlaw; in our time Figaro is the law-giver. Then he was poor and humiliated, snatching bits from the master's table and therefore feeling with the hungry; his laughter was full of fury. Now God has blessed him with all the gifts of the earth; he has grown flabby and sluggish; he detests the hungry and does not believe in poverty – he calls it idling and dawdling. The two Figaros have one thing in common – they are both flunkeys. . . . With us [i.e in Russia] this class is scarcely in sight; in Germany nothing else exists. . . .'

Paris in 1848 convinced Herzen that the model of the French revolution merely led to the dust and ashes of constitutionalism, democracy, majority rule, *laissez-faire* and the domination of the hated bourgeois. Bourgeois individualism is merely the right of greedy people to compete against each other; bourgeois freedom merely the right of rich men to keep and enlarge their property. The bourgeois pollutes everything he touches. 'Punch' is the highest form of art which he knows; he 'makes kitchen gardens instead of parks' and 'in place of palaces, hotels open to all, that is to all who have money'; he 'dresses up Robert Peel in a Roman toga' and inspires the proletarian with the hope that, if only he is sufficiently diligent and sober, he too can become just like his master. In his attack upon the bourgeois, Herzen not only clarified his own mind but he also left a lasting legacy to the Russian revolutionary movement.

The other development which 1848 provoked in Herzen was an increased belief in the value of the Russian peasant commune. From the streets of Paris and the outer suburbs of London (where Herzen took up his residence in 1850, explaining his choice unflatteringly: 'I had absolutely nowhere to go and no reason to go anywhere'), the Slavophil doctrines looked more inviting; at least the commune might save Russia from the blight of a middle-class society. The emancipated peasant, free to enjoy the material blessings of nineteenth-century scientific culture, free from the evils of private property and competition, cultivating his strips and living in brotherly communion with his equals, looking to the State only for defence against external enemies – it was this development of Russian society which Herzen advocated from his London home, first in the *Polar Star* and then in *The Bell*. His literary friends were hardly surprised that he had turned journalist; but many were amused by the fact that the aristocratic, sceptical, cosmopolitan, and Voltairian Herzen had placed all his eggs into the same peasant basket. Turgenev (one of the most consistent of the Westerners) complained that Herzen was worshipping the Russian sheepskin; Turgenev asserted that Russia did not differ at all from her Western European sisters 'except, perhaps, that her behind is a little larger'. But the Crimean War, Alexander II's determination to emancipate the serfs, suddenly made the exiled Herzen an important figure in Russian affairs. In 1857–61 *The Bell*, printed in London and smuggled into Russia, helped to mould the opinion which the Tsar needed as a temporary ally.

It seems at first strange that the rebellious and sceptical Herzen should, in the most effective period of his active life, be found working in co-operation with a Russian Tsar. But this was no retreat from the spirit of 1848. Herzen believed that emancipation would save Russia from a fate even worse than serfdom; he believed that if the Tsar were not assisted against the reactionary landlords, the commune would shortly be broken up by the invasion of the countryside by capitalism. Unity must be maintained among the friends of emancipation and therefore it was essential to keep the tone of *The Bell* fairly moderate. In spite of these tactical necessities, however, Herzen's tone retained some of the violence of 1848–9. He spoke ominously of the need to 'sharpen the axe'; he encouraged the students to desert the universities and to 'go to the people'; and he dealt contemptuously with the liberal officials like Chicherin who timidly supported the whole range of Alexander's reforms, even after 1861. Of them Herzen wrote: 'They will reconcile us with all that we hate and despise, and, by improving, consolidate all that should be thrown overboard.' The reality of Herzen's spirit of rebellion is proved by the growing distrust which he inspired in the Russian liberals.

But his compromising tactics during the struggle for emancipation earned him the distrust of the younger radicals. He was the father of Russian populism and yet the first generation of populists rejected him. This was partly because, as an exile, he was out of touch with Russian conditions, partly because as an aristocrat of an earlier generation he did not speak in the same language as the 'sons', and partly because as a rich man he did not appear to live the life of self-sacrifice which the radical leaders now demanded. To defend himself against the assaults of both the liberals and the radicals, Herzen wrote one of his most brilliant articles, 'The Superfluous Men and the Bilious Ones'. Of the latter he complained that they were 'all hypochondriacs and physically ill' and suffered from a 'special neurosis', 'a sort of devouring, irritable and distorted vanity'. These phrases were returned in kind by the 'sons'; and shortly before his death Herzen was viciously insulted by one of them, A. Serno-Solovevich. 'You our leader? Ha! Ha! Ha! . . . You the complement of Chernyshevsky! No, Mr Herzen. . . . Between you and Chernyshevsky there was not, and could not be, anything in common.' The Russian revolutionary movement always had a fissile tendency; its leaders seemed to be anxious to find grounds for disagreement rather than for co-operation. The quarrel between Herzen and the disciples of Chernyshevsky exemplifies the tragic waste involved in these controversies, for Herzen was obviously on the same side of the barricade as Chernyshevsky, obviously a 'bilious one' rather than a 'superfluous man'. The populist movement of the seventies was based upon ideas identical to those which Herzen had advocated during the fifties in *The Bell*. But the dynamic of the revolutionary movement seemed to make it inevitable that each generation should renounce the gods of the 'fathers'. Herzen rejected the Slavophils, the populists rejected Herzen, the Marxists rejected the populists, and the Bolsheviks rejected the more cautious Marxism of the Mensheviks.

CHERNYSHEVSKY AND THE SIXTIES

N. G. Chernyshevsky was born at Saratov. His father was a priest and many of his relations were peasants. He was destined at first for the priesthood but his father wisely decided that he was unsuited for it. He was consequently sent to the University of St Petersburg. There he rejected Christianity, became acquainted with the Petrashevsky circle, read voraciously and formed contemptuous opinions of official Russia, including the university officials. In 1850 he returned to Saratov as a schoolmaster. In spite of his squeaky voice and bizarre appearance he was a great success.

> He seemed to experience personally what he was teaching us. . . . The habitual cruelty of the teachers came to an end: they stopped beating the boys, and even the school Censor applied the birch less frequently. . . . It was sufficient for Nikolai Gavrilovich to make known his cordial relations with the helpless assistant teacher of Russian who could not control his class, to make the children stop tormenting him.

Outside the classroom, Chernyshevsky appeared to be equally fortunate. He wooed and won Olga Sokratovna Vasilieva, a doctor's daughter who was reckoned to be the finest catch which the little provincial town could offer. Chernyshevsky worshipped her for the rest of his life. He restricted his revolutionary activities so that he would bring her into no danger; he gave her that full freedom which, in radical circles, was considered to be the right of women. But all his gauche tenderness and devotion was poorly repaid. Olga Sokratovna made full use of her freedom in order to pursue her romantic interests elsewhere; she openly despised her husband's intellectual aspirations. 'How he bores me' she used to say. 'One can't ask a thing: instead of answering in two words he will start a whole conversation. Naturally, I don't listen. It's my only salvation. But he himself always listens to me, and does not get bored, simply because he likes my voice. He listens, but he does not hear what I say or forgets straight away. He couldn't possibly be interested in my pleasure trips, my turn-outs, dresses, dances, and chats with young men.'

Chernyshevsky's influence upon his pupils did not please the headmaster. He thought the young man subversive and he was probably right. 'I do and say things here, in class, which smell of penal servitude', Cherynshevsky noted in his diary. In 1853 he returned to St Petersburg with a wife and family to support. He turned to journalism and, until his arrest in 1862, his essays and articles in the *Sovremennik* (*The Contemporary*) made his one of the most influential voices in Russia.

At first his views were little different from those of Herzen. Especially after the imperial rescript of 1857, he was conscious of the need to unite the liberal and socialist opponents of serfdom. But as the shape which emancipation would take became clearer, *The Contemporary* and *The Bell* began to speak with different voices. Chernyshevsky saw that the provisions for compelling the peasants to pay a high redemption due, coupled with the failure to provide them with enough land, would make their economic condition worse than before. He also saw that the reform would not be accompanied by any political freedoms, and that the precarious freedom of the press which had existed since the beginning of Alexander II's reign would be removed as soon as the emancipation decree had been published. Herzen continued to give his unqualified support to the government and even attacked Chernyshevsky – fearing no doubt the effects of a breach in liberal–socialist unity. Their quarrel was never healed over in spite of a flying visit made by Chernyshevsky to Herzen in London (June 1859). It was further emphasised by a simultaneous quarrel between Chernyshevsky and the two aristocratic writers whose works had previously helped to make *The Contemporary* the most distinguished periodical in Russia – Tolstoy and Turgenev. Both had been stung by Chernyshevsky's utilitarian attitude towards imaginative literature. Both felt an aristocratic disdain for a man whom they felt to be their social inferior. Tolstoy was particularly scathing in his attacks. 'But what shall we do now with this gentleman who smells of bugs? I can still hear his thin voice, uttering disagreeable stupidities.' These 'disagreeable stupidities' were intelligible enough to the censorship. They were to cost Chernyshevsky twenty years of exile in Siberia while Tolstoy enjoyed a comfortable life and world renown.

The loss of his liberal allies, the realisation that the emancipation was the end and not the beginning of his hopes for the peasantry, was a crisis not only for Chernyshevsky but for the progressive forces in Russia. Chernyshevsky was personally a timid man, one, it was said, who could hardly bring himself to go into a shop to buy a bag of carrots. He was a thinker, a writer, not a revolutionary. But many of those around him, drawn to *The Contemporary* by his influence, were less cautious. Their reaction to the emancipation statute was to appeal directly to the peasants, to instigate a social war by revolutionary means. Such was Dobrolyubov, a young writer who died prematurely in 1861. In

1859 he wrote (in *The Bell*): 'You will soon see that Alexander II will show his teeth, as Nicholas I did. Don't be taken in by gossip about our progress. We are exactly where we were before.... No, our position is horrible, unbearable and only the peasants' axes can save us. Nothing apart from these axes is of any use.... Let your "Bell" sound not to prayer but for the charge. Summon Russia to arms.' Chernyshevsky could not easily echo such a crude summons. Neither could he join enthusiastically in the formation of the first secret society to exist since the liquidation of the Petrashevsky circle – Zemlya i Volya (Land and Liberty). This started to meet at the end of 1861 and most of the early members were associates of Chernyshevsky. They did not conceal their admiration for him; he was clearly the main intellectual influence behind their movement.

But Chernyshevsky moved cautiously, frightened for the safety of his wife, trying to find common ground with the liberals without betraying the cause of the peasants. He consequently welcomed the initiative of those nobles who – partly for economic reasons – criticised the 1861 statute and demanded more political freedom. They petitioned the Tsar in 1861: 'To summon delegates elected by the entire Russian nation is now the only way of solving satisfactorily the problems which have been raised but not solved by the manifesto of 19 February.' Chernyshevsky followed this with his own petition, an article called 'Letters without an Address'. In it, he warned Alexander of the danger of a peasant revolt. His words showed how far he had moved from the sentimental ideas of the Slavophils: 'The people are ignorant, dominated by primitive prejudices and by blind hatred for anything different from their own barbaric customs, make no distinction between one or the other in the class that wears different clothes from themselves. They will act against them all without exception and will spare neither our science, our poetry nor our arts, and will destroy our civilisation.' The only way to defend civilisation in Russia was for all educated men, conservative, liberal and socialist, to draw together in the exercise of power through a constitution.

But even before the article was written (the censor did not allow it to be published) the Third Section was busy collecting 'evidence' which would permit the arrest of Chernyshevsky. The government had been frightened by the disorders which had broken out after the emancipation edict. Peasants, students and nobles had shown that they wanted to go further than Alexander would permit. But Dolgoruky (Alexander II's police chief) found that his victim was too cautious for him. No evidence could be found linking him with the underground movements until Herzen unwittingly gave the opportunity. From London Herzen sent a letter inviting *The Contemporary* staff to continue their activities in England. An agent of the Third Section was present when Herzen despatched his messenger. Since *The Contemporary* had already been suspended, this gave the authorities the slender excuse which they were seeking: in July 1862 Chernyshevsky was imprisoned in the Peter and Paul fortress.

He had to wait two years before he was brought to trial; it took the Third Section that length of time to forge sufficient evidence against him. He first occupied himself with an ambitious intellectual project, a *History of the Material and Spiritual Life of Humanity*, in several volumes. But he was unable to get enough material and therefore turned to writing his novel, *What is to be Done?*. The authorities were so bewildered that they allowed the manuscript to get out of the fortress and to be published in *The Contemporary* (which resumed publication in 1863).

Through this novel Chernyshevsky influenced a whole generation of populists. It was not a literary masterpiece; its author said of himself: 'I have not a trace of artistic talent', and he was quite right. Yet it 'fired the minds of a whole generation; it was read with passion, in tattered, printed or hand-written copies. . . .' To have read it became the *sine qua non* of entry into the student 'circles'. Discussion of the novel 'ended in a brief ceremony. One had to answer three questions. First question: Do you renounce the old order? Answer: I do. Second question: Do you curse Katkov? Answer: I curse him. Third question: Do you believe in Vera Pavlovna's third dream? Answer: I do. A pair of sharp scissors then produced their harsh, energetic click and down went the luxuriant plait of hair.'

In this novel, Chernyshevsky expressed many of the ideas which he had confided to his Diary in earlier years. Its characters exemplified the life of personal freedom which the populist students in fact came to live. Both men and women enjoyed the pleasures of free love and complete equality. They lived and worked together in 'communes' in which all property was publicly owned. Their scientific interests were absorbed by lengthy discussion and the practical activity of harnessing technology to communal life. Chernyshevsky envisaged not only the happy agricultural community but also communally owned industrial plant. The State was to play a very small part in all this – it was restricted to making capital available to the communes. This would enable individuals to escape from poverty; but Chernyshevsky had no intention of overemphasising the economic side of his Utopia. Its main object was to enable people to find themselves – to escape from the necessity of being either slaves or masters, the exploited or the exploiters. The whole novel is inspired with a passionate (if clumsily expressed) belief in equality and brotherhood. This was Chernyshevsky's central affirmation. God, the State, getting rich, liberal economics, political freedom, nationalism, comfort, security, peace and happiness – all these things were to be rejected with varying degrees of ferocity. What remained was a passionate belief that man was noble, wise and creative if only society would permit him to be. To reach this goal would require revolution; the novel does not actually say so, but this is what its readers understood. But such revolution could never be the mere overthrow of the existing State. It must also include a personal rejection of all the evils of bourgeois society, of all the timidities of parental prejudices. This is why the novel appealed to young Russia. It encouraged young people to do what they already had a mind to do.

Its impact was made all the greater by the martyrisation of its author. In 1864 he was condemned to fourteen years forced labour and perpetual deportation to Siberia. His departure from St Petersburg was preceded by a special ceremony known as civil execution. He was put on a scaffold, forced to kneel and set in a pillory. A piece of cardboard on which was written 'State Criminal' was hung round his neck. A woman threw a bunch of flowers on to the scaffold and was promptly arrested. At a post-station on the way to Irkutsk, the custodian asked him what his trade had been. 'He looked at me, smiled, thought for a while and then said: "In the scribbling line, sort of clerical work".' His police escort was astonished by the gentleness of his behaviour. 'Such a fellow could be safely dispatched to forced labour on his own, without any guards. Just put him in a cart and tell him: "Drive to penal servitude".... Our orders were to escort a criminal, but we were escorting a saint.'

Chernyshevsky did little labour in the various Siberian camps where he spent the next twenty years. He lived for the most part in solitary confinement, although not altogether deprived of books. He was very ill for most of the time; he suffered from scurvy, stomach ulcers, rheumatism and malaria. Some of his companions were 'politicals' like himself. He valued their presence and through them the Russian radicals were kept informed about his fate. But for the most part he was surrounded by common criminals, men who 'would fight each other for a cabbage leaf and were treated as objects'. He remained kind, serene and courageous. His letters to his wife – who did not emulate the Decembrist women – are filled with a feeling of guilt for the distress which he had brought upon her. He sympathised with her (alleged) physical sufferings; he implored her to marry again. Of his fate he wrote to her in 1871: 'I am deeply sorry for your sake that it should be so. But I am content for my sake. And in thinking of others – of those tens of millions of paupers – I am glad that without will or merit on my part, my voice should have acquired more power and authority than before and should have enabled me to plead their cause in this way.' He never attempted to escape, although the revolutionaries tried to arrange it. The authorities at length became alarmed by the power which he continued to exert over young Russians. It was suggested that he should petition for the imperial mercy. 'I thank you', he replied, 'but, you see: exactly what am I supposed to petition for? . . . It seems to me that I have been deported because my head and that of the chef de gendarmes Shuvalov are shaped in different ways. Is this something about which one could ask for mercy?'

He was not forgotten. In 1881, the revolutionaries who murdered Alexander II made an extraordinary bargain with the government. They promised not to disturb the coronation of his successor if Chernyshevsky were allowed to return to European Russia. The bargain was grudgingly kept. Four months before his death he was allowed to return to Saratov. The perhaps excessive devotion this shy, gauche and gentle individual inspired is well exemplified by Ishutin's remark: 'There have been three great men in the world, Jesus Christ, Paul the Apostle and

Chernyshevsky'. Even Dostoyevsky, who bitterly opposed everything which he stood for, was moved to admire his character. Indeed, Dostoyevsky's fictitious world greatly resembles the real world so painfully occupied by Chernyshevsky.

PISAREV AND NIHILISM

A vacuum was created by the removal of Chernyshevsky. Young Russians continued to look for an intellectual guide, someone who would tell them what ought to be done. The need for such a figure was all the greater because, after 1862, the government restored the censorship and terrorised all those who refused to accept the official interpretation of the emancipation. The division between the world of the student and government circles became sharper. It was only with reluctance that the thinking young turned to open violence. Most continued to hope that society could be transformed peacefully; many foresaw that the adoption of capitalism would create a large number of new professional occupations which would offer the prospect of a profitable career to the educated. But the condition of Russia in the sixties did not easily permit a sensitive conscience to look to the future with optimism. The peasants were still impoverished, there was still no political liberty; it seemed very likely that Russia would absorb the worst elements of Western capitalism without even the benefit of the relatively free political structure which that capitalism had generated. Chernyshevsky's question remained pertinent to the situation of the sixties. But Pisarev, speaking directly to the needs of his generation, gave it a new twist. Instead of asking: What is to be done? he asked: Who is going to do it?

Pisarev was the son of an impoverished nobleman. His home was a typical 'nest of gentlefolk'. His mother spoiled him immoderately and he repaid her affection with uncritical obedience. 'He would never eat a sweet without permission, and when offered one by a stranger would keep it in his mouth without swallowing it until maman appeared on the scene.' When he became infatuated with his cousin, Raisa Koreneva, his mother soon chased the young lady away. But Pisarev's passion for her continued, and after she eventually married a German, he arrived at the station armed with a whip just as the couple were about to set out on their honeymoon. The German husband proved stronger than the Russian lover and Pisarev was humiliatingly thrashed before the departure of the train. As a student at St Petersburg he had decided to study philology because he liked the blue collar worn by the members of this faculty. But increasing poverty forced him to seek an income, and in 1858 he started to write for a women's magazine. His task was to describe for his young lady readers such books as would 'enrich their minds without damaging the spotless purity of their hearts' – a singularly inappropriate occupation for a revolutionary writer. The effort helped to undermine his mental stability. He tried to commit suicide and was confined in a mental institution. When he returned to St Petersburg in 1860 he became a contributor to the *Russian Word*, a journal which was even more radical than Chernyshevsky's *The Contemporary*. Pisarev entered enthusiastically into the envenomed exchanges between these two radical journals. He was much more effective with the pen than with a whip. It was not the last time that the conflicts of the radicals were to delight conservative Russia. But even Pisarev could not compete with the most violent of the *Contemporary* writers, Antonovich, who described his rivals as 'rascals, swine, rotten sandwiches, Blagosvetlov's abortions, Pisarevian lack-wits'. Much of this controversy had to be carried on from a cell in the Peter and Paul fortress, where Pisarev was lodged from 1862 to 1866. The immediate cause for his arrest was a pamphlet written in his usual bitter style:

> Not to seek revolution in the present state of affairs, one must be either infinitely dense or infinitely bribable in the service of the reigning evil. . . . Beside the people stand those who are young and willing to innovate. . . . As for the dead and the decaying, they will collapse into the grave of their own accord. Our business is merely to give the last push and throw mud on their putrid corpses.

Pisarev continued to write from prison. He smuggled his literary productions out in his mother's shoes. She bombarded those in high places with demands for her son's release; Pisarev himself dutifully insisted that he loved his sovereign and was a regular communicant of the Orthodox Church. At length he was released upon the occasion of the Tsarevich's marriage. But his nerves – never strong – were broken. In July 1868 he was drowned while on holiday at a Baltic resort. It was probably suicide.

Pisarev's influence upon the young puzzled the authorities. There was little to admire in his life; it was sad without being tragic. Nevertheless his works were not only read but studied; and 'every line serves as an occasion for heated and passionate debate'. His effect was partly the result of his literary gifts. He expressed himself with a vigour which Chernyshevsky altogether lacked. He was also a very capable purveyor of the ideas of others; for example, he introduced the reading public to the ideas of Darwin. He was above all a communicator of ideas to a generation hungry for them. Pisarev's readers did not all agree with him; but none could fail to see more clearly after reading Pisarev.

Pisarev's thought did not at first have much social content. He was never a populist, weeping over the sorrows of the peasant. Rather, he was concerned with his own sufferings which he only gradually came to associate with the shortcomings of the society in which he lived. Like Bazarov, the hero of Turgenev's *Fathers and Sons*, he preached a thorough egotism. 'To do what one pleases is to act in such a way as to be oneself at each and every moment, even while surrendering to others.' The 'social question' was primarily one of individual morality. If every man behaved according to his own temperament, this would strip away the cant and humbug which concealed men from themselves and from the injustices which their blindness permitted. This doctrine was particularly attractive to the young because it enabled them to cast away the allegedly outworn morality of their elders. To this view of ethics, Pisarev added a naïve but enthusiastic belief in the saving power of science. Bazarov's famous frog became even more celebrated because of Pisarev's treatment of this incident. There are no mysteries, nothing which cannot be explained by the scientific method, claimed Pisarev. All phenomena, from the twitching of a frog's leg to the ultimate meaning of the universe, are equally amenable to rational investigation. 'It is precisely here, in the dissected frog that lies the salvation and the renovation of the Russian people.' There was some truth in this paradox, a paradox which delighted his youthful admirers. A cool view of the structure of the Russian monarchy would certainly lead to revolution.

Views such as these earned Pisarev the attention and the

hostility of older men. They – and particularly Turgenev – hastened to give Pisarev's attitude a new name – nihilism. For the older men this meant a 'beatnik', a 'worldly, riotous, dissipated person'. The Third Section also disapproved. One report described a female nihilist: 'She has cropped hair, wears blue glasses, is slovenly in her dress, rejects the use of comb and soap and lives in civil matrimony with an equally repellent individual of the male sex or with several such.' For the old to attack the young on grounds of sexual laxity and the rejection of hygienic standards is a common enough phenomenon. But more perceptive spirits saw that nihilism went deeper than a mere fashion. It was a radical revolt against history and tradition, against all the accumulated 'wisdom' of the past, against the voices of the 'fathers', against all that seemed to be condemned by the voice of reason. It was a doctrine which the professional thinker could riddle with criticism; yet, by granting the intoxicating sense of launching out into a brave new world, it could enlist the sympathies of the young, the intelligent young for whom there was no place in post-reform Russia. In a land in which the only horizons seemed to be prison walls, Pisarev's words – which would have seemed merely absurd in Oxford or Paris – had a sinister echo: 'In a word, here is our ultimatum: what can be smashed, must be smashed. What stands the blow is good; what flies into smithereens is rubbish. In any case, hit out right and left; no harm will or can come of it.' Dostoyevsky did not make the mistake of underestimating the power of nihilism; in *Crime and Punishment* (1866) he showed what happened when a young man created his own moral world.

Only towards the end of his prison sentence did Pisarev turn to considering in detail the social applications of his ideas, chiefly in an essay significantly entitled 'The Thinking Proletariat'. Here he scornfully rejected the populist idea that the sole function of the intellectual was to wait upon the peasant, just as the midwife waited upon the termination of her client's pregnancy. He saw no fount of wisdom in the common people; he found no special destiny in the future of Russia; he did not think that the peasants would rise of their own accord and saw little good in their so doing. True to his egotistic individualism he saw, on the contrary, that the moving force in history is a conscious intelligentsia, an *élite* of persons who have cut themselves off from the ordinary concerns of ordinary men and who have devoted themselves to the cause of social regeneration. This might be achieved by political revolution; but Pisarev seems to have preferred the idea of propaganda and education. His idea of the 'superman' is not unlike that of Nietzsche. But there was one important distinction. Whereas Nietzsche foresaw his supermen wielding unlimited power over a cowed collection of inferior beings, Pisarev saw his *élite* playing the part of teachers and prophets – those marked by exclusion and suffering to accomplish the individual regeneration which he always preferred to mere social and economic change. The essential preliminary to becoming a 'superman' – or, as Pisarev would have put it, to becoming fully human – was to revolt against the past. Only the rebellious man could hope to grasp the truth. This was the trouble with the peasantry; they lived lives of dull routine, shackled by traditions and superstitions and consequently the victims of a social system which kept them at the level of rather inefficient machines. Pisarev did not descend to the vulgarities of political organisation. But his immediate successors took the hint; they transmuted his idea of a revolutionary *élite* into that of a secret revolutionary party. The step from there to the Leninist conception of party was a short one but one which Pisarev would have opposed. The future utopia did not much concern him. He welcomed capitalism not because it would give rise to a bourgeois class but because it would enable poverty to be eliminated. As a materialist he considered that poverty was the root of much evil. 'Our predicament consists in that we are poor because we are stupid and we are stupid because we are poor. To become prosperous we must become intelligent, but we cannot become intelligent because our poverty does not allow us to breathe.' He hoped that universal wealth would be achieved by the mastery of man over nature by scientific means. He failed to foresee that this would also entail the increasing power of the State over man.

But these economic and political questions were not central in Pisarev's mind. He was concerned with the quality of individuals rather than with the number of their possessions or the way in which they organised themselves. He brooded anxiously upon the heroes of the three great novels of the sixties – Bazarov, Rakhmetov and Raskolnikov. The last he rightly felt to be a fundamental criticism of his whole system of ideas. For

Raskolnikov found his salvation not through reason and rebellion but rather through suffering and supernatural grace.

RUSSIAN JACOBINISM: ZAICHNEVSKY, NECHAYEV, TKACHOV

Jacobinism was not much favoured by the radicals of the sixties and seventies because it was clearly opposed to the anarchist trend which so deeply influenced Russian thought at this time. In perspective, however, Russian jacobinism acquires some importance, for it was one of the currents which fertilised Marxism in Russia. The jacobins looked back to the French revolution. They claimed that a political revolution must precede social change. The revolutionaries must not destroy the State, as the anarchists insisted; rather they must preserve it, even strengthen it, in order to complete the revolution in the social and economic spheres. These writers claimed – like Marx – that the State would wither away once its task of social transformation was completed. But subsequent events have demonstrated that the anarchist–populists were right to doubt whether this would come to pass. In varying degrees, these jacobin writers applied themselves to a problem which could not be of interest to an anarchist: how to organise a political party which could both seize power and retain it after a successful revolution. However much the jacobins might protest their faith in democracy, their opponents rightly sensed that a party so carefully organised would be unwilling to relinquish power and would exercise it despotically when it was obtained. Distrust of the State was so deeply ingrained that the populists rejected even those who claimed that they were going to use its powers for the benefit of the hungry and the oppressed.

Zaichnevsky (1842–96) was the son of an impoverished nobleman. He went to Moscow to study mathematics but a course of extra-curricular reading in Herzen, Proudhon and the history of the French Revolution turned him into a rebel. From the first he concerned himself with the problem of communication. From 1859 he lithographed subversive works, sold them cheaply and used the money to buy a printing press. This was concealed in the provinces and was available for the publication of Zaichnevsky's most famous work, *Young Russia*, even after its author had been imprisoned in 1861. He and his companions also spent the summer months touring round the provinces and spreading their ideas among the peasants. They tried to impress them with the worthlessness of the emancipation, the complicity of the Tsar in this gigantic fraud and the need to seize by force the land which ought to be theirs by right. Numerous peasant revolts were in fact occurring at the time, and Zaichnevsky believed that a great explosion was about to annihilate the social bonds of rural Russia. 'We do not fear this revolution, even though we know that rivers of blood will flow and that perhaps even innocent victims will perish.' Zaichnevsky explained in some detail what would follow this spontaneous upheaval. Elections for a national assembly 'will have to take place under the influence of the government, which must at once make sure that the supporters of the present régime do not take part – that is, if any of them are still alive'. Universal suffrage would not prevent the government from carrying through – presumably by force – a series of socialist reforms. There would be no private property in land; it would be administered for the State by the communes. The nationalities – especially the Poles – would be given the chance to break away after a plebiscite. Education would be free, women emancipated, monasteries dissolved, industry nationalised, inheritance abolished, taxation rationalised and marriage forbidden. These plans were expressed in violent, uncompromising words. They assumed that Russia was already divided between the few 'Haves' and the many impoverished; Zaichnevsky bitterly assailed Herzen and Chernyshevsky for trying to substitute a liberal compromise for his own dramatic opposition of irreconcilable forces. Much of the rest of his life was spent in Siberia. He remained true to the ideas expressed in *Young Russia*; several of his disciples joined the Bolshevik party.

Nechayev (1847–82) was born in the industrial city of Ivanovo. His father was a craftsman and his mother a serf. In 1866 he went to work in St Petersburg as a teacher; but soon all his energies were taken up by subversive activities. He was much impressed by Karakozov's attempt to shoot the Tsar, which occurred a few weeks after his arrival. But it was not only the fact that Karakozov missed which demanded a fresh initiative from the students. Much more serious was the fact that the pistol shot was not followed by any peasant rising. In fact, it was claimed that Karakozov's aim was disturbed by a peasant who jolted his arm at the moment of firing. The authorities were delighted by this portent – the Tsar saved by the faithful peasant. The hero was fêted – until it was discovered that he was permanently drunk. But the incident posed a difficult problem for the revolutionary students. The peasants were not as easily inflammable as they had hoped; a revolutionary party would have to be formed in order to inspire and control the discontent of the people. The problem was one of organisation; and it was to this that Nechayev directed himself.

First, the revolutionary must be an absolutely dedicated man. 'Day and night he must have one single thought, one single purpose: merciless destruction.' Then he must reject the morality of the society in which he lives, he must reject love, honour and sentiment in order to further the revolutionary cause. 'For him everything that allows the triumph of the revolution is moral and everything that stands in its way is immoral.' He must be prepared for death and torture and he must in turn be willing to kill 'tirelessly and in cold blood'. Nechayev put his own principles into effect during his brief active career. He created a personal legend by relating how, with amazing skill and cunning, he had escaped from the dungeons of the Third Section. Armed with this reputation he went abroad and deceived the *émigrés* like Bakunin that he was the leader of a vast underground network. He encouraged his disciples to rob and to blackmail in order to raise funds for the party. He demanded complete obedience and when he failed to get it, or when he felt that there was a police spy in the ranks of his organisation, he was quite prepared to murder. In 1869 his suspicions fell upon Ivan Ivanovich Ivanov. The offender was invited to meet his fellow conspirators in the garden of the Moscow School of Agriculture. It was planned to disinter a typewriter which had been buried there at the time of the Karakozov affair. But in fact the conspirators murdered Ivanov – the first, but by no means the last to be executed in the name of revolutionary discipline.

Nechayev was arrested in Switzerland in 1872, and spent the rest of his life in the Peter and Paul fortress. The authorities were relieved when he eventually died (of scurvy). He had converted the soldiers of the guard into socialists – sixty-nine of them had to be arrested. Bakunin, whom Nechayev had tricked and betrayed, summarised his aims: 'For the body – only violence; for the soul – lies. Truth, mutual trust . . . exist only among a dozen people who make up the sancta sanctorum of the society. All the rest serve as a blind, soulless weapon in the hands of these dozen men.'

Peter Tkachov (1844–86) developed the trends exemplified by his jacobin predecessors. But his wide knowledge and acute intelligence enabled him to state the case more fully and more convincingly. He was an intellectual rather than a conspirator, the son of an impoverished nobleman. He spent only one long period in prison (1869–73). He escaped, went to Switzerland and died there in a lunatic asylum.

Tkachov was much influenced by the writings of Marx. He was excited by Marx's presentation of history as the conflict of rival economic forces, and applied the analysis to the situation in Russia. The result was that he found the current populist belief in a peasant-based utopia naïve and impractical. The peasantry were not a truly revolutionary force in the Marxist sense. A close examination of the economic structure of Russia revealed to Tkachov the fact that there was no revolutionary proletariat. The peasants – whom he claimed the populists idealised and misrepresented – were capable of a *jacquerie* (peasant rising) but were too ignorant, superstitious and shortsighted to carry through the political revolution which Tkachov and Marx considered necessary. If Tkachov had been nothing but a Marxist, the solution to this situation would have been simple – to encourage the growth of capitalism and to create a revolutionary urban proletariat in Russia. But the inevitable postponement of the revolution, combined with the fact that capitalist development would destroy the socialist possibilities of the peasant commune, drew Tkachov away from the application of the undiluted Marxist logic. He was impatient for revolution. He had a passion for equality, 'an equality which must by no means be confused with political and legal or even economic equality; but an organic physiological equality conditioned by the same education and common living conditions'. With such views, it was obviously intolerable to encourage Russia to adopt a course of capitalist development with the immediate effect of intensifying the inequalities between men, even though, in the end, these inequalities would give birth to a revolutionary situation.

In addition, Tkachov acutely analysed the effects of capitalist development upon the revolutionary intelligentsia.

> Life speaks to them. . . . I need agricultural foremen, technicians, industrialists, doctors, lawyers, etc. To each one of them I am prepared to give full freedom in the sphere of his own speciality and nothing more. You must help me. . . . And for all this I will give you a good and solid reward. . . . I will create conditions that correspond to your character and I will give you a feeling of satisfaction with your work and so do away with your melancholy.
>
> The demand for the intellectuals' work has increased. The 'ruined' and the 'enraged' now have the chance to escape from their ruin and to build for themselves a new, definite and secure social position.

Increasing prosperity might destroy the revolutionary ardour of the one class upon which Tkachov relied – the intelligentsia.

The part to be played by radical youth was crucial because Russia had to be jerked out of the slow march of the historical dialectic. Things could not be left to take their course; the dedicated and the intelligent had to take upon their own shoulders the responsibility for the future of the suffering and impotent masses. The revolutionary was not to regard himself as a superman; he was not to make the mistake of thinking that he created history himself; 'You go where you are driven, you are only the echo of life, the reflection of needs and dreams of practical action and daily routine.' But organisation, discipline and obedience would be necessary to accomplish the two objectives which Tkachov distinguished: first the accomplishment of the actual overthrow of the existing régime and secondly the creation of a socialist society. The members of the revolutionary party would require both the ability to destroy – if necessary by violence – and to create the new world afterwards. Such qualities would be possessed only by a very few individuals. The party must be kept small. A few individuals would have to substitute themselves not only for the masses but for the intelligentsia.

Such ideas naturally antagonised the anarchist element within the populist movement. Bakunin and his disciples expected that the revolution would lead to the destruction of the State and the dissolution of society into its 'natural' economic components. But, as Tkachov acutely pointed out, there was no guarantee that within such basic units the socialist ideas popular with the anarchists would take root. Would the peasants not continue to worship the Tsar, go to church, take the sacraments, and practise despotism over their wives and daughters? Would the anarchist intelligentsia find that, in the end, it was necessary to seize power by force in each of the peasant communes? If so, they might just as well accept the need for a single initial seizure of power. He tried to point out that, although anarchy was a phase of any successful revolution, as a goal it did not in itself solve any of the social problems. 'By not wanting the State, the anarchists would end by creating a million States.' In reply, the anarchist writers drew attention to a part of Tkachov's teaching which, as events were to show, made his plans highly dangerous to the future happiness of the human race. They asked whether the revolutionary minority would, in fact, relinquish power when their task was done. Tkachov defended himself vigorously but erroneously. 'You say: Any power corrupts men. But on what do you base such a strange idea? On the examples of history? Read biographies and you will be convinced of the contrary.' He then quoted the cases of Robespierre, Danton, Cromwell and Washington, all men whom power had not corrupted. All would be well, he concluded, so long as the minority genuinely had the interests of the people at heart. He would have thought differently had he lived through the period 1917–21.

THE POPULISTS AND THE MOVEMENT TO THE PEOPLE

Populism was the most pervasive, the most widely influential and the vaguest of the intellectual fashions which gripped educated Russians during the reign of Alexander II. It required less courage and less intelligence to be a populist than to be a nihilist, anarchist or jacobin. Anyone who felt that the countryside was beautiful, the peasant noble, and the town degrading and immoral, was in danger of becoming a populist. When Tolstoy wrote about the nobility of mowing hay, when he donned his smock and started to repair his daughter's birchbark shoes, he was adopting attitudes which appealed to the populist. At a time when industrialisation was making slow but perceptible progress, when the railways were spreading over Russia and the towns developing ugly industrial suburbs, it was natural that the members of the landowning class, economically undermined as they were, should look back with nostalgia to the happy days of rural Russia. Improved transport made the contrast between town and country all the sharper; trains brought more people into contact with the towns. The rapid urbanisation of Western Europe seemed menacing to the security of Russia. Populism had some appeal to the many unthinking people who regretted the passing of pre-reform Russia.

But the populist passion did not stop at the admiration of natural sublimities. Like Rousseau, their chief interest was in the humanity which occupied the landscape. The peasant was at the centre of their attention. This preoccupation gave them many links with conservative Russia, for the Slavophils and even the Panslavs had the same preoccupation. Some of the leading populist writers (like Mikhailovsky, 1842–1904) continued to write unhampered by the censor. Many populists gave up their revolutionary activities during the Turkish War of 1877–8.

Like the conservative Panslavs they went off with enthusiasm to fight for their Serbian brothers. Tikhomirov (1852–1923), having taken part in the assassination of Alexander II, repented in 1888 and was allowed to return to Russia. There he was received back into the Orthodox Church, and became so absorbed by his religious duties that he took to anointing his toothbrush with holy oil. His zeal was such that he was given a job in the Ministry of Internal Affairs on the recommendation of Pobedonostsev, the leading conservative. He rewarded his new masters by writing *Monarchical Government*, a work in which he praised the intelligent and enlightened manner in which Nicholas II conducted affairs. These incidents show that Marx may have been correct when he wrote: 'Indeed a *narodnik* [i.e. populist] once he has given up terrorism is easily transformed into a Tsarist'.

But although it is necessary to point out the links between populism and conservative Russia, it must be remembered that populism was a revolutionary doctrine. This was because the populists saw that emancipation, far from improving the lot of the peasant, was in fact turning him into a pauper. The commune was being undermined by the development of rural capitalism; the much vaunted egalitarianism of village life was being poisoned by the appearance of the *kulak*, or rich peasant. The traditional wisdom, serenity, stability of this unique institution was being threatened by a poison imported from Western Europe. This situation faced the populists with a dilemma. Who would defend the commune? At first, a conservative populist like Mikhailovsky thought that the salvation of the commune could be achieved by the Tsar. 'We must avert the proletarian plague which now rages in the West, and threatens to infect Russia in the immediate future.' Such a programme would have completely prevented the economic development of Russia, an objective pursued (admittedly without much energy) by the government of Alexander II. But later on in the seventies, neither Mikhailovsky nor the bulk of the populists looked to the throne to preserve their beloved commune. They had perceived 'growing up under the shelter of the Tsar's ermine robe' the foxy features of the bourgeois entrepreneur. In other words, the autocracy was already sold out to the capitalist; 'the more despotic the policeman is, the more easily can the *kulak* plunder'.

The discovery that the autocrat was no longer 'conservative' (at least in their eyes) was a crucial moment for the populists. Their objective remained the same; but they now had to adopt illegal or revolutionary means to achieve it. At the beginning of the seventies it was still possible for the movement to attract people of fairly conservative temperament, people who believed that the success of populism would merely restore the happy state of affairs which had existed before the reforms of Peter the Great, people who thought that this objective could be gained by the gradual transformation of society rather than by revolutionary violence. But after the great disappointment of the summer of 1874, these hopes were no longer tenable, and during the later seventies, populism, combined with more radical tendencies, entered upon a more violent phase.

The most influential 'gradualist' was Peter Lavrov (1823–1900). He was an artillery officer of noble birth who spent much of his professional life teaching mathematics at the Artillery Academy. He came into close contact with the underground movement only when he was banished to the Vologda region (1866–70). There he was horrified to find that his young companions were deeply influenced by the works of Pisarev; and in his most influential work, *Historical Letters* (1868–9), he set out to combat Pisarev's nihilism. This dry, abstract work apparently caused its young readers to shed hot nocturnal tears. One wrote: 'To the devil with enlightened self-interest and critical realism – to the devil with frogs and experimental science!'

Lavrov's message (further elaborated in the periodical *Vpered* (*Forward*), published in Zürich and London from 1873), was typical of a mind nurtured on mathematics. He claimed to have discovered the law which governed the development of human societies. This was that the State represented an archaic, outmoded and unprogressive stage of human development. It was destined to be replaced by the 'natural' economic groups like the peasant commune and the *artel*. The transition to this golden age could only be assured by the moral improvement of individuals. The enlightened must prepare themselves and others by individual moral improvement, study and peaceful propaganda. The students must not enter underground conspiracies, they must not plot violence, they must not leave their studies in order to play games with the police, they must not fall victim to nihilistic despair and world-rejection. They must proclaim the truth of Lavrov's law of development which, once heard, would, like a Euclidean theorem, recommend itself to the listener as an absolute truth. They must study the economic structure of society as it then existed with the purpose of showing the connection between its injustices and inefficiencies and the state organisation which corrupted it. 'The history of the Russian State is the history of the systematic economic looting, intellectual oppression and moral corruption of our country. Every progressive thing that has been done in Russia has been done against the State and everything that has come from that source has been harmful to society.' He warned his young readers against jacobinism. 'We do not want a new constraining authority to take the place of that which already exists, whatever the origin of this new authority may be.' But, against the absolute anarchy of Bakunin (with whom, like everyone else, he quarrelled), he admitted that the power of the State would only gradually be superseded by social organisation. Above all, he was anxious that his young disciples should guard the purity of their motives and morals.He frequently proclaimed that the success of the 'new order' would depend upon the moral quality of those who built it. Populism must have its saints.

These ideas suited the situation in which discontented and idealistic youth found itself at the beginning of the seventies. The prospect of an immediate peasant revolution had faded. The career of Nechayev had discredited extremism. Only a gradual transformation of society, combined with puritanical self-improvement, seemed to offer an answer to the question: 'What is to be done?'

Loosely organised groups, of which the largest was that which gathered around Chaikovsky in St Petersburg, were formed to put these ideals into practice. They satisfied the desire of Russian students to live communally. The lack of a defined programme was an advantage; it enabled the impatient anarchists like Kropotkin to work in harmony with the more cautious adherents of Lavrov. The absence of a powerful leadership, the lack of any urgent political task (like assassination), enabled the devoted students to concentrate upon the development of the beautiful moral character deemed so important in populist circles. 'We must be as clean and clear as a mirror', wrote Chaikovsky. In all the university towns, little groups of students attempted to achieve this laudable objective, one which could hardly be criticised even by the Third Section. All property was held in common; asceticism was enthusiastically practised. 'One day . . . Bardina admitted that she liked strawberries and cream and was teased by the group to which she belonged. From that day on [the group] looked upon her as bourgeois.' Women played a very large part in the groups. Some, like Sophya Perovskaya, were to pass on to sterner things. But others were more concerned with such problems as whether their ascetic code would allow them to marry. 'When . . . the programme of the new revolutionary organisation came under discussion, the girls proposed that it should include a renunciation of marriage. The men protested. . . .' Some of the male populists entered into 'platonic' marriages with nobly born girls in order to save them from aristocratic suitors chosen by their parents. Most of the students learned a humble trade both for the purpose of supporting themselves and in order to draw closer to the peasant and workman.

> The workrooms are all of the same type. At the same time they act as 'communes'. Let us go into one of them: a small wooden

> house with three rooms and a kitchen, in the district of Vyborg in St Petersburg. Little furniture, spartan beds. A smell of leather. It is a workshop for cobblers. Three young students are working there with the greatest concentration. At the window is a young girl. She too is absorbed by her work. She is sewing shirts for her comrades who for days have been preparing to go to the people. Haste is essential. Their faces are young, serious, decided and clear. They talk little because there is no time. And what is there to talk about? Everything has been decided. Everything is as clear as day.

In addition to living humbly and preparing beautiful souls, the populist circles were also much concerned with propaganda. This might take the form of merely going to live and work with the 'humble and the oppressed', not merely to inspire them to improve their economic condition but also to enable the students to develop the splendid moral virtues which 'the people' were alleged to possess. In this way a lot of information about humble life was collected – especially by Flerovsky in his *Situation of the Working Class in Russia* (1869). This very intelligent book concluded not with an appeal for revolution but with a demand that the government should abolish the redemption dues. It was hoped that such a moderate attitude would encourage the liberal middle classes to support the populists. The distribution of books also absorbed much energy. Some of them were printed on secret presses in Russia and in addition to the stock textbooks of socialist literature – Lassalle, Marx and Chernyshevsky – included some pamphlets especially composed for the peasants. The villagers were provided with free copies of such works as: *The Story of a French Peasant*, *Tale of a Kopeck* and a collection of subversive songs. They were also given brief lives of Stenka Razin and Pugachov. They thought little of these works – although there is evidence that they enjoyed 'the choral singing of one of the revolutionary hymns'. The Third Section, however, acted firmly against the amateur publishers. By the end of 1873, most of the leaders of the student circles were in prison.

This inspired the gradualist populists to a final effort. Already the extremists were arguing strongly that peaceful propaganda would achieve nothing. Some were even starting to lose faith in the revolutionary potential of the peasantry. So far much effort had produced no tangible result. In the summer of 1874 the greatest effort was made to establish a firm link with the peasantry. It was an unplanned and unco-ordinated movement, more like a conversion campaign than a revolutionary movement. Men were trying not just to reach a certain practical end, but also to satisfy a deeply felt duty, an aspiration for moral perfection. Some even joined the Orthodox Church before setting out on their mission to the people. The students set out in their 'communes'. They adopted the dress of the *muzhik* and often insisted upon working like peasants – a physical strain which was too much for many of them. They went chiefly to the traditional areas of peasant revolt, the Volga, the Don and the Dnieper. Some concentrated upon educational and medical work; others encouraged the peasants to seize the land and refuse to pay the redemption dues. They addressed the peasants as their brothers, but the peasants still failed to trust or to understand them. Beneath the assumed *muzhik* clothes they recognised the features of the educated, the urban intellectuals, members of the other Russia. When they did take fire from the numerous harangues directed at them, they tended to mistake the point made by the populists. One peasant interrupted a speaker who was describing what things would be like when the people 'owns its own land, woods and waters'; he shouted, 'That's grand! We'll divide the land and I'll take two workers and then I'll be in a fine position.' But there was hardly time for close contact with village life to sap the ardour of the students. The police moved very swiftly; four thousand were arrested or questioned. The peasants did not help them to resist; in fact the police had more trouble in extracting the fugitive students from the local manor houses. Liberal and even conservative Russia was much more impressed by the students' sacrifices.

THE SECRET SOCIETIES

The fiasco of 1874 caused the survivors to reconsider not the general aims of populism but the means by which they might be attained. Some began to see that the revolutionary possibilities of the industrial workers had been neglected. Factory life was attracting large numbers of peasants to the towns and most of them still retained strong links with the villages, to which they returned each summer. Low wages, very often long in arrears, unhealthy working and living conditions, made agitation fairly easy. The more intelligent workmen were readier than the peasants to join the active revolutionaries. At first the populists tried to make use of the workers for their own ends. The workers could carry propaganda to the villages; and the 'spontaneous' strikes, inspired with increasing frequency in the seventies by intolerable conditions, could be turned into revolutions. Consequently the populist organisers of the south Russian workers' unions which emerged in the seventies, encouraged their followers to murder their employers rather than to remain satisfied with mere wage negotiations. But by the end of the decade, especially after the union movement had made some headway in St Petersburg, it became clear that the workers, like the peasants, were not satisfied with populist leadership. The wave of strikes of 1877–9, the most extensive which had yet occurred in Russia, was led by men of working-class origin like Khalturin and Obnorsky. The programme drawn up by their organisation, the Northern Union of Russian Workers, shows the gradual emergence of a specifically working-class policy. This document continued to proclaim the old populist objectives – the destruction of the State and the creation of socialist communes. But it also demanded political rights – freedom of speech and of the press – and factory legislation which could only be the work of a powerful State. The emergence of a working-class policy was a turning point in the populist movement. It showed that a policy which was entirely directed at the peasant and the countryside was irrelevant to the needs of the 'hungry and the oppressed' in the towns. This was a reason for speedy action before the town proletariat became too powerful and numerous. The lesson of the seventies was that the town worker was more revolutionary than the peasant, but that he wanted the wrong sort of revolution.

More congenial to the populists was an attempt to arouse a peasant revolt carried out with considerable ingenuity at Chigirin, near Kiev, in 1876–7. This was an area of rising poverty and land-hunger. Troops had been used on a large scale to put down disorders. Hundreds of peasants had been sent to prison in Kiev. There they were contacted by the populist Stefanovich; he discovered that they were convinced that the Tsar knew nothing of their plight and would improve it if he did know. Stefanovich told them that he would go to St Petersburg on their behalf. He returned with a 'Secret Imperial Charter' (addressed to 'our loyal peasants') and a book called *The Statutes of the Secret Militia*. Both had been printed on a populist press in Kiev (where the organisation and fighting spirit was particularly strong) but the peasants believed that they had come from the Tsar. These documents 'proved' that although Alexander intended all the land to go to the peasants, to his distress, the nobles had kept most of it themselves. He had done his best for twenty years but now owned himself defeated. The peasants must help themselves. They must form secret bands under the command of the leaders mentioned; then followed the names of Stefanovich and his friends from the Kiev student circle. For about a year Stefanovich drilled his peasant bands but the police penetrated the organisation before a peasant rising could be accomplished. Even so, the Chigirin conspiracy was the most effective piece of populist agitation. Many populists regretted that its success depended upon the gullibility and ignorant loyalty of the peasants.

The failure of the movement to the people was attributed by the populists to the wickedness of the government rather than to the indifference of the peasants. New tactics were needed; unwillingly it was conceded that the jacobins may have been partly right. The fight against an absolute State required the weapons of conspiracy and terrorism. Comrades must be rescued from the hands of the police, the student settlements in the countryside must be protected, police informers must be eliminated, secret printing presses kept secret. The new tendency towards organisation was seen in the second Zemlya i Volya (Land and Liberty), a secret society with its headquarters in St Petersburg, which came into existence in the winter of 1876–7. Its leading figures were Natanson, A. D. Mikhailov and a southerner, Frolenko. All had passed through the phase of 'going to the people'; all were disillusioned by the results; all retained a typical dislike of powerful, centralised organisation. The structure of their society reflected their ideas. It was a federation of groups, each retaining some autonomy. But there was a central or directing group of about twenty-five members who were neither propagandists nor agitators but more like professional revolutionaries. There was a division of labour within the central group. Some, like Zundelevich, managed the underground press; others specialised in the rescue of captured comrades. But there was no dictatorial 'leader'; and the populist programme remained unchanged. The purpose of Zemlya i Volya was not to seize political power; it was so to weaken the State that the peasants would be able to carry out the economic revolution which had been so long awaited.

Evidence of a firmer spirit among the revolutionaries, evidence that they were turning away from the gradualism of Lavrov towards the jacobinism of Tkachov, was clearly shown by the demonstration in the square of Kazan Cathedral, St Petersburg (December 1876). There the red banner of 'Land and Liberty' was publicly unfurled for the first time. Only a few hundred demonstrators stood around while Plekhanov made a speech. The police treated many of them with extreme brutality. But the incident frightened the government. Nobody had publicly demonstrated in the capital since the Decembrists. It was found that the populist cause aroused sympathy among the educated and propertied. The police complained that they got no help in their anti-populist activities. The government consequently decided to enlist public opinion against these 'beatniks'. The patriotic fervour induced by the war against Turkey in 1877 was a good opportunity. The means chosen was a series of public trials.

THE ASSASSINATION OF ALEXANDER II

Official Russia was caught in a dilemma. On the one hand, bureaucrats and monied men were repelled by the militant conservatism which in the late seventies began to make itself heard from the circle surrounding the Tsarevich and of which the spokesman was Pobedonostsev. Such conservatives claimed that the reforms had been a mistake and that any advance towards a constitution would be a disaster. On the other hand, the liberals could not approve of the socialist and anarchical ideas of the populists; and any move on their part against the existing order of society was likely to be of indirect aid to the populists. Liberal Russia was consequently compelled to stand by while officialdom fought it out with unofficial Russia: the Tsar and his Third Section versus a handful of students and workmen.

The populist movement was not crushed by the fiasco of 1874.

The students continued to 'go to the people' and the more active leaders continued to seek a way to gain their objectives. It was with the object of enlisting educated opinion against the populist nuisance that the government launched the three great public trials of 1877. Officials were convinced that it was only ignorance which prevented the government from obtaining the support of public opinion. The Committee of Ministers reported: '(The populists) themselves say that torrents, rivers – a deluge of blood – are necessary to bring about their ends. . . . But for public opinion to break away from those who hold such doctrines, their principles must no longer remain unknown.' In fact the government's attempt to arouse public opinion was as futile as the populists' effort to arouse the peasantry. This was partly because the individuals put on trial were so noble and unselfish that they inspired sympathy rather than disgust; the populist cult of the beautiful soul was of some practical advantage after all. But public indifference had a deeper motive. The liberals hoped that the struggle with the populists would so weaken the government that in order to gain allies the autocrat would have to grant a constitution. Having been admitted to power, the liberals would of course turn against the socialist aims of the populists. The fear that the State would be greatly strengthened by the grant of a constitution, the fear that their own terrorist activities might help the liberals, haunted the populists during the last three years of Alexander's reign. Only a few drew the conclusion that it might be better to take a step towards socialism by way of liberalism – in other words to support the liberal demand for political freedom. But most of the populists felt that added urgency was given to their programme by the imminence of a liberal victory. The peasant revolution must be hastened so that it came before the liberals could seize power. As the periodical *Zemlya i Volya* put it: 'If we carry out a peasant policy, freedom will not be a fundamental aim but an inevitable result. . . . Who wants freedom must defend the harmless peasants. That is the way to get freedom in Russia.' The liberal merely wanted political freedom so that he could exploit the 'humble and the oppressed' the more efficiently. From 1877 to 1881, official Russia was trying to gain the sympathy of the liberals, if possible without giving them what they wanted – a constitution; the populists were trying with increasing desperation, to prevent the liberals from gaining what the autocrat had no intention of giving them – a constitution.

The 1877 trials gave the populists a public platform and a sympathetic audience. They also provided the setting for an act of terrorism, a pistol-shot as dramatic as any in a play by Chekhov, which initiated the widespread violence of 1878–81. In a prison which contained a large number of populists who were about to be brought to trial, the Governor of St Petersburg, General Trepov, who already had a long record of brutality, was insulted by one of the prisoners, Bogolyubov, who refused to take off his cap when the General passed. Trepov got authority from the Minister of Justice, Pahlen, to have Bogolyubov flogged. The sentence was carried out publicly and with such brutality that Bogolyubov went mad. Revenge came in January 1878 when Trepov was shot in his office by Vera Zasulich who had started her revolutionary career in the Nechayev circle. She made no attempt to escape and the government hoped that her trial would accomplish what all the others had failed to do. It was thought in official circles that Zasulich had been Bogolyubov's mistress, and that it would be possible to prove that her motive had been merely personal. Trial was consequently in open court with a jury. The plan completely miscarried. The jury was so strongly influenced by public sympathy for the attractive assassin that it refused to admit a fact which was hardly in question – that Zasulich had committed murder. She was acquitted; the police attempted to arrest her outside the courtroom but she escaped and fled abroad.

This was the moment when populist and liberal Russia came closest together; if the former had been less extreme and the latter less timid, a useful union of forces might have been achieved. But the honeymoon was deliberately broken by the more determined revolutionaries. In August 1878 two of them, Kravchinsky and A. D. Mikhailov, carried out the most perfectly executed of the terrorist coups. They stabbed Mezentsev, head of the Third Section, in broad daylight in the heart of St Petersburg and escaped in a carriage. The efficient clandestine press enabled Kravchinsky to explain his motives. In his leaflet *A Death for a Death* he showed that Mezentsev had not died so that the liberals should enjoy a constitution. He had been killed as an act ot revenge.

These events constituted a declaration of war. On one side the populists tightened up their organisation; the teaching of Nechayev, once so hated, was now openly followed. On the other, the government suspended trial by jury, infiltrated spies into 'Land and Liberty', got the conservative journalist Katkov to launch a press campaign against the populists, and organised bands of hooligans to disrupt student meetings and parades. The police tactics were quite successful. In October 1878 many of the leading members of 'Land and Liberty' were arrested. But Mikhailov escaped and kept his secret press going (it was hidden in a flat in the middle of St Petersburg). In addition, he got one of his disciples, Kletochnikov, into a job at police headquarters. From there Kletochnikov was able to provide him with lists of police spies and details of movements and arrests.

In spite of the disruption of the central organisation, assassinations continued. In February 1879 the Governor of Kharkov, Kropotkin, cousin of the anarchist, was murdered. The motive was again revenge. But in April a more significant attempt was made; Solovyov fired five shots at the Tsar while the latter was walking in the garden of the Winter Palace. They missed, and the poison which Solovyov took immediately afterwards failed to work. In his deposition, he explained that he had at first tried to help the people by becoming a schoolmaster, but he found that his school was attended only by the sons of the bourgeoisie. He consequently lived among the people as a railway carpenter. His experiences of the helpless poverty of the great bands of migratory workmen aroused first his pity and then his anger. He explained his motive for shooting at the Tsar as follows: 'I could think of no more powerful means of bringing the economic crisis to a head. . . . We revolutionary socialists have declared war on the government.' Clearly Solovyov's attempt was not merely revenge. It opened a new phase in the struggle. The government resorted to military rule in six of the most turbulent provinces; a special commission was set up to inquire into the reasons for the rapid spread of 'subversive doctrines'; and Alexander went off for a holiday in the Crimea. On the populist side, some tender consciences were shocked. How would the people benefit from the death of a Tsar? Was it not better to continue the work of propaganda and to reserve terrorism for purposes of revenge? Would political assassination not entail the seizing of political power by the revolutionary party? To settle these doubts 'Land and Liberty' summoned a conference to meet at Voronezh in June 1879. The fate of this congress was determined, however, by a meeting of the extremists at Lipetsk a few days earlier. There, disguised as a bathing party, they created the organisation which was shortly to murder the Tsar.

The ideological conflicts among the populists were foreshadowed by the whole development of the revolutionary movement since the beginning of Alexander's reign. The adherents of 'Land and Liberty' were not divided on the question of terrorism; all agreed that bloodshed was justified. They were not even divided about the ultimate objective of their activities: it was to secure an economic revolution for the benefit of the peasantry. All the populists continued to reject the jacobin line taken by Tkachov; they were not working for a political revolution in which state power would merely be taken over by the revolutionary party. But the Lipetsk group – who found a particularly forceful and lucid spokesman in the southerner, Zhelyabov – moved much closer to the position formerly taken by Tkachov. They argued that experience had shown the impossibility of promoting an economic revolution while the autocracy remained intact. Success could only be achieved in a freer political atmosphere. The first task was to create political liberty by launching terror against the Tsar and others in high places. For some at Lipetsk, this would be the end of the political activities of the group. The murder of the Tsar would be followed by one of two consequences: either the people would take the chance to carry through the populist economic programme, or conditions would be created in which the populists could return to propaganda in the villages. But others of the Lipetsk group – and many of their opponents – recognised that a third alternative must be faced. This was that the assassination of the Tsar would be followed by popular apathy, and that by either a conservative counter-revolution or the seizure of political power by liberals hostile to the populist aims. In this case, should the revolutionary party retain its political purity and merely stand by and permit others to enjoy the revolution which it had created? The answer given to this was really a quibble: it was, that the revolutionary party must continue to hold political power until the people had organised themselves for the economic transformation of society.

This idea was very close to Tkachov's notion of the revolutionary State using political power to force through an economic revolution. It was this possibility which divided the populists in 1879. The deepest tradition of populism was hatred of the State, even of a revolutionary State. 'Land and Liberty' was disbanded; the traditionalists organised themselves in the 'Black Partition' – a group which carried on the propaganda work in the villages. The impatient ones formed 'The People's Will' – a tiny, highly organised group of professional revolutionaries. In August 1879 they solemnly sentenced Alexander to death. The division of funds and material between the two groups was carried out quite swiftly and amicably. There was none of the lengthy wrangling and bitter denunciation which attended the strangely similar division between Bolsheviks and Mensheviks twenty-five years later.

The refusal of the 'Black Partition' to adopt the conspiratorial and authoritarian internal organisation favoured by 'The People's Will' brought quick disaster. A police spy led his colleagues to the group's printing press early in 1880. The leaders – among whom were Vera Zasulich and the future Marxists, Plekhanov and Akselrod – fled abroad.

For nearly two years, the executive committee of 'The People's Will' carried on a daring campaign in which they came increasingly closer to their single objective – the Tsar's death. Three separate efforts were made to blow up the train in which Alexander returned to St Petersburg from the Crimea in November 1879. One failed because the Tsar took an unexpected route; another because Zhelyabov made a technical mistake in assembling an electric battery; the third, which involved digging a long tunnel under the line, blew up the wrong train. But the Tsar's safe return to the Winter Palace did not discourage the plotters. Khalturin, the labour organiser and factory worker, had an advantage over his more educated companions; he was a working man and he knew a trade. He had worked as a carpenter on the imperial yacht and gained such a good reputation that, later on, he was employed in the Winter Palace. He wanted to kill Alexander with an axe; but Zhelyabov persuaded him to adopt a different method. He provided him with dynamite which Khalturin carried in past the guards in his tool basket. For weeks he slept with the dangerous material under his pillow, half-poisoned by the fumes. At length he hid it in the ceiling of the room beneath the imperial dining room. He exploded it in February 1880, killed eleven people, but left Alexander untouched.

This explosion presented the government with a fresh crisis. The capture of Zundelevich and the press which belonged to 'The People's Will' (January 1880) encouraged the government to believe that they were making some progress against the revolutionaries. The hope was proved false by Khalturin's explosion. The attempts against the Tsar's life had not aroused any patriotic reaction against the revolutionaries. A passionate debate ensued. On the one hand, the conservative Pobedonostsev argued that no concession whatever should be made to liberal wishes; 'le mot "constitution" ne devait même pas être prononcé'. On the other hand, some officials (notably Valuyev) argued that mere appeal to liberal opinion, such as had been tried in 1877, was no longer enough. The autocrat would actually have to call a constitutional assembly and would actually have to give it some power. Khalturin's explosion had once again brought forward the central problem of the reign. The tactics chosen were a compromise between these positions. Loris-Melikov, Governor-General of Kharkov, a bureaucrat with a mildly liberal reputation, was given 'dictatorial' powers to co-ordinate the rather diffuse powers of the government. The various police forces, the Third Section, the Ministry of the Interior, all passed under his control. Simultaneously, he was instructed to make some concessions to liberal opinion. He dismissed the reactionary Minister of Education, Tolstoy; he made it easier for the *zemstvos* to carry out their functions; he slightly relaxed the press censorship. But, above all, he started to talk about the possibility of a constitution – talk which was intended to reach liberal ears. Such talk had been going on since the sixties and Loris-Melikov added nothing original to the debate. But by a dramatic twist of fate – and one which gave rise to the legend that Alexander was a true liberal at heart – he got Alexander to sign a 'constitutional' document on the very morning of his murder: 1 March 1881. This contained an elaborate plan for the creation of three 'provisional preparatory commissions' which would contain a few members elected by the *zemstvos*. Their proposals for legislation would then pass to the Council of State – which would be entitled to ignore them altogether. This timid plan, which treated constitutionalism as if it were as explosive as Zhelyabov's dynamite, elicited the following comment from the Tsar: '. . . I do not hide from myself that it is the first step towards a constitution'. It was no such thing. It was propaganda for liberal consumption. It would have been repudiated as quickly as Nicholas II repudiated the constitution of 1905 if Alexander had lived. His heir repudiated it a few days after his father's assassination.

The new line taken by the government led to a temporary deviation of tactics in 'The People's Will'. Attempts were made in 1880 to counter-attack the 'liberalism' of Loris-Melikov by finding new allies among the educated classes. Considerable progress was made among the army officers, a group which had been so far neglected by the populists but which had played a leading part in the Decembrist rising. During 1880, Zhelyabov made friendly contact with several officers; some put their expert knowledge at the disposal of the conspirators and one, Sukhanov, promised that he could organise a rising at Kronstadt and get units of the fleet to bombard St Petersburg. Less successful was an attempt to arouse student anger against the new Minister of Education, Saburov. At a noisy meeting, Zhelyabov and others tried to 'heat up the atmosphere' from the back of the hall. The minister got his face slapped, but student feeling remained cool. The students, like the liberals, were keeping out of the contest; neither the blandishments of the government nor the encouragement of the revolutionaries tempted them out of neutrality.

The leaders of 'The People's Will' realised that they had neither the time nor the resources for further propaganda. Police activity started to grope towards the centre of the organisation. In November 1880 the most daring organiser, Mikhailov, was caught in a photographer's shop; in a moment of recklessness he had ordered prints of two comrades who had just been hanged. Good progress had been made in organising working-class groups in most of the big towns; it seemed likely that a successful coup would be followed by urban risings. Thus a 'now or never' mentality led to the most ambitious plans being laid.

It had been observed that the Tsar followed a fixed routine on Sundays. He walked or drove to the Catherine Canal where his former mistress, Princess Catherine lived. Although his route was often changed it nearly always took him along the Malaya

Sadova. Consequently, the executive committee rented a cheese shop there and began to dig a tunnel under the street. The suspicions of neighbouring shopkeepers were aroused; the new arrivals seemed to show so little interest in trade and they did not even know the retail price of cheese. The police were informed and they searched the house. They found nothing suspicious except a pile of fresh earth; this was explained as a means for keeping dairy produce fresh – an explanation which the police accepted. The work went ahead and explosive placed under the street. In case the Tsar failed to pass that way plans were made to bomb him elsewhere on his route. Four throwers were chosen; number one, Rysakov, was a factory worker aged nineteen. It was held that his humble birth made him particularly apt for the honour. Each thrower was equipped with a bomb newly designed by Kibalchich; after much research he had invented a bomb which was certain to explode, no matter at what angle it hit its target. When everything was already prepared Zhelyabov – who was at the centre of the conspiracy – was captured by the police; but the assassins pressed on with the attempt under the command of Zhelyabov's mistress, Sophya Perovskaya. On Sunday 1 March she played a key rôle. By waving a white handkerchief she warned the throwers that the Tsar's sledge was about to pass along the Catherine Canal. This allowed time for the four men, each carrying a bulky parcel, to station themselves. Rysakov's bomb stopped the procession without killing the Tsar; but the second bomb (thrown by a nobleman, Grinevitsky) killed both the thrower and the autocrat.

The success of the plot was in sharp contrast to its complete failure to elicit any popular reaction. The peasants thought that Alexander had been murdered by a group of noblemen who wanted the government to restore serfdom. There were no demonstrations (or counter-demonstrations) in the cities. Even the students would do little more than refuse to contribute towards a wreath for the dead Tsar. The public was numb, passive, shocked and inactive. The police recovered rapidly from their defeat. Within a few days of the assassination they had captured all the ringleaders but not before the revolutionaries had published and extensively circulated their *Letter from the Executive Committee to Alexander III*. This caused a great sensation. It was written (by Tikhomirov) as from one independent government to another. It outlined the developments of the previous ten years, and showed how a movement which had started peacefully as a campaign for social improvement had been forced into an underground, partisan war. It claimed that Russia had no 'authentic government'. But its conclusion was quite moderate. The revolutionary leaders still wanted to attract the support of the liberals. They offered to call off their warfare against the government if Alexander III would only pronounce a democratic constitution. They called only for a freely elected parliament accompanied by the usual liberal freedoms, of the press, of speech and of assembly. The *Letter* did not mention socialism. The revolutionaries had realised that socialism must lie on the other side of democracy. Although they did not know it, the *Letter* was published at the moment when Loris-Melikov was attempting to rally the liberal bureaucrats against the reactionaries who came into power with the new Tsar. Loris-Melikov wanted to press forward with the constitutional proposals which had been authorised by Alexander II on the morning of the assassination. But he lost his fight: Alexander III returned to a policy of Russian reaction.

The conspirators were brought to trial. With the exception of Rysakov (who repented of the part which he had played), they all made an impression of nobility. They were condemned to be hanged and the sentence was carried out on 3 April 1881. Kibalchich whiled away the last few days of his life by designing a flying machine. As he mounted the scaffold, his final regret was that he would not be able to see whether it worked. But at least he had the satisfaction of knowing that his bombs were effective. Sophya Perovskaya and Zhelyabov died with the same defiance which had animated their lives.

These executions broke up the directing force of 'The People's Will'. But its ideas and its adherents lived on. Its ideas were revived in organised form at the beginning of the twentieth century in the Socialist Revolutionary party. Its adherents, however, tended to drift away from populism towards Marxism. These developments are well illustrated by what happened in the Ulyanov family. Alexander Ulyanov fell in with one of the remaining groups of 'The People's Will'. He resolved to throw a bomb at Alexander III. But the plot was discovered and the elder Ulyanov was executed (1887). Alexander's younger brother, Vladimir, decided to follow the Marxist line. He wanted to revenge his brother but populism struck him as being an unscientific and reactionary doctrine. The younger Ulyanov was not, however, a Marxist like other Marxists. He took with him a dedication, an idealism and a passion which were wholly Russian and wholly of the revolutionary movement of the sixties and seventies. Lenin (as Vladimir Ulyanov later called himself) eventually avenged the death of his elder brother in the completest imaginable manner. He did so with the aid of a party which owed much to the extraordinary vitality of underground Russia in the nineteenth century.

6

The Era of Witte: Reaction, Reform, Revolution and Recovery: 1881-1907

A PERIOD OF NEW POSSIBILITIES

The Russian revolution of 1917 is an event so portentous as to cast its shadow back over the two or three decades before. It is easy, and mistaken, to examine the events of the period of the last two Romanovs (Alexander III, 1881–94 and Nicholas II, 1894–1917) as if they were nothing but the prelude to the revolution. To do so is to accept without a struggle the determinist element in Marxist historiography. It is also a mistake to be so horrified by the cruelty, obstinacy and irrational stupidity of the old régime as to think that anything – even Bolshevism – would be better than that. It is true that the last years of the Romanovs are filled with personalities and events which, with the advantage of hindsight, can be fitted into a pattern of decadence and incipient social dissolution. Most societies at most times can provide plenty of such examples. But to concentrate upon this side of the picture is to ignore the signs of rational progress towards the solution of some of Russia's most pressing problems. For during these last years of its existence, the old régime inaugurated an industrial revolution of its own upon a scale and with a tempo which, in time and without a war, might have enabled a peaceful transformation of Russian society.

The puzzling truth is that during these years Russian government and society developed rapidly and powerfully along contrasting and conflicting lines. On the one hand Western influences became much more powerful. Such influences were discernible in the decision to industrialise, a decision which is rightly linked with the name of Witte but which was shared by large sections of the ruling bureaucracy and the educated classes.

Again, Western influences played upon the wealthy and the educated in the form of a sharpened desire for a constitution, for a broadening of the basis of power and for the institution of the rule of law. Another aspect of Western influence was the spread of Marxism among the revolutionaries. As Russia became in some ways like the West, so the West became more interested in things Russian. In the years before the War, Russian literature, music, painting, poetry and ballet became known in the West and Russian artists took the lead in many forms of twentieth-century art. The great Russian cities became closely linked with their counterparts in the West in the cosmopolitan culture of the twentieth century. But this movement, in such sharp contrast with the passionate Slavophilism of an earlier generation, met with violent opposition in conservative circles. Bureaucrats, landowners and churchmen continued to insist that Russia had a special destiny for which Western Europe provided no model. Russia, in their view, did not need industry, banks, railways, middle classes, large towns or constitutions. The Russian Empire and its many nationalities would be subdued not by a common pursuit of wealth but by common subjection to one Tsar, one faith and one language. Such was the bewildering nature of political ideas in Russia that this conservative view was shared in many respects by the left-wing heirs of the populists, the Socialist Revolutionaries. They too thought that Russia had a special destiny but in their case it was a destiny in which the Holy Russian peasant occupied the centre of the stage.

Between these two extreme positions there were many intervening compromises. There was also much genuine bewilderment. In which direction should Russia go? A simple utilitarian calculation of profit and loss was not enough for the educated Russian of the age of Chekhov. Who was the heir to the Cherry Orchard? It seemed inconceivable that it should be the rough, ignorant merchant yet it was plain that some heir was needed. The historian who is puzzled by the contradictions of late Tsarist Russia can take encouragement from the fact that the most enlightened Russians of the period were equally puzzled. When so many courses of action seemed to be open, it needed a rare heroism to stick to one. Witte was one such hero, Lenin was another, and the spokesman of the right was Pobedonostsev. But such heroism required the taking up of an extreme position.

The inevitable conflict between such extremes might have been contained in time of peace; it might have been assuaged by the growing prosperity of Russia. But to subject a society undergoing such tensions to the strain of war was a dangerous gamble; a gamble which the successors of Peter the Great were, however, bound to attempt. For an unchanging element of the Petrine State was its inability to keep out of any major struggle between the Great Powers.

Up to the end of the nineteenth century, the arena of this struggle had been Europe. But during the last years of the Romanovs, the stage widened, more actors appeared. The destiny of Russia was to be settled not only in Germany and Poland but also in the Far East. Russia could become a world power and compensate herself in Asia for the humiliations endured since the middle of the nineteenth century. Was Russia to be a world power? An industrial power? An autocracy? A peasant utopia? A liberal constitutional State? Clearly, not all at once. Seen in this perspective, the problems of twentieth-century Russia are visible long before the 1917 revolutions, and those revolutions were merely a stage towards their solution.

FORCED INDUSTRIALISATION

One aspect of the new forces at work in late Tsarist Russia is clearly shown in the career and ideas of Sergei Yulyevich Witte (1849–1915). He was the last of the great statesmen of the old régime, a man of superabundant talent who found that he could act effectively even through the creaking and corrupt mechanism of the Tsarist bureaucracy. He was born into the highest circles of the administrative class – his mother was a Fadeyev, the daughter of a provincial governor – and his early career was in government railway administration. Throughout his life the railway remained near the centre of his economic ideas. In many ways he was an unorthodox bureaucrat; he was energetic and efficient and he forged close links with the growing commercial community. He married a Jewish wife and he was closely linked with many of the Jewish railway magnates. This required courage at a time when anti-semitism was becoming an increasingly powerful force in court and government circles. It helps to explain the many enemies who obstructed his plans. Witte approached the centre of power in 1889 when the Minister of Finance, Vyshnegradsky, brought him to St Petersburg as a railway expert. In 1892 Witte became by rapid promotion first Minister of Communications and then Minister of Finance. In the latter capacity he remained until 1903, making out of this Ministry a self-contained 'empire', opposed by the more conservative bureaucratic elements entrenched in the Ministry of the Interior. This 'civil war' within the bureaucracy is in itself testimony to the bewildering nature of the problems facing Russia.

Witte was not a liberal. He did not believe that Russia's destiny was to be decided by the clash of political parties in a free society. He was a loyal servant of the Tsar and without the strong support of Alexander III it is unlikely that he would have reached the top. Witte always claimed to admire Alexander who was an inflexible autocrat but a much more admirable man than either his father or his son. He was slow-witted, candid, sexually puritan and enormously strong. His death was appropriate for a successor of Peter the Great. His train was involved in a crash and the monarch fatally ruptured himself by holding up the roof of the twisted coach until his family could be extricated. Alexander gave Russia thirteen years of peace but he allowed his subjects to know that he was not intimidated by foreign powers. His remarks about Queen Victoria ('a gossiping old hag') and the Kaiser ('a lunatic') endeared him even to those who disapproved of his political inflexibility. He restored the monarchical office to some degree of respect and affection. It seemed perfectly in accordance with the old Russian ways that the Guards officers had to give up one vice (fornication) which the Tsar disapproved of and to solace themselves with another, drunkenness, with which, it was known, Alexander was in sympathy.

Alexander III was only forty-nine when his muscular feats prematurely ended his days. It is possible that his death was a turning-point. He would have backed Witte more strongly than Nicholas II did, for Alexander, in spite of his old-style Slavophil image, was more realistic than his romantic subjects realised. During the decade before Witte's arrival in St Petersburg, the first hesitant steps had already been taken towards the policy of industrialisation which Witte was to push forward so ruthlessly. Bunge and Vyshnegradsky, Witte's predecessors in the Ministry of Finance, had already resorted to foreign (and particularly

French) loans, raised a high tariff barrier against imported manufactured goods (particularly that of 1891) and increased the amount of indirect taxation designed to soak up the tiny profits of the huge peasant population. The main outline of the Witte plan – that the State should become richer and stronger with the aid of foreign capital and at the expense of its poorest subjects – was already in being before 1892. The terrible famine of 1891, which was directly caused by the government policy of taxing the peasant until he had no more food reserves, forced the dismissal of Vyshnegradsky. A stronger hand was needed, one that would push through industrialisation quickly, before the peasants were goaded beyond endurance and before conservative society should have been whipped up into a frenzied hatred of urbanisation.

The central problem of what may be called 'the Witte policy' was that of providing a market for the products of the new Russian industry. Since the peasants had to be deliberately impoverished in order to provide capital for industrialisation, they could not be expected to purchase industrial products. In part, it was hoped that improved communications with China by way of the Trans-Siberian railway would provide Russian industry with a foreign outlet. But this would take time to develop; during the critical phase of industrial 'take-off', the only possible purchaser of goods was the State itself. Consequently, capitalism developed under Witte in a manner unlike its growth in Western Europe. Although Witte was anxious to encourage habits of commercial enterprise in the Russian people, he realised that this development must be encouraged from the top. Industry must be stimulated to provide what the State wanted and what would in turn help to create a vast market in the future. The obvious answer was railways. These were desirable for both strategic and economic reasons and their building and maintenance would provide work for a native metallurgical industry. Under Witte's direction, the length of track in the Empire increased from about 30,000 kilometres in 1890 to about 60,000 in 1904. Much of this mileage went into the Trans-Siberian railway: but most of it consisted of lines in European Russia designed to link together the new industrial areas and to connect the grain-producing areas with the ports. Preferential tariffs gave the government dictatorial economic powers. By their use, for example, it was possible to make it more profitable to send grain to a distant port than to a nearby town. Under Witte, all the new lines were state-owned and most of the old ones were taken out of private ownership. He provided Russia with a railway system which was not greatly altered until his successor, Stalin, had worked it to death during the first five-year plan. The enormous scale of railway building may be judged from the fact that by 1900 a half of the total state debt of 6000 million roubles was invested in railways.

The effect upon the metallurgical industries, protected behind the 1891 tariff, was dramatic. Although the rich deposits of Ukrainian iron and coal had been exploited since the reign of Alexander II, they had lacked a lively demand, and output had not expanded rapidly. During the nineties, the coal of the Donets basin was linked by efficient rail transport with the iron ore of Krivoy Rog. Coal output in the Ukraine increased from 183 million *poods* in 1890 to about 672 million in 1900 – by far the largest source of coal in the Empire. Iron and steel production increased from about 9 million *poods* in 1890 to about 76 million in 1900. This far outstripped the total production of the Urals and made the southern Ukraine one of the great industrial areas of Europe. New towns sprang up, like Ekaterinoslav and Rostov-on-Don, populated with hordes of raw peasant workers. The new industries were generously paid at prices far above the costs of the most efficient firms. Witte was determined to make the investment of capital in heavy industry a profitable business. By such means the State could both initiate an industrial revolution and shape its progress.

Other industries which also surged ahead during the Witte era included oil extraction and textiles. Oil was made available in the Moscow region and for export to world markets by an improvement in rail and river communications from the Azerbaidjan wells around Baku. Cotton textiles (both thread and cloth) nearly doubled production during the nineties and remained, in terms of employment, the largest industry in the Empire (about 600,000 workers in 1904). Moscow, St Petersburg, Ivanovo-Voznesensk and Lodz in Poland were the largest centres. The new tariff barrier encouraged the production of a sharply increased proportion of raw cotton from Russian Central Asia.

In crude statistical form, the results of this industrial revolution were impressive. Between 1892 and 1901, the index of industrial production rose from about 31 to about 61 (1913 = 100).

The annual rate of growth was about 9 per cent, a rate surpassed only by Japan, Sweden and the United States. Much of the new industry was based upon the most modern European machinery; in Russia the rule was to encourage huge industrial units, far bigger than in other industrial nations. The reason for this was the lack of skilled management: what managers were available could be better employed when in charge of large units. In theory the Russia of Witte's time had unlimited cheap labour – it should be remembered that in the demographic upsurge of 1860–1914 the Empire's population increased from 80 to 175 million – but, in fact, not much use was made of the idle hands in the villages. High capital investment in modern machinery made the unskilled labour of the villager a drawback rather than an advantage. Besides, Witte had to take account of the still dominant landlord class, loath to see its cheap labour removed to the towns. Although Witte's policy undoubtedly led to more rapid urbanisation, this side of it should not be exaggerated. Many of the new industries, particularly mining and smelting, were carried on outside the towns. Many of the new workers were not really proletarians: they were still peasants whose closest links were with the village. It should be remembered that in spite of Witte's efforts there were still at least twice as many peasant craftsmen (about 7 million) as industrial workers at the end of the nineteenth century. Russian industry had made great strides but Russia remained predominantly a peasant land. The very old and the very new co-existed side by side (as they still do).

The money for industrial development was wrung from the peasant and charmed out of the foreign capitalist. The former had to pay both the redemption dues and sharply increasing indirect taxes upon tobacco, sugar, paraffin and matches. In addition, Witte established a most profitable state monopoly of spirits in 1894. His system made the peasant so miserable that drunkenness was the only way out: even this was turned to the profit of the State. Probably about one-fifth of peasant income was taken in taxation – a far higher proportion than in the case of any other social class. In addition, the grain exports had to be kept up in order to maintain the balance of payments and thus to keep the foreign creditor confident. During the nineties the peasant share of the harvest dwindled rapidly. Widespread starvation returned in 1899. In the meanwhile the foreign investor received very large profits. His confidence was maintained by the favourable balance of trade (broken in only one year in the period 1882–1913) and by placing the rouble on the gold standard (1897). This measure was carried through in the face of bitter opposition from the bureaucracy and its landlord allies. Witte's victory was a sign of his continuing influence at the court of the new Tsar. Nicholas II probably failed to understand the vital importance of the reform but, for the time being, his loyalty to his father's memory persuaded him to continue his support of the Witte system.

In spite of the success of his policy, Witte still needed a plentiful supply of foreign capital. Domestic resources were insufficient, especially as he was unable to prevent large increases in the expenditure upon armaments and foreign travel. The social structure of Russia forbade the use of the heroic measures later employed by Stalin: it was impossible to curtail the consumption of those classes which could exert pressure upon the government. In spite of skilful and probably dishonest budgetary devices, Witte had to borrow money abroad to maintain the favourable balance of payments. Much of the borrowed money had to be used to pay the interest on existing loans. The bulk of the foreign investment was French and it was very helpful that France had strong political reasons for investing money in Russia. French government policy, however, was to steer investment into enterprises which would show a military return, particularly strategic railways. Such lines were often of little economic value in themselves and it was the profitability of investment which chiefly interested Witte. Besides, France was politically unstable: the Dreyfus Affair might well lead to a government which could no longer afford to underwrite the anti-semitic régime in Russia. It was consequently essential to attract the private French investor whose savings were marshalled through the French banks. Such investors needed strong inducements in the form of security and high interest rates. The disadvantages of investing in Russia were numerous. Local authorities were antagonistic to foreign concerns, and an autocratic régime provided no legal safeguards against arbitrary seizure of property. To some extent Witte could overcome these difficulties by his skilful management of the French money market. Bribes were freely distributed and the French press adroitly squared. But such measures were merely tactical. In the long-term planning of foreign borrowing, it was essential to give the investor the virtual certainty that his money would not depreciate in value and that his profits would not be paid in a devalued currency. The gold standard was necessary to achieve this indispensable confidence-trick.

This aspect of Witte's policy was highly successful. French capital poured into the new Russian industries, especially the metallurgical enterprises of the Ukraine. The interest of the French investor was decisively secured. So much French money was invested that more was always forthcoming in spite of defeat in war, revolution and social anarchy. Enormous French industrial monopolies, operated from Paris by the banking houses, were created. Behind the Russian tariff, encouraged by high rates of interest guaranteed by the Russian government, French capital was able to make profits which it could not make at home. Even during the period after Witte's fall, when industrialisation proceeded at a considerably slower tempo, foreign loans could not be avoided. By 1914 foreigners owned a large slice of Russian industry – exactly how much has since been a matter of controversy. Marxist historians, anxious to prove that Tsarist Russia was little more, in economic terms, than an imperialist colony exploited by European capital, have tended to put Russia's foreign indebtedness at the highest figure. One such estimate is that during the last years of the Empire, foreigners owned 90 per cent of Russian mines, 50 per cent of her chemical industry, 40 per cent of her engineering plants and 42 per cent of her banking capital. Russia was, it seems, midway between an industrial power – by 1914 she was the fifth largest producer of iron and steel in the world – and a colonial territory like China. The dependence upon foreign capital worried patriotic Russians but Witte was unconcerned. His plans, unlike Stalin's, had never included the idea of economic autarchy. He planned for peace and for an era of international capitalism. Given peace he knew that the foreign capitalists would never permit large investments to go to ruin for lack of fresh supplies of money. The American industrial revolution had also been fertilised by huge supplies of European money, but America had luckily got far beyond the 'take-off' point before her industrial economy was tested by foreign war.

WITTE AND HIS CRITICS

Witte's system prospered and he retained his ascendancy over the Tsar until about 1900. His enemies were vociferous, influential and well-informed. They included those persons and groups whose material interests were obviously injured by the Witte system, and also most of the academic economists. Their arguments turned upon two rather different viewpoints. The more conservative critics – those who were associated with Konstantin Fedorovich Golovin – argued that Witte had forcefully developed the Russian economy along disharmonious lines. Russia was essentially an agricultural country and the way forward to industrialisation was by way of a prosperous agriculture. Cheaper metal for more agricultural machinery, less indirect taxation – this would make possible the gradual accumulation of capital in the countryside and eventually provide a market for Russian industry. Exactly the same arguments were to be heard from the right wing of the Communist party when Stalin undertook industrialisation thirty years later. Other critics – obviously more urban, more Westernised and more liberal than the Golovin circle – attacked the Witte system on the ground that it smothered capitalist enterprise. The State did too much, absorbed too much of the available capital, failed to trust the vigour of the native entrepreneur. Economic development should be along the *laissez-faire* lines which had already proved so successful elsewhere.

While Witte was able to show nothing but success in his annual budgets – always produced with a great flourish and allowed to be the subject of free debate in the press – he could afford to ignore these critics and their highly-placed friends, the Grand Dukes and Guards officers who adorned the imperial court. But at the turn of the century a series of set-backs defied even Witte's skill in public relations. A series of bad harvests (1897, 1898, 1901) underlined the problem of the hungry countryside. Peasant risings became common, especially in the overcrowded central agricultural regions. The much advertised Siberian railway, the most dramatic feature of the Witte system, was opened in 1901 but failed to attract the profitable traffic which had been forecast. Its completion involved complications in foreign policy, complications which Witte had not foreseen.

They were, in essence, whether Russia was prepared to defend, by war against Japan, that domination of Manchuria which the Siberian line made possible. Witte spoke for peace but failed to explain how peace could be had while at the same time totally excluding Japan from Manchuria. But most serious of all was a sharp industrial recession, common to all Europe, but which struck Russian industry with particular severity. In spite of heavy subsidies from the state banks, industrial enterprises were ruined and closed and thousands of workers became unemployed. This raised the spectre of urban strikes – the most serious known in Russian history occurred in 1903 – and encouraged the revolutionary parties. Student riots and political assassination again became common: Bogolepov, Minister of Education, and Sipyagin, Minister of the Interior, were murdered in 1901–2. Once again, the State and the revolutionaries were face to face and it was easy enough to cast the blame upon Witte.

The future of the Witte system now turned upon the attitude of Nicholas II. This simple, honourable, industrious, pious, obstinate and bigoted family man did not possess the intellectual equipment necessary for the analysis of the choices facing him. On the one hand, there were the voices of conservative Russia, always at his elbow. Not the least influential was that of his wife, Alexandra Fyodorovna. Like many other consorts of weak monarchs, she wanted her husband to be a man: 'Don't let others be put first and you left out.' She hated Witte because he had absorbed so much power and because he showed scant respect for the Orthodox Church which she had ardently embraced upon her marriage. She cocooned Nicholas in the protective but exclusive happiness of a strenuous family life. Even in the imperial lavatory photographs of the happy couple's numerous children reminded the Tsar of his domestic bliss. Alexandra encouraged the darker elements in her husband's character, his anti-semitism and his superstition. Under the strain of being forced to consider a radical change of policy, Nicholas turned to the consolations of religion. Unfortunately, these were administered by persons of little holiness. About 1900 one Monsieur Philippe, a French adventurer and 'spiritualist', turned up at court. He specialised in procuring the emanation of the spirit of Alexander III, a consoling presence for his puzzled son. In 1903, Rasputin took Philippe's place as charlatan by imperial appointment. In the face of growing unrest in Russia, the voice of the obscurantist court told Nicholas to discard the Witte system and resort to old-fashioned repression and nationalism.

The voice of the court was backed by that of the bureaucracy. Witte's enemies were strongly reinforced by the promotion of Plehve to the Ministry of the Interior in 1902. He rallied the bureaucratic forces which were opposed to Witte's foreign and economic policies, and advised that the mounting revolutionary pressure should be met by repression and a victorious war against Japan. Witte fought hard for his system. He was unwilling to abandon any of the essentials, tariff, gold standard and foreign loans, neither was he more ready to compromise on the peasant question. As chairman of a powerful interdepartmental committee on agriculture, Witte pronounced his view on the peasant in 1902: the commune must go and capitalist agriculture be introduced. This could be done without reducing the capacity of the peasant to pay his taxes – the aspect of the question which most interested Witte. These ideas pleased neither Nicholas nor the conservative ministers. There was no means of forming a united ministerial front. Until Nicholas decided, the bureaucrats had to wrangle and wait. It is a striking illustration of Witte's fundamental loyalty to bureaucratic tradition, even after so many years of great power, that he was willing to accept the rules of this medieval political system.

Nicholas wavered until August 1903. On the one hand he appointed ministers notably hostile to Witte; on the other he publicly congratulated Witte upon the success of his ten years in office. At length he seemed to gather confidence from participation in a great pilgrimage to a monastery lately graced by the presence of the newly canonised Seraphim of Sarov. Three hundred thousand of his pious subjects accompanied him as he helped to carry the saint's coffin. Flickering candles, massed choirs, the acclaim of the pilgrims, the enthusiasm of the clergy – all these helped to persuade Nicholas that he was in touch with the soul of his people, and that this soul was essentially conservative and religious. On 16 August 1903, at the end of his usual weekly meeting with Witte, he kindly remarked that, in view of the strain imposed by the Ministry of Finance, he intended to promote Witte to the chairmanship of the Council of Ministers. This was, it is true, promotion according to the official code, but in fact it deprived Witte of real power by removing from him a minister's right of direct personal contact with the Emperor. Although Witte himself retained much influence and was to be called upon to save Nicholas II in 1905, and although much of his economic system remained in force, the Witte era was over. Forced industrialisation, the effort to catch up with the West before the West got too far ahead, the ruthless pursuit of the good of the State at the expense of the well-being of the individual – these things had to await a sterner despot than Witte and a more centralised, fanatical, rational and optimistic totalitarianism than the medieval-style autocrat could provide.

POBEDONOSTSEV AND CONSERVATIVE RUSSIA

Witte represented one side of the Romanov state tradition: dynamic, revolutionary, modernising and materialistic. During the years of his ascendancy another side of the same tradition was sustained with equal confidence and ability by Konstantin Petrovich Pobedonostsev (1827–1907). While Witte encouraged railways and factories, Pobedonostsev emphasised the religious and moral aspects of human society. He was not, as Witte was, a worshipper of state power; rather he saw in state power a divinely appointed means for the conduct of fallen man through a transitory life. It is a striking illustration of the powerful tensions within Russian society that two such very different traditions could nurture such vigorous protagonists at the same time.

Pobedonostsev made his career as a professor of law. He had an exceptionally clear mind and wrote with authority and wide knowledge. Alexander II invited him to tutor his sons and he had a profound influence upon both Alexander III and Nicholas II. In 1880 he was promoted to be Chief Procurator of the Holy Synod, a post which he held until 1905. This gave him complete secular control of the Orthodox Church and kept him at the centre of the bureaucratic machine. Thence his personal influence radiated over the Tsars, over their choice of ministers, over the educational system, and over the political ideology of the governing classes.

Pobedonostsev's political thought was sincerely felt, profoundly learned and expressed with force and clarity. Since his time his views have become, so it seems, totally irrelevant. The contemporary student can sympathise with Witte's plans: they foreshadow what a score of governments in the modern world are trying to accomplish. It is, however, very easy to dismiss Pobedonostsev as an arch-reactionary who for merely selfish reasons defended a social system which was obviously unjust. Such an attitude would be wrong. Pobedonostsev represented a cultural tradition with deep roots in the Russia of his day. It touched upon the Christian conservatism of Dostoyevsky, the nationalism of the Slavophils, and that persistent anti-westernism which had led so many Russian thinkers of all political complexions to react against any attempt to follow the path taken by the West. Although Pobedonostsev's thinking was similar to that Official Nationality expressed earlier in the century by Uvarov, the religious side received more emphasis. Man was naturally bad and power was given by God to the autocrat to overcome, to some extent, the results of natural evil. Only worse evil could come from sharing this divine authority with elected representatives of the people. The autocrat needed two qualities, firmness and the ability to choose the right ministers. Firmness was needed because there would be constant pressure from selfish individuals for representative government. Pobedonostsev described parliamentary government as 'the great lie of our time'. With considerable skill he showed that political parties and elections are an elaborate charade to persuade the populace that it enjoys a share of political power while in fact that power is autocratically manipulated by the party leader.

> By the theory of parliamentarianism, the rational majority must rule; in practice, the party is ruled by five or six of its leaders who exercise all power. In theory, decisions are controlled by clear arguments in the course of parliamentary debates; in practice, they in no wise depend from debates, but are determined by the wills of the leaders and the promptings of personal interest. . . . It is sad to think that even in Russia there are men who aspire to the establishment of this falsehood among us. . . . We may not see but our children and grandchildren assuredly will see, the overthrow of this idol. . . .

Whoever has witnessed a democratic election might feel some sympathy with Pobedonostsev's analysis. In brief, he argued that while parliamentary government was nothing but another sort of autocracy, only eloquent, forceful and dishonest men could ascend to the top of the parliamentary ladder. In the Russian autocracy, however, the wielder of power did not have to struggle for it. He could be given a philosophic character by his tutor and could be taught to surround himself with unselfish ministers. Together with parliamentary government, Pobedonostsev rejected the idea of law, a free press, the jury system and popular education beyond the primary stage. He was a strong nationalist and approved of the Russification which was applied to the national minorities of the Empire during his period.

Pobedonostsev's influence became paramount in 1881. Some liberal members of the bureaucracy wanted to meet the assassination of Alexander II with slight concessions to liberal demands. Such were the new Tsar's first two Ministers of the Interior, Loris-Melikov and Ignatiev. In 1881–2 these men, relics of the happier days of the sixties, tried to persuade Alexander III to accept the mildest forms of representation. Ignatiev, much influenced by the Slavophils, intended nothing more than a revival of the *zemskii sobor*, the feudal representative system of the pre-Romanov era. The assembly he envisaged would have been dominated by the Church and the landowners. It would have been about 2000 strong and would have acted in a purely consultative capacity at such times as the Tsar was pleased to call it. But Pobedonostsev was horrified and hastened to put his imperial pupil on the right path: 'If the direction of policy passes to some kind of an assembly it will mean revolution and the end, not only of government but of Russia itself.' Alexander obliged his tutor by dismissing both Loris-Melikov and Ignatiev and by appointing Pobedonostsev's nominee Dmitri Tolstoy to the Ministry of the Interior in 1882. On only one other occasion before 1905 did Pobedonostsev have to exert himself to defend Russia against the horrors of a constitution. That was in 1895 when some highly respectable *zemstvo* representatives hoped to extract concessions from the new Tsar. Pobedonostsev wrote the imperial reply to these 'senseless dreams'. 'Let it be known to all that I devote all my strength to the good of my people, but that I shall uphold the principle of autocracy as firmly and unflinchingly as did my ever-lamented father.'

Pobedonostsev's hatred of representative institutions was so great that he even wanted to abolish them where they existed only in local government. He would have liked to abolish the *zemstvos* altogether but feared the results. Instead he abolished (1889) the most important official elected by the *zemstvos*, the justice of the peace and replaced him by a new official, the 'land captain'. The 'land captain' must be a member of the local nobility, must have certain property and service qualifications and, in the end, must have the approval of the Ministry of the Interior. His powers were both judicial and administrative and were sufficient to prevent the growth of that peasant self-government which had been one of the aims of the reformers of the sixties and which was essential if the peasants were to emerge from their medieval habits of thought. Although Pobedonostsev claimed that the reform of 1889 was designed to restore the landed nobility to its rightful place as the guardian of the peasant and the support of the throne, the bureaucrat gained very much more than the landowner. In spite of elaborate efforts to shore him up, like the Nobles' Land Bank (1885), the economic position of the landowner was declining beyond the power of the State to revive him. This was one of the weaknesses of the Pobedonostsev system: the autocrat, however firm, could not reverse economic trends. In spite of official hostility, the *zemstvos* actually greatly increased their activities during this period. They employed a growing number of experts – teachers, doctors and engineers – whose idealism was proof against the poor wages which the *zemstvos* offered. Such people were generally reformers who, in their professional organisations, developed a corporate strength which was turned against the conservative ideals of the bureaucracy.

Pobedonostsev was particularly concerned with education and censorship, government activities which he controlled through his nominees up to the revolution of 1905. In the field of education, he successfully procured the abolition of the University Statute of 1863. The universities were placed under the direct control of the Ministry of Education. This was followed by an effort to restrict university entrance to the children of the politically reliable classes. Neither secondary nor higher education was to be available to the 'children of coachmen, servants, cooks, washerwomen, small shopkeepers and persons of a similar type'. National minorities, and especially Jews, were also excluded from higher education. Under Pobedonostsev's guidance the number of secondary and university students increased little, if at all. The census of 1897 showed that only about 100,000 persons in the whole Empire had received a university education. This was not a good preparation for Witte's policy of industrialisation; it illustrates the contradictory effects of the divided opinions about Russia's future. In the field of elementary education, Pobedonostsev was much more active. Church schools, which provided a bare minimum of reading, writing, scripture and the study of Old Slavonic for liturgical purposes, were founded in large numbers. Together with the contemporary effort being made by the *zemstvos*, a real impression was made upon illiteracy. By 1904, about a quarter of Russian children attended an elementary school. At the same time 93 per cent of Japanese children were being educated – a measure of the superior speed at which a nation with problems similar to those of Russia was progressing.

Although the censorship laws were tightened up in 1882 – any paper which had been warned three times ran the risk of suppression – and although Pobedonostsev made no secret of his detestation of journalists and the press, this was in fact a period of surprisingly free expression. Only one paper suffered the final penalty under the 1882 law. It was possible, with caution, to discuss freely the important issues of the day. Censorship was unable to prevent the emergence of two major writers, Chekhov (1860–1904) and Gorky (1868–1936), neither of whom wrote in a manner destined to delight the autocrat. Tolstoy continued to write about Christianity and the relations of Church and State. His views were so displeasing to the censors that he was excommunicated and persons found in possession of his works were banished. Russian science moved into its greatest age in spite of the censor's activity. A period in which Mendeleyev (1834–1907), Popov (1859–1905) and Pavlov (1849–1936) were at the height of their powers can hardly be said to have been a cultural blank. It must be allowed that Pobedonostsev's censorship was not a very onerous affair. No doubt it was a misfortune for the subjects of the autocrat that they could not enjoy the amenities being currently offered by the British press (like *Tit Bits* and *The Daily Mail*): but this was a misfortune which diminishes in historical perspective.

RUSSIFICATION OF THE NATIONALITIES

Probably the most original twist given to Uvarov's political formula by Pobedonostsev and his conservative contemporaries, was the emphasis laid upon Great Russian nationalism. In the absence of other satisfactions, the dominant race was invited to enjoy its superiority over the numerous lesser breeds which inhabited the vast Empire. In part, this was merely a matter of bureaucratic convenience: it was tidier to have a single language and a single culture dominating the whole. In part, Pobedonostsev and his nationalist friends like the writer Katkov were merely reproducing in a Russian form the imperialist sympathies which swept all the advanced nations at the time. Neither Britain nor even the United States were free from the racialist prejudices which Pobedonostsev proclaimed. There were, however, two special factors in Russian nationalism. First, it was closely connected with the Orthodox Church; this combination resulted in the appearance of a particularly virulent form of anti-semitism. The late Tsarist government enjoys the melancholy distinction of being the first in modern Europe to use anti-semitism as a deliberate instrument of policy. Secondly, several of the 'lesser breeds' who felt the whip of Russification were, like the Poles and the Finns, ancient races with a long tradition and a powerful culture. They were able to react forcefully against their oppressors and to add their voices to the growing chorus of revolt against the official conservatives.

Poland still remembered the repression which had followed the revolt of 1863. Perhaps this goes some way to explain why there was not a more violent reaction to the Russification decrees of the eighties. As a result of these, Polish was almost completely displaced as the language of instruction in all Polish schools even at primary level. It had already disappeared from the administration and the law courts. But there were many other reasons why Poland remained comparatively quiet. International considerations counted for much. Although Russia was probably the most oppressive of the occupying powers, few Poles wished to take action which might lead to the increased power of Germany or Austria. After the alliance between France and Russia (1894) there was no encouragement from the traditional quarter. Besides, the industrialisation of Russia offered a profitable market to the rapidly growing Polish industries. To workers and capitalists alike, insurrection threatened a way of life which seemed to offer a comfortable future. The Polish intelligentsia was bitterly divided in its attitude towards Polish nationalism. Some Polish socialists were nationalist before they were socialists: others, like Rosa Luxemburg, believed that nationalism was an outmoded creed and that Poland was destined to become a part of a larger multi-national workers' state. It was indicative of the changed conditions since 1863 that Roman Dmowski, the leader of non-socialist Polish radicalism, advised his countrymen to seek the alliance of Russia. He thought that Germany was the main enemy of Polish nationalism. The growing rift between the three occupying powers was a key factor in Polish thought. Independence, it was thought, would come as a result of an international war – that is, so long as the Polish nationalists put themselves on the winning side. Insurrection would merely serve to reunite Russia, Austria and Germany.

Finland still enjoyed the separate status granted by Alexander I. This annoyed the Russian nationalist bureaucrats. Certain commercial interests were frightened by the competition from Finnish timber and textiles. The programme of Russification was launched in 1898 by a decree aligning Finnish military service with that of the rest of the Empire. This was followed in 1899 by another decree which declared that the Finnish Diet was merely ornamental and that Finland was nothing but a Russian province. The Governor, Bobrikov, pressed on with the usual measures of Russification in spite of growing resistance. Russian officials were introduced and Russian made the language of administration. But Finnish resistance, unlike that of Poland, was single-minded. All classes and parties, even the old Swedish aristocracy, united against the Russians. Sibelius wrote nationalistic music. In 1904 Bobrikov was assassinated and Finland had to be held down by force. Both in 1905 and 1917 it provided a highly convenient hide-out for revolutionaries who sought a brief rest from their troubles in the imperial capital.

Elsewhere in the Empire, the Armenians, Ukrainians, Lithuanians, Latvians, Estonians, Baltic Germans, and Tatars felt, in varying degrees, the hand of Pobedonostsev. The Armenians had always been the most loyal of the Caucasian peoples. Now they suffered because they proffered aid to their co-nationals in Turkey. The Russian government would not tolerate any form of insurrection, even one outside its own borders. The centre of Armenian national life, the Gregorian Church, was suppressed (1903) and the inevitable Armenian reaction crushed by a novel sort of pogrom. The natural enemies of the Armenians, the Tatars of Azerbaidjan, were encouraged by the Christian imperial government to slaughter the Christian Armenians. This they did to some effect in Baku (1905). Another race notably loyal to the dynasty, the Baltic Germans, was also diverted from its allegiance by attacks upon its religion (Protestant) and language. Pobedonostsev took particular interest in a great new Orthodox cathedral built under his direction in Riga.

ANTI-SEMITISM

It was in its attitude towards the Jews that the imperial government showed the greatest perversity. There was already a powerful anti-semitic tradition in Russia before the reign of Alexander III. From the beginning of the century the Jews had been forced to live in the fifteen provinces of the Pale – that is, in south-western Russia. They had been subjected to humiliating conditions designed to separate them from the rest of the community. Only the richest merchants could travel to the capitals: the Jews could not employ Christian servants: they could not use the Hebrew language; they were not supposed to own land or to work on it. They were constantly faced by the hostility of local officials, landowners and peasants. It was fatally easy to blame poverty, starvation or misfortune – and such things were common enough in Tsarist Russia – on the Jews. These laws forced a lopsided social development on the Jewish communities. There was no peasantry. Instead, there was a swollen number of shopkeepers and moneylenders and a growing number of discontented intellectuals for whom nothing could be worse than the Russian State. The situation deteriorated still further during the predominance of Pobedonostsev. The murder of Alexander II was widely and falsely attributed to the Jews, and the government at first did little to protect them from the pogroms which broke out in Kiev, Odessa and elsewhere in 1881–2. Fresh restrictive laws were added to the 650 already in force. Industrial cities like Rostov-on-Don were excluded from the Pale; quotas prevented Jews from getting higher education; they were excluded from the Bar and from *zemstvo* elections; large numbers of Jewish artisans were expelled from Moscow. But still more sinister was the growth of a new sort of virulent anti-semitism in part religious but, to an increasing extent, racial in tone. The old-fashioned anti-semite, like Alexander III, had been concerned mostly by the fact that 'it was the Jews who crucified our Lord and spilled his precious blood'. For such people, conversion to Christianity would mean that a Jew ceased to be a Jew. But during the last years of the century certain extreme nationalists, well supported both at court and in the bureaucracy, began to move towards a new formulation of anti-semitism. Just as the ignorant peasant had for a long time blamed the misery of his life upon the Jews, so the nationalists would blame national disasters on the same cause. Attention could be diverted from such disasters by officially inspired pogroms.

In this combination of anti-semitism, nationalism and mass appeal, lies the birth of modern fascism – and its birthplace was Holy Russia. These ideas were first given effective political form in the Union of the Russian People, otherwise known as the Black Hundred, founded in 1905 through the energy of a former official, Vladimir Mitrofanovich Purishkevich. Nicholas II was for a time proud to wear the button badge of this organisation which openly baited Jews, organised pogroms and whipped up the hatred of the masses by arranging 'ritual murders'. Purishkevich and his associates were equipped with a piece of anti-semitic propaganda which later became the centrepiece of German Fascist anti-semitism. This was *The Protocols of the Elders of Zion*, a work which probably appeared first in 1895 and which may well have been written by, or with the connivance of, the Russian political police. The book is a farrago of nonsense which would have needed an ignorant, uncritical and fanatical mind to believe it: there were many such in late imperial Russia. The book explains that a vast Jewish machination is about to promote a Jewish world government, presided over by a King of the seed of David. To secure their aim, the Elders of Zion work through any organisation likely to cause disorder – trade unions, socialist parties, a free press, and revolutionary activity. If the gradual evolution of the plan is interrupted by its accidental discovery by some non-Jew, then the Elders of Zion will spring their master plan. All the large cities in the world have been mined with underground railways and, at the touch of a button, the Elders will blow up all the paraphernalia of civilisation. To such depths had the official conservatism of Pobedonostsev sunk. In this fantasy world, it was hardly possible for sane political decision to be taken. Faced by the accumulation of so much ignorant hatred, the Russian Jews had two alternatives. They emigrated if they could and, if this was not possible, they joined one or other of the revolutionary groups. Such groups contained a large proportion of Jews. To the new wave of pogroms which began at Kishinyov in 1903 with the connivance if not the support of the government, the Jewish intelligentsia replied not only through their own Social Democratic (Marxist) organisation, the Bund, but also through passionate membership of the other underground movements.

EMERGENCE OF POLITICAL PARTIES: THE LIBERALS AND THE SOCIALIST REVOLUTIONARIES

The growing resistance to the autocratic system of which Witte and Pobedonostsev were the pillars, took two main forms, economic and political. Economic discontents fanned the poor peasant and the oppressed urban worker into demands for an immediate alleviation of their lot. But although the protest of the masses caused the government much annoyance and even, in 1905, sparked off a revolutionary enthusiasm which went far beyond the basically economic demands of the poor, such opponents, however numerous and however incensed, could never be a match for the resources of the Tsarist State. Far more serious was the growing hostility of educated Russia, the demand from the professions and from a segment of the gentry for a share in political power. Russian liberalism was already half a century old by 1900. But at the turn of the century it had taken on a spirit quite different from that of the 'fathers' of the sixties. Then, the liberals had hoped for reform by means of the Russian tradition of reform from the top. When Alexander II stopped reforming, the liberals were lost. But in the new century the liberals, or at least some of them, were in a less passive frame of mind. They were willing to provoke revolution, or certainly to make use of it should it occur, in order to force the autocrat to disgorge his power. The ultimate pathetic fate of Russian liberalism in 1917 should not disguise the fact that in 1905 it was the most formidable opponent of the régime.

Russian liberalism in the era of Milyukov (1849–1943) and Maklakov (1870–1957), two of its most distinguished practitioners, was composed of three groups. First, and in the Russian scene of great social weight, the *zemstvo* gentry. This was a class which had seen its power eroded by the bureaucratic centralism of Pobedonostsev and which disliked Witte's industrialisation. But as a class the gentry were the most conservative of the liberals. They were a country party, demanding local independence and a return to those economic conditions in which, it was hoped, landlord agriculture could flourish. Secondly, the most radical element in Russian liberalism was the group of technical experts employed by the *zemstvos*. These teachers, doctors and statisticians were young, badly paid and idealistic. They were the heirs of those populists who had 'gone to the people' during the reign of Alexander II. These people saw that their work was hindered at every turn by a hostile, obscurantist and corrupt bureaucracy. Thirdly, the group which provided the leadership of the movement was the urban professional class, frequently men of great wealth and culture. The advance of urban and industrial life and the increase in wealth was making such men more numerous. They were used to meeting their less exalted *zemstvo* colleagues at the numerous professional congresses which during the nineties helped to take the place of political life in Russia. Such bodies aroused the suspicion of Pobedonostsev who, as an example, closed one of the most illustrious, the Moscow Law Society, in 1899. It was from among the urban liberals that there came the first move towards setting up a liberal party in Russia. It had to be underground. It was called the Union of Liberation and its first clandestine meeting took place in 1904. Its aims were strongly radical – far too radical for some of the vaguely liberal *zemstvo* gentry. It demanded a democratic constitution based upon universal suffrage and national self-determination: clearly the Union was seeking to make friends on the left.

At the same time as Russian liberalism was preparing for action, the left wing was organising itself upon even more revolutionary lines. The descendants of the populists organised themselves as the Socialist Revolutionaries (1901). Their ideology (ably expressed by Victor Chernov) had changed little. The peasant would be encouraged to revolt, society would collapse, private landholding would be abolished and the land would be held in common by those who tilled it. This was a creed which commanded great enthusiasm among certain uprooted intellectuals (including many women) and, to some extent, appealed to the peasants too. Chernov and his allies propagandised the areas of strongest peasant discontent, the central and Volga provinces, and reported much success. A special secret section of the party, descended from 'The People's Will', specialised in political assassination, a romantic occupation which attracted a number of gifted individuals. The S.R. combat unit claimed several distinguished victims – Plehve and the Grand Duke Serge being the choicest. The assassins were usually caught and revealed themselves to be persons of incoherent idealism, obsessed by the notion that state power against which they fought was so absolutely evil that all methods were justified. These S.R. fanatics, Maria Spiridonova, Kalyayev, Balmashev, are of great interest as individuals; among their ranks they contained one of the most successful Judas figures, Yevno Azeff (1869–1918), who for many years was privy to the secrets of both the police and the terrorists. So skilfully did he conceal his tracks that it is still impossible to say which side he mainly betrayed. Although the biographer can find much interest in these people and although the Tsar's ministers had good reason to fear them, the S.R.s were not an important political force. In spite of their numbers and fanaticism, their doctrine was too archaic and their organisation too undisciplined to constitute a serious threat to the State in 1905.

The other left-wing party, the Social Democrats, is described in Chapter 10. Its leaders were mostly in exile and mainly concerned with doctrinal wrangles at this time. In spite of its small numbers, it produced individuals who played an important part in the events of 1905. The urban workers, among whom it was strongest, owed little to its leadership at this time. For the S.D.s, 1905 was a time to learn rather than a time for effective action.

For the liberals, the existence of these left-wing parties was an embarrassment. The already timid conservative liberals were paralysed by the fear that their actions might weaken the autocracy. Power would then fall, not to them but to the leaders of the masses who held socialist doctrines far removed from liberalism. For such timid spirits, progress could only be made with the co-operation of the autocrat. Only by a process of evolution could the revolution which would sweep away everything that the liberals valued, be avoided. On the other hand, if the liberals made no attempt to ally with the forces of the left, their threat to the autocrat would be so slight that he could afford to ignore them. This was the liberal dilemma: how to force the autocrat to yield without destroying the whole basis of society. Caught between the rival rigidities of left and right, the liberals were in an impossible position. The need to compromise, either towards the left or towards the right, dominated liberal policies during the revolutionary period 1904–7.

THE REVOLUTION OF 1905

Plehve's 'little war' against Japan (and against the revolution) started in January 1904. It aroused no patriotic emotion: it was too far away and too obviously unconcerned with vital national interests. The first defeats (April–May 1904) served to encourage the hordes of discontented. The Poles began a series of strikes and other disorders which, during the whole revolutionary period, kept at least 250,000 Russian troops in garrison. But it was the murder of Plehve in July which touched off the wave of political activity which reached a climax in October 1905. In the face of mounting disorder in the non-Russian borderlands, in the countryside and on the city streets, the autocrat was forced to move very slowly towards an accommodation with his most moderate enemies, the liberals. But Pobedonostsev had done his work well. Nicholas showed a powerful resistance to any concession. When one was made, it was either worthless or insincere. The latent power of the autocrat was clearly illustrated in 1905 even when society seemed to be united against him.

Plehve's death was followed by an official effort to wean the conservative *zemstvo* liberals away from their more radical allies. Mirsky, Plehve's less reactionary successor, permitted the *zemstvo* men to meet in a national conference and to present their mild requests to the sovereign. These included the traditional liberal freedoms but stopped short of the political machinery which would guarantee them. That is, the *zemstvo* liberals did not demand a constituent assembly but merely a representative body which could advise the Tsar if he so wished. Had he taken the milder liberals into his confidence at this point, Nicholas might have avoided much further trouble. But, after granting a few worthless concessions to *zemstvo* activities, he told them to mind their own business. This helped to keep the conservative liberals in line with their more radical friends for the time being.

The loss of Port Arthur followed by 'Bloody Sunday' (December 1904 and January 1905), provided the emotional drive necessary to fuse together most sections of the rebellious in Russia. The first revealed the incompetence of the autocracy: the second its heartless cruelty. Faced by this double blow, the autocrat hastened to make peace with the external enemy so that he could the better eliminate the more dangerous enemy at home.

The economic conditions of the Russian proletariat which burst so dramatically on to the political scene in January 1905 were terrible. Wages were low, accidents were frequent, trade unions were forbidden and life in the huge factory barracks was overcrowded, insanitary and brutish. But to all these evils the Russian worker, for the most part a peasant by birth, ought to have been inured. It seems, however, that in Russia, as elsewhere, a rudimentary political consciousness was fostered by the disciplines of factory and urban living. The opportunities of organisation were greater; the contrasts between rich and poor were stronger; the prospect of success was brighter. Large-scale strikes had been common for about ten years before 1905. They had, on the whole, been brutally suppressed. Witte did not intend to jeopardise the success of his industrial revolution by allowing the workers to gain higher wages. Neither was he willing to go to the expense of a Bismarckian code of state socialism. Apart from repression and a few labour laws (which were chiefly concerned with children), the government also tried an experiment peculiarly Russian. That was to set up, under police control, trade unions in which the workers would be given a basic education and encouraged to turn their grievances away from politics and towards the improvement of their economic status. It was to this police–unionist movement that Father Gapon, the leader of the workers' movement on 'Bloody Sunday', belonged. Gapon formed a remarkably successful police union in 1904. At the end of the year he suddenly found himself the uncrowned king of the St Petersburg workers. He, not the left-wing parties, controlled the 100,000 strikers who proposed, with their families, to march peacefully upon the Winter Palace in order to present their grievances to the autocrat. Gapon warned the Minister of the Interior that the demonstration was imminent. Presumably the authorities intended from the start to teach the workers a lesson. The orderly crowds, dressed in their Sunday clothes, streamed across the Neva carrying portraits of the Tsar and religious banners. Some were singing hymns. This was quite in the tradition of Old Russia. The good Tsar would relieve his humble people from the oppressors, of whose wickedness he, in his remote glory, knew nothing. But the soldiers of the good Tsar met the demonstrators on the respectable side of the Neva and, without warning, charged them with drawn sabres. About 1000, including women and children, were killed and wounded.

This cruel slaughter quickened the intensity of disorder throughout the Empire. In the towns there were strikes, in the country peasant revolts, and in the borderlands national risings. The liberals moved leftwards. In the Union of Unions the more radical professional men sanctioned the use of terrorism to achieve constitutional reform; and even the *zemstvo* men moved away from their original idea of a merely legislative assembly towards the more radical notion of a constituent assembly. In the Far East, the naval disaster of Tsushima followed the defeat at Mukden. The government hesitated to bring back the demoralised troops for fear that they would infect the European population. In the face of incipient anarchy, Nicholas hastened to make peace with Japan (August 1905) and attempted to drive a wedge between the conservative and radical liberals by the Bulygin rescript (the details of which were made public in August). This promised a consultative assembly elected on class lines without any vote being given to urban workers or professional men: it could be summoned and dismissed by the Tsar at will; its deliberations could be ignored if the Tsar so desired.

But the Bulygin rescript did not have the desired effect. It was rejected even by the conservative liberals. The government had obviously ceased to govern. In the universities, in the press and in the streets, political opinions were being freely expressed for the first time in Russian history. The next push came again from the urban workers. In October 1905 the railwaymen started a national strike and this rapidly became general throughout Russian industry. Witte, who had hurried home from peacemaking with Japan, now advised the Tsar that he must either concede the liberal demands or set up a military dictatorship. No candidate could be found for this office. The Grand Duke Nikolay Nikolayevich told the Tsar that he would shoot himself if forced to undertake it. Eventually in October the Tsar capitulated. In the manifesto of 17 October he granted the fundamental personal liberties, promised to enfranchise the social groups excluded in the Bulygin rescript and to give the Duma (parliament) legislative powers.

These concessions still left him with plenty of elbow-room. But they were sufficient to split the liberals. With varying degrees of enthusiasm they agreed to work within the lines laid down by the October manifesto. To the right wing, this was as far as they wanted to go: to the left, it provided a fulcrum with which other concessions could be levered. But for the working men of Moscow and St Petersburg, the October manifesto was not far-reaching enough. It made no promise of social reform and they had no hope of dominating the legislative assembly. It seemed to them that after the workers had breached the walls of the autocracy, the liberals had slipped into the citadel of power and had sealed it up against their former allies. They reacted with surprising effectiveness through an institution destined to make its mark in the world: the Soviet.

This was first heard of in the textile town of Ivanovo-Voznesensk. It owed much to the Russian peasants' habituation to communal action and very little to party political direction. In St Petersburg the Mensheviks and Trotsky quickly grasped its possibilities. It enabled working men to elect, by direct democracy on a factory basis, representatives of their own class whom they could trust and whom they could remove at once if unsatisfactory. The delegates to the Soviet were like shop-stewards with a political mandate. They were elected on the factory floor and were certain to be militant. The Soviet was characteristic of a country in which the working men felt totally excluded from society. They would trust only an institution which was dominated by their own class. By mid-October, the St Petersburg Soviet represented all the large factories in the city and had an executive committee in session in the Technological Institute. It spoke with authority. It would have nothing to do with the formed left-wing parties: it demanded a constituent assembly, the end of martial law, an amnesty and the eight-hour day, and proclaimed a general strike to achieve these ends.

But the October manifesto had deprived the workers of their liberal allies. Without money, the strike could not be carried on. In December, the government moved against the St Petersburg and Moscow Soviets. In Moscow great brutality was used against the working-class quarters. This was, for the autocracy, the turning point. Its enemies were divided and one was crushed. Now the Tsar could afford to withdraw the concessions which he had made to the liberals in October. With Witte about to conclude a new French loan, Nicholas could show the liberals that a Russian autocrat was not bound by mere promises.

The first signs of counter-revolution had been seen during the days after the October manifesto. While most of Russia rejoiced at the apparent victory gained over the autocrat, the government organised a nation-wide series of pogroms. This showed its continuing power even at the moment of apparent defeat – a sign which was correctly read by the Soviet leaders as meaning that the October promises were illusory. The accuracy of this prediction was shown in the spring of 1906 when the government published the detailed provisions of the new constitution. Although all classes received some representation in the Duma, the electoral system was heavily weighted in favour of the propertied classes. Two thousand landowners and ninety thousand urban workers each received the right to be represented by one deputy. Further, the Duma was balanced by an equally powerful upper house, the Council of State. Its 196 members were to be chosen by the Church, the nobility and the universities. The military estimates of the budget were to be removed from the competence of the Duma. By the fundamental laws, issued in the name of 'Nicholas the Second, Emperor and Autocrat of all the Russias', it was made clear that the Duma could not initiate legislation, that it could not amend the fundamental laws, that ministers were in no way responsible to it, that it had no control over foreign policy, that it could not examine the government's loans, that it could be summoned and dismissed at the will of the Emperor, that by Article 87 the Emperor could rule by decree when it was not in session. Its members were elected for five years, but the period between elections could be less if the Emperor desired. With the publication of this document it became clear that the constitution, such as it was, came to the people by gift from above. The recuperation of the autocracy had been so swift that it was difficult to believe that Russia had been in dissolution only a few months before.

THE COUNTER-REVOLUTION: STOLYPIN TAMES THE DUMA

The first national election in Russian history was held (1906) in the shadow of the ferocious punitive expeditions by which the government restored its authority. Russia had a constitution but it was obvious that its government still paid little heed to legality. In spite of the renewed official terror anti-government parties won a great success. The Radical liberal Kadets won the largest number of seats (179); the Labour Group, closely linked with the S.R.s, won 94; there were 18 S.D.s and large numbers of national representatives (e.g. 51 Poles). On the right wing, there were 5 extreme monarchists and 17 Octobrists – that is, a party of conservative liberals pledged to work within the limits of the October manifesto. In spite of his detestation of the Duma, the Tsar punctiliously opened its first session in the Tauride Palace. From the first, relations between government and Duma were impossible. The Duma majority objected to the fundamental laws and wanted to introduce a radical land reform by which the landowners would have been expropriated. Nicholas and his new minister Stolypin also intended land reform but of a very different sort. The two sides were so far apart that it was even impossible to persuade a few of the more conservative liberals to become ministers. The Kadet leaders, like Milyukov, were perhaps over-confident about their strength. They reckoned that they had merely to remain firm and the government would have to concede their demand for a Kadet-dominated ministry. If so, they were badly mistaken. In July 1906 troops occupied the Tauride Palace. Energetically managed by Stolypin, counter-revolution in the form of pogroms and field courts-martial enabled the government to keep the initiative. The Kadets and their allies fled to Finland where from the comparative security of Vyborg they advised the populace to refuse to pay taxes and to serve in the army. This appeal to civil disobedience had no response. The Kadets found themselves isolated between the autocrat and the indifferent masses. The party which had commanded the loyalty of the voters in April 1906 received no sympathy from those same voters in the summer. The reason for this was, in part, that revolutionary-minded persons had voted Kadet in April simply because the left-wing parties had not taken part in the election.

It might seem that at this point the obvious course for Nicholas and Stolypin to take was to complete the discomfiture of Russian liberalism by abolishing the Duma altogether. That they did not do so may be attributed to two reasons, neither of them very flattering to the influence of liberalism. First, foreign and particularly French opinion was soothed by the existence of parliamentary institutions in Russia. French governments found it much easier to procure money and sympathy from the French people for an ally with a parliament. Secondly, Stolypin, the lineal heir of Witte, found, like his predecessor, how useful it was to have a channel open between the government and the public. Witte had used the press: Stolypin intended to use a parliament through which he could propagandise the scheme which lay nearest his heart – land reform. But for the achievement of this purpose, which was virtually to make the Duma another wheel of the bureaucracy, it was necessary to ensure a subservient assembly.

The second Duma met in February 1907 and in spite of intense government pressure on the elections was a disappointment to Stolypin. True, the number of Kadets had fallen to 92 and the number of right-wingers (who noisily demanded the abolition of the Duma from its floor) had increased. But revolutionary opinions were strongly represented by a vocal contingent of S.D.s (who like the right-wingers had little use for Dumas). The result of the presence of such violently conflicting opinions was that the Duma members took to throwing inkpots. The right-wingers sang the national anthem and asked the S.R.s if they happened to have any bombs in their pockets. Peasant deputies demanded radical land reform: one caused considerable effect by stating that in pre-emancipation days his uncle had been exchanged for a greyhound. The S.D.s stirred up the pot, the Bolsheviks among them openly using the Duma as a revolutionary tribunal. In brief, the second Duma merely brought to light all the irreconcilable hatreds, hopes, memories and aspirations of a land in which classes and institutions co-existed in time although separated by hundreds of years of development. Such scenes hardly suited Stolypin's purpose. He dissolved the Duma (June 1907) and, acting under Article 87, by a process which was illegal even by the elastic standards of the 1906 constitution, he sweepingly reformed the electoral law. The new law was of a bewildering complexity but had the effect of almost eliminating the national minorities, the urban intelligentsia, the urban worker and the peasant from representation. Henceforth the Duma would be the preserve of the landowners and the richest city dwellers. As such it could (and did) obediently speak for the ruling bureaucracy to the people. The counter-revolution was virtually complete. It seemed to Nicholas that the teaching of Pobedonostsev was excellent. Patient firmness would overcome all difficulties.

During the age of Witte Russia had adopted a decisively new course of development. Witte had given first place to economic development. State power was not to be used merely for repression and the preservation of the old but for a radical change in the economic structure of society. In the purely economic field he had achieved great success but he had failed to prevent revolution. Was this failure merely the result of the enmity of the conservatives in governing circles? Could Witte have succeeded if Pobedonostsev had not been so powerful? Or was it impossible to solve the problem of Russian backwardness without first completely changing the political and social structure by revolution? Was the dismissal of Witte by Nicholas II the final and fatal error of the dynasty? It could be argued that the Witte system, for all its success in bringing Russian industrial strength up to Western European standards, actually increased the chance of revolution in Russia. Industrialisation under Witte was based upon capitalism. The foreign capitalist was bound to demand his pound of flesh. He demanded high profits and these could only be paid at the expense of the Russian worker. The foreign capitalist also demanded a pro-French alignment of Russian foreign policy.

This brought war nearer and with war came revolution. The Russian capitalist (and managerial) class was made much more powerful by the Witte system, yet Witte did not propose to give it political power. This was bound to produce antagonism to the régime among the rich and skilful, an antagonism which is shown in the hard line taken in 1905 by the radical liberals. Industrialisation brought more peasants into the towns. They were certain to be poor, and there was no attempt to make them feel that they had a stake in society. In the villages they would have been merely anarchical: in the towns they were potentially revolutionary, as is shown by the Soviet movement which, in the autumn of 1905, gave the revolutionary intelligentsia its first taste of success. Forced industrialisation on the Witte model was bound to sharpen discontent both in the village and in the town. It might make the State stronger but it was certain to widen the distance between the State and its subjects. The latter were for the most part indifferent to Witte's main objective, which was to make Russia strong. Only the liberals shared this hope and they were either denied power or cheated out of it. For this, Witte must take his share of the blame. He may have been impeded by the Slavophil conservatism of Pobedonostsev but much more serious was his own bureaucratic temperament. He believed that autocratic Russia could gain all the advantages of the Industrial Revolution without paying the price of political readjustment. He impatiently strove to undo the neglect of the past in a short time. He would not let capitalism grow slowly by means of a gradually improving agriculture, but sought to force the pace by daring innovations. Here was an inevitable contradiction: those who were his economic allies were bound to become his political enemies. The better capitalism worked the more dangerous became the capitalist enemies of the autocratic régime. In the end, the discontented capitalists brought down the Romanovs in 1917. But perhaps Witte was right to despise them: within a few months they too had been brought down by the Bolsheviks.

7

The Strange Death of Official Russia: Murder or Suicide? 1907-17

STOLYPIN AND THE DUMAS

A decade elapsed between the counter-revolution of 1907 and the final elimination of the Romanov autocracy in the revolution of February 1917. This period was marked by the apparent revival of state power and initiative, most clearly shown in its ability to pursue a thirty-month war against the most powerful military force in the world. It was marked by the decline of revolutionary tendencies on the political left and the growth of purely economic demands among the industrial workers. It was marked by the continuing struggle between the autocrat and the liberal intelligentsia, a struggle which was intensified by wartime conditions and which ended, so it seemed, with the victory of the liberals in February 1917. But this victory was illusory. The revolution was stronger than the liberals and their tenure of power lasted for only eight months. The most constant political feature of the decade was the obstinate refusal of Nicholas II to share his autocratic power.

This steadfastness can be seen in different lights. The more popular view has been that Nicholas 'flouted the spirit of the age' by refusing to compromise with popular demands. His obstinacy is seen as the reverse side of his weakness. Dominated by his wife, he is alleged to have been motivated by the sole desire to transmit to his sickly heir the full plenitude of power. But another interpretation can be placed upon the Tsar's actions – and it is one which his own words support. It is, that as a deeply religious man he took literally the words which he had spoken at his coronation. He believed that even were he to agree to a constitution he would still, in the eyes of God, be responsible for what happened to his subjects. But by surrendering to the politicians he would have deprived himself of the power to fulfil his responsibility. Such an outlook upon power is hardly likely to appeal to the modern mind, to which the voice of the people is the source of all authority (or so it is alleged). But when the actions and words of Nicholas are weighed against the words and actions of his liberal and revolutionary opponents, it is tempting to salute his dignity if not his acumen.

Until his murder in 1911, P. A. Stolypin was an outstandingly energetic prime minister (a title in general use after 1905 but which did not imply that its holder enjoyed the powers of, say, the British Prime Minister). He was a landowner rather than a bureaucrat, efficient, lucid and ruthless. He had been particularly successful as the governor of Saratov province during the peasant risings of 1904–6, in suppressing which he earned the special hatred of the S.R.s. As Minister of the Interior, with the aid of especially constituted courts, he quelled the last stirrings of revolution. To the bomb and the revolver of the S.R.s – weapons which killed more than 2500 people in 1907, most of them officials – he opposed a police firmly and intelligently led. He was nearly killed himself when a bomb exploded in his house – an incident in which thirty-two people died. But his outstanding personal courage was proof against terrorism. He continued to place his agents in the terrorist organisations and to mete out summary justice. He showed that, if it came to terror, the resources of the State were superior to those of the terrorists. He thoroughly discouraged his adversaries.

His distinguished career ended in the Kiev opera where he was shot by Dmitry Bogrov, a man who was both revolutionary and police agent. It is still not known in which capacity Bogrov was acting on this occasion. Stolypin's high-handed actions had made him enemies at court as well as in revolutionary circles. Some thought that they detected an irony in his actions during the moments which followed the pistol shot. As he sank back into his seat, he raised his hand in the sign of the cross towards the box occupied by Nicholas and the royal family.

No one of his calibre served the dynasty after his death. Perhaps Nicholas resented such capacity so close to the throne: if so, he deserved to perish. The Empress Alexandra made little effort to conceal her joy at Stolypin's death, for not only had the minister overshadowed her husband but he had also dared to order her favourite Rasputin out of St Petersburg. The man of God from Tobolsk annoyed Stolypin by trying to hypnotise him. Rasputin was in Kiev as the imperial party drove through. It is alleged that he called out as Stolypin's carriage passed by: 'Death is after him! Death is driving behind him!' This incident has given rise to the suspicion – which still remains no more than that – that Bogrov acted with the connivance of Rasputin. The defence of a régime in which a court favourite was suspected of being privy to the murder of a great minister requires a powerful apologist.

Stolypin's constructive policy was centred around his peasant reforms. Elsewhere it has been seen with what vigour and success he carried them through. Under the impetus of a more efficient agriculture the Industrial Revolution moved forward again after the check at the beginning of the century. By 1913 the value of industrial production was more than 50 per cent greater than it had been in 1909: steel production increased five times between 1900 and 1913 and the age of Stolypin saw a similar rate of growth in oil and coal production and in the manufacture of machinery. As in the time of Witte, taxation, and especially the indirect taxes, remained very high. Tax receipts doubled between 1900 and 1914. After 1905 less state money went into railways and much more into rearmament. By 1914 one-fifth of state expenditure went into the army and navy.

Stolypin was a constitutionalist but not a parliamentarian; a nationalist but not a reactionary. He led from near the centre of the political spectrum, with violent enemies both to the left and the right. In his treatment of the Duma he had considerable success. This was partly because he was, unlike his bureaucratic colleagues, a good public speaker who was able to sway the assembly almost in the style of a British minister. His success is also partly due to the close links which he forged with Guchkov, the leader of the Octobrist majority in the Duma. Both men had a common interest in protecting the constitution against the reactionaries (especially those at court) who were trying to persuade Nicholas to withdraw what little power remained with the Duma.

Both Guchkov and Stolypin were much concerned with military efficiency and Stolypin found Guchkov useful as an ally against the entrenched conservatism of the military and naval hierarchy. Guchkov was the chairman of the Duma Committee for Imperial Defence. Such a man, speaking publicly in the Duma, fully informed about his subject, enjoying the respect and support of the Prime Minister, calling attention to the ineptitude of the Grand Ducal incumbents of high military office, was a new portent in Russian political life. Although the Duma was far from being a legislative assembly, it had at least become the ally of the liberal bureaucrats, a fifth wheel of the administrative chariot, a means whereby a powerful minister could make public his programme. Not only matters military benefited from this flow of fresh air. The Duma budget commission forced the Minister of Finance to account in public for his department – perhaps even to move slowly away from such close dependence upon indirect taxation; the education commission persuaded the government to spend more than ever before upon primary education; and the tolerant religious views of the Duma majority certainly helped to reduce the persecution to which the non-Orthodox were subjected and almost persuaded Nicholas to undertake a major church reform. Rasputin prevented this by pointing out to Nicholas that if the Church were headed by an elected Patriarch rather than by an appointed Synod, there would be a rival autocrat in the land.

Although there were, during these years of Stolypin's ascendancy, remarkable signs of growing co-operation between bureaucracy and non-official conservative liberalism, there was no doubt as to who was the ultimate master. Stolypin behaved, on occasion, with autocratic indifference towards the susceptibilities of the parliamentarians. Some of his most important legislation was pushed through by the cynical use of Article 87. The great land reform was started in this way in 1906. Although the Duma had in theory the right to refuse to accept an edict issued under Article 87, it was in practice almost impossible to baulk an important measure which was already being put into effect. Such behaviour merely invited immediate dissolution and it was one of the chief objects of the leaders of the third Duma to make the sessions as long as possible. With perhaps pardonable vanity they believed that if the nation became used to reading their speeches, the idea of free debate and informed criticism – so strange in Russia – would take root. In fact the standard of public speaking was very high and the criticism of the government sensible and constructive.

But the deliberate restraint of the Duma was not proof against the impetuously high-handed actions of Stolypin in 1910. He put through a law by which the *zemstvo* institutions were to be introduced into the western provinces. The electoral provisions were so arranged that the Polish landowners and the Jewish townsmen would be dominated by the Russian peasants. Stolypin's motives were openly nationalist. He had already shown his great-Russian chauvinism in his treatment of the Finns and the Ukrainians, and this policy had gone down very well with the Duma majority. Consequently a Duma dominated by Russian landowners gave their assent to Stolypin's 1910 law – a clear indication that in this assembly national feelings were, at this time, stronger than class sentiment. But in the Council of State (the upper house) this was not so. Its conservative majority, led by two of Stolypin's bureaucratic colleagues, Trepov and Durnovo, refused to accept the measure. This public indication of divisions within the Council of Ministers was in itself a novel, and some might say healthy, indication of major changes in Russian political life. Stolypin reacted with energy. He demanded that the Tsar should prorogue the Duma for three days: in this period he proposed to push through his legislation by Article 87. He also demanded that the ministerial rebels should be ordered out of St Petersburg and that Nicholas should give a written agreement to these actions. This last demand infuriated the Tsar: as if the word of an autocrat were not sufficient for one of his servants! The Duma was also annoyed by the cynical use of Article 87. Astute persons, like Prince Meshchersky, who started at this time to advise Nicholas behind the scenes, saw the great significance of this crisis. Suppose that Stolypin, instead of acting on Article 87, had chosen to put himself at the head of the Duma majority? As it was, he had exacted concessions from the Tsar almost as if he were a Western minister backed by a parliamentary majority. Having tasted such success, even at the expense of losing his popularity with the Duma (something which could always be repaired with a couple of skilful speeches) where would Stolypin stop? The 1911 crisis showed Nicholas the danger of having a popular minister in touch with at least a section of the public while he, the Tsar, was in touch with nobody except his family and a handful of courtiers. If Stolypin had not been murdered he would have been dismissed in 1912. His successors were, for the most part, servile nonentities.

LITERARY MOVEMENTS

There were other signs during these years that the liberal intelligentsia were learning to adopt a less revolutionary attitude towards the problems of society and government. A group of distinguished intellectuals who published *Vyekhi* (*Landmarks*) in 1909 pointed the way. Many of the contributors were Christians – in itself a novel feature in the ranks of the intelligentsia. They blamed their predecessors for adopting a negative attitude towards the problems of power, for failing to move towards a position in which a working compromise between government and society would be possible. They alleged that concentration upon the idea of abstract justice had killed both charity and the love of truth. They claimed that the celebrated 'love of the people' so prominent in populist circles was in part merely a sentimental self-glorification. This book was widely read and violently attacked by the left-wing leaders. Its significance is great. For the first time since the rise of the intelligentsia, since the days of Belinsky and Chernyshevsky, a number of leading thinkers had written strongly against the tradition of revolution. In time, such an attitude would have permeated by way of the universities to the idealistic but disorientated youth who threw the bombs and fired the pistols. To them, the men of *Vyekhi* said in effect: work with the State towards a gradually improving future, not against it towards an immediate revolutionary utopia. There was just enough evidence in pre-war Russia that social problems were moving towards amelioration to make this appeal reasonable.

The tone of *Vyekhi* towards social and artistic questions was similar to that of Chekhov – progressive, optimistic, humane. But it must be remembered that not all the writers of this era (one extraordinarily rich in talent) shared this middle ground. Tolstoy's voice continued to be heard until his strange death in 1910. With increasing age, Tolstoy became ever more the bitter, sceptical critic of his contemporaries. His latest works contained a comprehensive denial of the utility of science, government,

Orthodox Christianity and family life. His amazing literary power was employed in the service of sexual abstinence, a non-Christian form of Christianity, non-resistance and the repudiation of alcohol and tobacco. This seems a strange programme to be advocated by the author of *War and Peace*. Tolstoy, in fact, had little faith in the willingness of his contemporaries to learn from him. With much gloomy pleasure, he foretold in his later works some of the terrible things which actually did come to pass during the war. But too much should not be made of this. In his old age Tolstoy resembled an Old Testament prophet, foretelling doom for his wicked contemporaries. Such prophecies have usually come true at all times. It may be that more credit is due to those who try to prevent disaster than to those who merely prophesy it (with whatever degree of literary skill).

On the side opposite to Tolstoy, the elderly millionaire anarchist, stood the group of brilliant young poets like Blok and Yesenin. Although their techniques were modern their ideas were old. They saw the revolution – which they foresaw with varying degrees of ambiguity – as a sort of rebirth of the spirit of Christ which, they asserted, would provoke a universal spiritual rebirth. Blok's masterpiece *The Twelve* is the most celebrated expression of this idea which must, in retrospect, appear to be rather eccentric.

THE URBAN MASSES

The era of Stolypin, tantalisingly brief though it was, also witnessed some significant changes in the activities of the urban masses. The number of urban industrial workers increased rapidly – from about 1¾ million in 1910 to 2½ million in 1914. Trade unions had been legalised in 1906 but made little progress, partly because both the government and the capitalists were hostile to them. It is unlikely that working conditions improved very much, although the government made some move towards social security. Accident insurance and sickness benefit schemes were introduced but they applied to only a small section of the workers. Up to about 1912 there were comparatively few strikes but then the brutal treatment of the strikers in the Lena goldfields (170 were killed) sparked off a wave of strikes which culminated in the St Petersburg general strike of July 1914. There is no doubt that, especially in large industrial centres like St Petersburg, Moscow and Baku, the influence of Bolshevik and Menshevik agitators was considerable. But the significant point is that even the professional revolutionaries had to conceal their political aims behind economic demands. The angry workers were mainly interested in wages, hours, housing conditions and the truck system. A feeling of solidarity grew up between the widely separated industrial centres, shown by the fact that when the St Petersburg general strike began in July 1914 it was in sympathy with a strike which had been going on for some weeks at Baku.

AFTER STOLYPIN AND BEFORE THE WAR

It would be an error to see this outbreak of strikes simply in terms of bad economic conditions and left-wing agitators. It also reflected the worsening political situation which followed Stolypin's death. The fourth (and last) Duma was elected on the franchise of 1907 and its membership was similar to that of the third. But its liberal majority became slightly more obstinate in face of the changing conditions at court. There Rasputin and the reactionaries had things all their own way. After the dismissal of Kokovtsev (February 1914), who was at least a sound financier, the aged Goremykin became Prime Minister. Maklakov was a notoriously reactionary Minister of the Interior: he and several other ministers advised Nicholas either to get rid of the Duma altogether or to reduce it to a purely consultative assembly. Nicholas was obviously inclined to follow this advice and was prevented from doing so mainly by his fear of what might be thought in France and Britain, his potential allies. But it was nevertheless highly distasteful for the Tsar to have to listen to the strictures of the president of the Duma, Rodzyanko. The latter, the spokesman in some sort of liberal Russia, a middle-class gentleman with instincts about as revolutionary as those of the average British peer, told his monarch that his government was an anarchy. None of the ministers knew what the others were doing and none knew what the Tsar wanted. Rodzyanko implied that a government responsible to the Duma majority (that is, to him and his friends) would at least be more efficient. Had Nicholas been astuter (more dishonest?) he would have seen the advantage of taking Rodzyanko and his friends into the Ministry. They would have worked hard to maintain the system which gave them power. But even at this stage, Nicholas could not stomach the idea that a man like Rodzyanko could rise to political power by the voice of the people. In the Duma, at elections, in the press, such a man would tap sources of power which were closed to the Tsar himself. The coming of the war found the Tsar and liberal Russia in much the same positions as they had been in 1907. In spite of some progress towards understanding during the Stolypin years, the two sides were still divided by mistrust, contempt and even hatred. A war of the sort which began in 1914 was bound to create national unity: the question was, would Russia consent to be united under the Tsarist autocracy?

THE WAR: AT THE FRONT

There is some evidence that Nicholas and a few of his advisers believed that foreign war was necessary in 1914 in order to regain control over the revolutionary forces in Russia. The same calculation had proved to be disastrously mistaken in 1904, but the war against Germany did at first prove to be universally popular. The only influential voices raised against it were those of Lenin, Rasputin and Witte, and it was impossible that these three should unite. The strike movement faded away: most of the S.D. leaders, with the significant exception of the Bolsheviks, believed that the victory of Russia would be a lesser evil than that of Germany; and the liberal groups in the Duma gave enthusiastic support to the war effort. Russia went to war in a spirit of exalted unity which delighted both her rulers and her allies. But the patriotism which inspired this unity meant different things to different groups. The Tsar hoped to strengthen his position by victory; the liberals hoped to win a constitution; the workers and peasants hoped for a life less harsh; and the nationalities hoped for independence. The Western Allies, of course, hoped for a powerful military effort and this they got, accompanied by much effusion of Russian blood.

Russian mobilisation proceeded rather faster than anyone had expected. By December 1914 six-and-a-half million men had been mobilised; by the end of 1917, about fifteen-and-a-half million. Russia used her manpower lavishly: about eight million men were killed, wounded or missing. Millions of others died as a result of the hasty evacuation of the western provinces. However, in its three major campaigning seasons before the February revolution, the Tsarist army did little worse against the Germans than did its Western allies, despite its much weaker industrial base and poor communications. It would be rash to assume that a nation which was capable of sustaining such a gigantic effort was notably corrupt or 'ripe for revolution'. It is too simple to argue from the fact that a revolution did occur to the assertion that it was bound to occur. To argue from the disasters which befell the army that only a revolution could be the outcome, is the same as saying that since the British army failed so completely at the Battle of the Somme, revolution was inevitable in Britain. Revolutions are only inevitable once they have happened.

In 1914, the Russian invasion of East Prussia persuaded the Germans to shift two army corps from the western front. This allowed the French to win the battle of the Marne and saved France as an independent nation. The initial Russian advantage in the north was soon lost, however. At the battle of Tannenberg (August 1914), incompetent Russian commanders threw away their armies and the German territory gained. But in the south, most of Galicia was captured from the Austrians in a victorious campaign which was halted only by a German counter-attack in Poland. In 1915 the Russian army, like the French and British, met with a series of disasters. The failure of the Anglo-French Dardanelles campaign, followed by the entry of Bulgaria on the German side, meant that Russia was cut off from her allies. The lack of sufficient war material began to tell. After a break-through at Gorlice (May 1915) the German and Austrian armies continued to advance until the autumn. Confusion but not despair reigned in the Russian armies. By the end of the year, without any reported mutinies, the line had been stabilised. It ran from just outside Riga, east of Vilna and Pinsk, to the mouth of the Dniester. The great retreat had profound political and social repercussions, just as similar disasters shook the British politicians. The survival of Russian morale was demonstrated by the 1916 campaigns. Fierce attacks in the north prevented the Germans from concentrating all their troops in France at Verdun, as they would have liked. In the south, Brusilov was so successful against the Austrians that he broke through to the Carpathians and persuaded Romania to enter the war on the side of the Allies (August 1916). Romania rapidly succumbed to a German attack: neither the Russian armies in the north nor the allied armies at Salonika were able to give any help. Nevertheless, by the end of 1916 the Russian military position was not desperate. Plans were going ahead for close co-operation with the Allies in 1917. As usual, the rôle of the Russian army was subordinated to the military needs of the Western Allies. But victory was expected and with it, no matter where the Russian army stood at the moment of German surrender, enormous gains were to be expected. Nicholas trusted his allies considerably more than Stalin trusted his at the end of the Second World War. At the end of 1916 nobody expected revolution although many hoped for it. Of these optimists, most hoped for changes which would make the Russian war effort more efficient: only a few looked to a revolution which would take Russia out of the war.

THE WAR: BEHIND THE LINES

Behind the lines, the war accentuated the political and social stresses already apparent in peacetime. The bureaucracy was quite unable to cope with the new problems of supply and organisation. By its traditions it was so much a class apart from the rest of the population that it was unable to achieve the rapid expansion which, for example, eventually enabled the British administration to cope with problems of supply. The Tsarist bureaucracy was also quite unable to tap the enthusiastic patriotism which was powerful in 1914. All governments during this war discovered that it was essential to make the war popular, to draw in the willing consent of the masses whose labour and bloodshed made the struggle possible. In this respect, Russia was at a grave disadvantage. The other combatants made extensive use of newspaper propaganda to arouse the martial spirit of literate but semi-educated masses. The Russian government was unable to do this: literacy was not widespread and there was no popular press. It was partly for this reason that the Russian masses eventually reacted in a normal way to the horrors of war – that is, by shooting their officers and going home.

In the absence of any other links between the ruling bureaucracy and the masses, the Tsarist government was forced to rely upon the patriotic efforts of the liberal-dominated *zemstvos*. After half a century in which their efforts had been jealously circumscribed and limited to petty local affairs, the *zemstvos* were allowed to form an All-Russian Union (chairman, Prince Lvov) and to undertake those tasks which the bureaucracy was unable to perform. From the first weeks of the war the *zemstvos* became indispensable in the provision of hospitals, trains, nurses and doctors, food, fodder, and even arms. Russia already had two separate and often conflicting administrative systems. In the perspective of the long struggle for power between Tsar and liberals, these concessions swung the balance strongly in favour of the latter. With the benefit of hindsight it can be seen that Nicholas might as well have granted gracefully what he was bound to have to grant anyway: that is, a measure of political power to match the new-found importance of the *zemstvos*.

The military disasters of 1915 naturally aroused the political ambitions of the liberal leaders. These men were patriots, extreme nationalists even, and they could not afford to let their opposition to the régime weaken the war effort, yet they could not utter too loudly the classical cry, 'The Fatherland is in danger', because they might arouse the ungovernable fury of the masses. They proceeded cautiously. Under Milyukov's leadership, Kadets and Octobrists came closer together and demanded a government of public confidence. This was, of course, the familiar liberal demand for an executive responsible to the legislature, dominated by the liberals. Nicholas refused, but he did move some way towards meeting the less extreme liberal demands. Several of the most unpopular and reactionary ministers were dismissed, including Sukhomlinov, the Minister of War. A Special Council of National Defence, which contained representatives both of the state departments and of the Duma, was created to co-ordinate the official and the unofficial war efforts. A War Industries Committee (chairman, Guchkov) was empowered to undertake the central direction of production. It contained representatives of the bureaucracy, of the industrialists and of the workers. Emboldened by these successes, which they attributed to the pressure they were bringing to bear, the liberal leaders, acting through a number of ministers who were by this time favourable to their programme, tried to get Nicholas to dismiss Goremykin. Nicholas had rarely, if ever, had to face a collective demand from

The Spassky Cathedral of the Andronikov Monastery, Moscow

Michael Speransky

Alexander II

Count Sergei Witte

P. A. Stolypin

Above: *Tolstoy while he worked on* War and Peace, *1868*

Above right: *A portrait of Dostoyevsky by V. Perov*

Right: *N. G. Chernyshevsky*

'The Wanderer': an evocative figure of pre-revolutionary Russia

Red Square, Moscow, in the 1900s

Women pulling a raft upstream, 1910

Compulsory labour for the bourgeois, 1918

Red Army men reading revolutionary newspapers, 1918

A. P. Chekhov and M. Gorky, 1900

Nazi hangmen at work in Minsk, October, 1941

The Third Ukrainian Front, 1944: a tank battle near Odessa

a group of ministers. Alexandra advised him to comb his hair with Rasputin's comb before the interview (September 1915). Perhaps it was this that permitted Nicholas to say 'No', charmingly yet firmly. For although in the face of the 1915 disasters Nicholas had been willing to apply the liberal nostrum up to a point, his deepest instincts led him towards an altogether different solution to his problems – to take up the command of his armies personally. By this move he hoped to restore that mystical unity between Tsar and people (*sobornost*) which, in his view, was more likely to win the war than any political or administrative reorganisation. It was his own decision to go to G.H.Q. at Mogilev, not a sinister plot by Rasputin and Alexandra to get him out of Petrograd. It was a decision quite in harmony with his simple religious nature. He took with him as Chief of Staff, that is, as effective commander of the army, Alekseyev, probably the best of the Russian generals. Between them, as the events of 1916 showed, they improved the situation at the front considerably.

The tactics of the liberal leaders changed significantly after their defeat in 1915. From open opposition they moved over to conspiracy: from charges of mere incompetence they moved to insinuations of treachery. It must be said that Nicholas gave them a large target. It was easy for the liberals to make out that Alexandra and Rasputin really ruled Russia, that it was they who advised Nicholas to make the bad ministerial appointments of 1916, and that their object was to make a separate peace with Germany. 'The German Woman' was an easy target for the whispering campaign undertaken by Guchkov and his friends. She was isolated, hysterical and imperious; there was no means for her to answer the charges, which were generally believed. In fact, there is no evidence at all in the German sources that Alexandra and Rasputin were contemplating a separate peace: the extent of their influence upon appointments is also in doubt. Guchkov, Lvov, Kerensky and others who were to play a large part in the overthrow of the Tsar, had many advantages. They were linked by the bonds of Masonry, they exercised great power through the voluntary organisations: above all, they had some ready disciples at G.H.Q. and among the enormously swollen officer corps. Their appeal was essentially patriotic and conservative. To save Russia, they said, it was necessary that

Nicholas should abdicate in favour of another member of his family and that the new Tsar should take an oath of loyalty to a parliamentary constitution. Guchkov was certainly preparing such a *coup dètat* timed for March 1917. His method was to undermine the loyalty of the officers and it is likely that Alekseyev was among the conspirators. Without these preparations for the revolution which did not happen, it is unlikely that the revolution which did occur in February 1917 could have succeeded. It should also be remembered that had Nicholas been less obstinate in resisting the liberal demands there would have been no need for Guchkov and his friends to spread the rumours which undermined the loyalty of the educated classes.

By the end of 1916 both the strength and the weakness of the liberal position were clearly apparent. Its strength was best shown in the celebrated speech made before the Duma by Milyukov in November. He accused the government of various shortcomings in war administration and ended each point with the question: 'Is this treason or folly?' The speech, along with other pieces of liberal propaganda, reached a wide public by means of thousands of mimeographed copies. The effectiveness of liberal propaganda was demonstrated by the murder of Rasputin in December. This was carried out by an aristocratic clique led by Prince Felix Yusupov with at least the knowledge of the liberal leaders. Its effect was small. Nicholas replaced the incompetent premier Trepov with the still more incompetent Golitsyn. The populace was apparently unmoved. The weakness of the liberal position was the time factor. Duma elections were due in 1917. Notable allied victories might well accompany these elections and the ensuing Duma could, in the flush of Tsarist victory, be dominated by the right wing. Since the liberals, as well as the Tsar, banked upon an allied victory it was essential for them to be actually in power when that victory was gained. Defeat by Japan in 1905 merely weakened the autocrat: defeat by Germany would dissolve both the State and society too.

THE URBAN MASSES ON THE EVE OF THE FEBRUARY REVOLUTION

But the liberals were forestalled at the last moment by a popular rising in Petrograd which they may have desired but which they were unable to control. The war had had contradictory effects upon the masses. Wages and prices had both increased sharply, with prices for the most part in the lead. City populations had been rapidly expanded by the needs of war industry: the population of Petrograd, for example increased from 2,100,000 in 1914 to 2,465,000 in 1917. Food had been relatively plentiful up to the end of 1916. The export of grain was impossible and for the first time Russians ate the produce of their own land. But the lack of labour in the countryside, the breakdown of railway transport and mere bureaucratic inefficiency led to a temporary food shortage in the cities at the end of 1916. The existence of bread queues added another grievance to the discontent which had already during the previous eighteen months brought the Petrograd workers out on many strikes.

The problem of who led the industrial masses in and before 1917 is not easily solved. Communist orthodoxy demands the assertion of the legend that the workers were inspired by the Bolshevik committees who in turn obeyed the exiled leaders. This is a myth which has not appealed to the infidels. Marxist but non-Bolshevik writers have rightly pointed to the considerable number of Mensheviks in the trade union movement. Both sections of the faithful have united against the view that the most influential workers' organisation in Petrograd in 1916–17 was a group which was neither Bolshevik nor Menshevik, a group called the Mezhrayonka. This was a committee of working men who retained the essentially non-political revolutionary fervour of the 1905 Soviet.

There is much evidence about the history of the working-class movements in pre-revolutionary Russia which has been suppressed or distorted by Soviet historians. Non-Bolshevik foreign writers, puzzled by the lack of evidence and unwilling to admit ignorance, have fallen back upon the historian's stock response to an unresolved problem: 'spontaneity'. In this view, the Petrograd workers' rising was a spontaneous reaction to intolerable conditions. A newer view which has recently received much support is that German agents armed with plentiful supplies of money and organised by the colourful millionaire international pedlar, Alexander Helphand ('Parvus'), infiltrated the Petrograd working-men's organisations in order to provoke crippling strikes. Thanks to the lucky capture at the end of the Second

World War of much German documentary material, this view has received some plausible confirmation. It appears that Helphand's agents were successful in provoking simultaneous strikes in Petrograd and Nikolayev in January 1916: but it has yet to be proved that the Germans were behind the Petrograd risings of February 1917.

In view of all these attractive possibilities, the reader would be safe in concluding that the working men of the Russian cities had good reason to be dissatisfied with their lot and that especially in Petrograd there were active groups attempting to turn economic discontents into revolutionary action.

THE FEBRUARY REVOLUTION AND THE ABDICATION OF NICHOLAS II

The wave of strikes which broke out in Petrograd in the first weeks of 1917 seemed little different in scope or violence from a dozen such affairs in previous years. The strikers drifted in and out of work, apparently taking little notice of the appeals either of the Bolsheviks or of the Workers' group of the Central War Industries Committee. On 22 February, convinced that the disorders were dying down, Nicholas returned to G.H.Q. at Mogilev. He left the Minister of the Interior, Protopopov, to disperse the last of the strikers. This may well be a key to much of what followed: for Protopopov, who was not a bureaucrat but had risen to the notice of the Tsar by loyal service in the Duma, was a monster of ineptitude even by the unexacting standards of the Russian bureaucracy. He commanded 6000 police and 150,000 troops for the maintenance of order in Petrograd. Normally the Okhrana (secret police) was well abreast of events in revolutionary circles. Police agents penetrated every secret group and long experience had shown them just how to lead on the revolutionary leaders into a compromising position. But in February 1917 the Okhrana was in low spirits. It had no use for Protopopov, who expressed virtuous horror at its cunning ways (probably because he was unable to understand them), and its morale had been shaken by the murder of Rasputin (whom it was supposed to protect). But the police would probably have been able to control Petrograd had it not been for the presence of the enormous and demoralised garrison.

The dangers of a military mutiny had already been pointed out to Protopopov. He took no notice and for this he can hardly be blamed. Morale in the Russian army at this time was not noticeably low although few welcomed the prospect of a fourth bloody campaign. But it may be surmised that few British, French or German soldiers looked forward with pleasure to the campaigning season of 1917. The Petrograd garrison was particularly inflammable because it was largely composed of holding units, that is, units which were greatly swollen in size, in which the officers hardly knew the men and in which nobody had anything to do. A tenth of the number of disciplined troops – and there were still plenty such in the army – would have been a much more efficient force for quelling disturbances. In this demoralised rabble, revolutionary agitators had been active in many units and the officers were much influenced by the liberals' propaganda.

On the day of the Tsar's departure to the front, the Putilov factory workers went on strike. Their demand was for a 50 per cent pay rise – a demand which was obviously intended to provoke a refusal. The troops, when at length brought out against the strikers, were incompetently handled: their officers showed little vigour and the men were unwilling to fire upon crowds which contained many women and children. The turning point came on Monday 27 February, when four regiments mutinied and shot some of their officers. Thereafter the task of the police was hopeless. They could not control the crowds of mingled workers and disorientated soldiers. The latter were an important factor in pushing the revolution still further. They knew that if the old régime were to be re-established they would be shot. Out of the turgid masses, on the very day of the garrison mutiny, a primitive centre of authority crystallised. This was the Petrograd Soviet which met, amid scenes of revolutionary ardour and picturesque confusion, in the Tauride Palace – the very building in which the Duma also held its sessions. The Soviet was directly elected by the workers and soldiers. As in 1905 it was not dominated by any one political party although the majority of its executive council were Mensheviks and S.R.s. It was powerful because it was armed, a key factor when most expected a determined attempt by the Tsar to recapture Petrograd. By its Order No. 1, the Soviet claimed the primary allegiance of the Petrograd garrison; the soldiers were to obey no orders unless these had been countersigned by the Soviet. The political line followed by the Soviet was more liberal than socialist. It called for the usual freedoms (press, speech, assembly, etc.), for an amnesty, for democratic elections, and above all for a constituent assembly to decide upon the political future of the nation. Although the Petrograd Soviet did not claim the powers of a national government, its lead was rapidly followed by the workers and soldiers in other large towns. The working men drew themselves apart into their own organisations and waited to see how the new government of Russia would carry out their will.

Faced by uncontrollable disorders in the capital, the Tsarist ministers implored their master to dismiss them. Constitutional convention prevented them from resigning or even from forcing one of their number to resign. Protopopov, who by this time had completely lost his head, offered to simplify things for his colleagues by shooting himself. Nicholas refused to grant his ministers' craven request and announced his intention of returning to Petrograd himself. The ministers left their offices and went into hiding in order to evade the threatening crowds. Thus perished ingloriously a bureaucracy which had misgoverned Russia for two centuries.

On 26 February, the day before the government had lost control of the streets of Petrograd, Nicholas responded to the news of continued disorders by the prorogation of the Duma. He presumably thought that the Duma politicians were still his most dangerous enemies and that the task of keeping order would be simplified if they were out of the way. Thus on 27 February the Duma was in a strong position to take over the leadership of the revolution. By meeting in spite of the prorogation edict, it could clearly show its independence of the Tsar and it could claim to be the one remaining valid authority, deriving its authority from the will of those who had elected it. Such a course of action would have been supported by the Soviet which was at this time anxious above all for united action against a punitive force. But at this moment of destiny, the Duma acted indecisively. Only a few members spoke in favour of revolutionary disobedience. Milyukov, the leader of a minority of radical liberals, did not want this Duma, elected on the narrow franchise of 1907, to make itself a sovereign power. His plan was for a constitutional monarchy based upon a Duma elected on a wide franchise – something like the first Duma, in which his party, the Kadets, had been so powerful. But the majority of Duma members shared the doubts and hesitations of the Octobrist president of the Duma, Rodzyanko. For the next few days this worthy nonentity – contemptuously dismissed by Nicholas as 'that fatty Rodzyanko' – played a central part in the political drama.

Rodzyanko was no revolutionary. The disorders of 27 February inspired him to nothing more than a sheaf of hysterical telegrams to Mogilev. He hoped at first that Nicholas would ride the crisis by appointing a government of national confidence, responsible to the Duma – the old liberal nostrum. But the crowds of soldiers and workers' deputies who cluttered up the corridors of the Tauride Palace, in which he and his friends were trying to decide the destiny of Russia, soon convinced him that such a solution was impossible. The Tsar must abdicate, possibly in favour of his brother Michael, possibly in favour of his son. The act of abdication must be accompanied by a promise of constitutional government. But while Rodzyanko was dithering (28 February–1 March), Nicholas was approaching Petrograd. If he arrived, together with a force of loyal troops, there would be a long and bloody civil war. The war effort would be (at the least) impeded, and even Rodzyanko and his friends might not be safe from the Tsar's revenge. They had already disobeyed the prorogation edict and set up a Provisional Committee of the Duma to take over the functions of the vanished ministers. In spite of his conservative nature, Rodzyanko had already been forced beyond the bounds of legality.

Terrified of the Soviet, aware that Milyukov was intriguing against him, faced by the possibility of the imminent arrival of the angry Tsar in Petrograd, Rodzyanko turned to the one power which might save Russia from civil war: the generals. The attitude of Alekseyev and most of the other high-ranking officers was that order must immediately be restored in Petrograd but that firmness must be accompanied by political concession. Alekseyev agreed with Rodzyanko that the imperial trains should not be allowed to reach the capital: and with his connivance Nicholas was turned back at Malaya Vishera and shunted off to the H.Q. of the northern armies at Pskov. There (1–2 March), occurred the final scenes in the history of the Romanov dynasty. It appears that Ruzsky, the general commanding at Pskov, at first tried to persuade Nicholas to make political concessions. He imagined that Nicholas would be willing and that such concessions would pacify the Petrograd mob. In both ideas he was wrong. Misinformed by Rodzyanko, Ruzsky and the generals were not aware that the Petrograd mob by this time wanted much more than a mere government of national confidence. But Nicholas in fact resolved the generals' difficulties. He would not share power: he would abdicate, both for himself and his son. This he did (2 March). In the instrument of abdication he named Prince Lvov Prime Minister. It was hoped by some of the members of the Provisional Government (formed as the result of an agreement between the Soviet and the Provisional Committee of the Duma) that the Grand Duke Michael would accept the vacant throne. Milyukov thought that this would help the new government to protect itself from the demands of the more radical members of the Soviet. With admirable foresight, he perceived that the new government would lack authority unless it was supported by a symbolical link with the past. But Michael was persuaded that if he accepted power it would provoke a civil war, and consequently refused (3 March). The revolution had already carried the unwilling liberals past constitutional monarchy to a democratic republic.

In such circumstances the first liberal government in Russian history achieved power – of a sort. Its origins were inauspicious. 'Who chose you?', the mob astutely shouted when Milyukov announced the names of his colleagues. With a rhetoric typical of those days, Milyukov replied: 'We were chosen by the Russian revolution'. But the question remained a dangerous one. Lvov and his government already shared power with the Petrograd Soviet: they would remain in power only so long as they obeyed the Soviet. They had not been chosen by the people: it was merely that they had been around at a moment of counter-revolutionary danger. They had received the blessing of the last Tsar – but this was a liability in March 1917. They had received the backing of the generals. This would remain their strongest bulwark until such time as the generals had no more soldiers to command. On the credit side, it could be said that the liberals had been strong enough to step in and prevent a civil war. But on the debit side, so far from being strengthened by their long struggle for power with the autocrat, they had been debilitated by frequent compromise. Not the least grave of the charges against Nicholas II is that by concentrating too much upon the succession of his natural heir, the Tsarevich Alexis, he so weakened his political heir, the liberals, that they were unable to retain the power which he bequeathed them. In the Bolshevik revolution eight months later that power was to be seized by persons whom both Nicholas and the liberals would have opposed together had they realised what was going to happen during the next few months.

The fall of the last Romanov was not really procured in the streets of Petrograd. Perhaps too much time can be spent wondering about whether the Germans or the Bolsheviks were more influential in the working-class suburbs of the capital. The February days were merely a repetition of what had occurred a score of times before. The Petrograd 'street' had gone further than it usually did: it had secured the aid of a section of the army. But there were plenty of loyal soldiers left. Nicholas could have restored order in Petrograd if he had been allowed to get there. The first Russian revolution of 1917 really took place at Malaya Vishera, where the imperial trains were diverted. This was not the work of the Petrograd 'street': it was done by a combination of liberal politicians and nationalistic generals. The only thing which united these men was a common desire to prosecute the war to a successful conclusion. It was because he too believed that Russian war aims could be better gained without him that Nicholas abdicated. The war itself was First Murderer at the political assassination of Nicholas II. This was a fitting end for the Romanov dynasty. From the time of Peter the Great, the dynasty had regarded the defence and extension of the frontiers of Russia as their chief objective. Russia did not exist for Russians but for the dynasty and its external ambitions. Immense success had been gained but in the end the effort was too great for a backward and divided people to support. Peter had started by merely opening a window on to the West; his successors knocked through many more windows and doors. But in 1917 the prize was much greater than it had ever been before. The defeat of Germany and Austria would lead to Russian domination of the whole of eastern and central Europe. No wonder that the generals and the liberals subjected everything (even the dynasty) to winning this war. The more intelligent knew that they were gambling. Which would come first – victory or the social revolution? In 1917, revolution preceded victory, but in 1945 the hopes of 1917 were fully realised by politicians and generals quite different from those of 1917.

8

Russia, Turkey and Europe: 1812-1917

THE DEFEAT OF NAPOLEON AND THE 1815 TREATIES

When the remnants of Napoleon's army struggled out of Russia, Kutuzov wanted to stop short at the old Russian frontier. He believed that Europe could be left to look after Napoleon. Alexander I, however, had very different ideas. He thought that the Russian army should pursue Napoleon until he had been utterly defeated. He considered that it was his duty (and in the interest of his country) to preside at the remaking of Europe which was bound to follow the dismantling of the Napoleonic system. This was a fateful decision, of a kind such as no future ruler of Russia was able to take until Stalin at the end of the Second World War when he too laid claim to his share of Hitler's Europe. Alexander, like Stalin, found that his allies were almost as frightened of him as they were of the nation which they were supposed to be fighting. Yet Alexander in 1815 did better for himself than Stalin in 1945. He refrained from taking so much as to drive his former allies into immediate enmity.

The Russian advance into Europe was slow: Napoleon was still a formidable general who could still inflict defeats upon the lumbering Russian armies. But the main problems of 1813 were diplomatic, not military. The Great Powers circled each other warily, each suspicious of what the others hoped to gain from the dismemberment of the Napoleonic Empire. First to join Russia was King Frederick William of Prussia. At Kalisz (February 1813) he undertook to declare war on France, thus restoring the friendship between the ruling houses of Hohenzollern and Romanov which was to last until the fall of Bismarck (1890). But Austria still hestitated. Its ruling dynasty (the Habsburgs) was linked to Napoleon by marriage; Metternich (the effective maker of its foreign policy until 1848) had no wish to exchange a Europe

Moldavia and Wallachia were united as Romania 1859-62
S. Bessarabia (*) ceded 1856, regained 1878
Danube Delta (✳) became Russian 1829-56

Russia's western border, 1815–1918

dominated by France for one dominated by Russia. He feared that a powerful Russia would damage Austrian interests in Poland and the Balkans. For some months he attempted to act as an honest broker between Alexander and Napoleon. Both were inflexible and so were the British. Realising at length that Napoleon's day was over, Metternich made haste to get on to the winning side (August 1813). He still hoped that Napoleon would see sense and make peace before he had been utterly defeated, but he was disappointed. Although Napoleon still continued to win battles against the enormous but poorly led allied forces, he was driven back into France. In March 1814 Alexander entered Paris. He had not only saved his own country: he had played a major part in the defeat of Napoleon.

With equal skill Alexander now proceeded to make a peace treaty highly advantageous to Russia. He arrived at the Congress of Vienna like a conquering hero. He had been fêted in Paris and London – so popular had he been in the British capital that the jealous feelings of the Prince Regent (who was rightly very much hated by his subjects) had been inflamed. At Vienna Alexander pushed his European peace plans vigorously, his main objective being to procure the whole of Poland for Russia. To console the Poles for their lack of independence he offered them unity. In 1795, their country had been carved up between Russia, Prussia and Austria. They had not liked it and, failing independence, most of them in 1815 accepted the prospect of national unity under Russian domination. Prussia fell in with the plan since Alexander offered the Prussians attractive compensation for giving up the slice of Poland which they had held in 1795. He proposed that they should take the whole of Saxony, a rich German State whose ruler had committed the imprudence of joining the victorious alliance too late. This acquisition would give Prussia a dominating position in Germany. Metternich strongly objected to the plan. It would give Russia a free hand in Eastern Europe; it would make Russia's client, Prussia, too strong in Germany; moreover, he had no intention of giving up Austrian Poland (Galicia). Metternich persuaded the British and French governments to join him in opposition to Alexander's plan. In January 1815 this strangely assorted trio made a secret treaty to oppose Alexander by force if need arose. But Alexander knew when to retreat, and the next month he agreed to a compromise. Poland would remain disunited. Austria would keep Galicia, Prussia a part of her share of Poland; in compensation, Prussia would take only a part of Saxony. In spite of this withdrawal Alexander still got the lion's share of Poland. But much more important than this territorial gain was the settlement which he had helped to impose upon Europe. Prussia was a grateful client state, well aware of Austrian (and French) enmity and of the consequent necessity to stand well with Russia. Austria was in a different position, but one which was highly advantageous to Russia. The 1815 treaties had given Austria enormous territories to govern. South of the Alps, she dominated Italy; north of the Carpathians she ruled Galicia. With the rising tide of nationalism she was weakened still further. To survive at all she too would need to stand well with Russia.

Alexander did so well in 1815 that he was bound to become a conservative. Any change in the 1815 system would be to his disadvantage. He realised that changes in the balance of power come about not merely by war between States; they could also be provoked by internal revolution within States. After 1815 he set himself up not only as the guardian of frontiers but also as the chief antagonist of anybody in Europe who wanted to alter the existing political constitution of the State in which he lived. He called this plan the Holy Alliance and secured the co-operation of both France and Austria in its execution. The British were less enthusiastic; by 1820 they had deserted the alliance system. But France under her restored Bourbon rulers took the place of Britain as Russia's Western ally. In the name of the Holy Alliance French troops invaded Spain in 1822 in order to force the Spanish people to restore the King whom they had rightly ejected. This was a strange but potent illustration of the post-war power of Alexander I. He cracked the whip in St Petersburg and Spanish liberals were executed in Madrid. No Russian ruler was half so powerful in Europe until the time of Stalin.

RUSSIA, BRITAIN AND TURKEY 1821–41

The position gained in 1815 in northern and central Europe was so strong that no Russian ruler seriously considered changing it during the century up to the First World War. Further advance in northern and central Europe would merely pose impossible international problems. In the south, however, relations with Turkey were not so agreeably stable. It was here that Russia had uncompleted ambitions. Great progress had been made in this direction during the second half of the eighteenth century. Russia had secured a foothold on the Black Sea. By the Treaty of Bucharest (1812) she had gained Bessarabia and reached the north bank of the Danube. She had found powerful allies among the Orthodox Christian peoples who lived, as Turkish subjects, in the Balkans. In 1812 the Turks had been forced to concede self-government to the most warlike of these peoples, the Serbs. But in spite of several defeats the Turkish Empire refused to collapse. While it remained in being, however weak, geography unfairly assured it a great advantage over the Russian Empire, pressing in from the north. Russian control of the Black Sea was useless so long as the Turks continued to control the exits from that sea. This they effectively did at the Bosporus and, a little further south, at the Dardanelles. The control of these narrow waterways became one of the dreams of Russian rulers, but as soon as the dream looked like becoming a practical policy it aroused fundamental difficulties. Austria also had interests in the Turkish Empire. The Danube valley was her main outlet to the sea and any further Russian movement southward would place the Danube under Russian influence. This prospect was intolerable at Vienna. For Russian rulers during most of the nineteenth century it was a cardinal point of policy to maintain good relations with Austria. European conservatism (and consequently Russian interests) depended upon the St Petersburg–Vienna *entente*. A forward policy in Turkey would be certain to break this essential friendship.

Unfortunately, the problem of Turkey after 1815 was complicated by novel factors. However much Alexander I and his conservative successors wanted to leave Turkey alone, however resolved they were to abandon Catherine's southern policy, events over which they had little control forced them into unwilling action. The Turkish government was unable to rule its own subjects. From time to time (with little or no prompting from Russia) subject nations in the Turkish Empire broke out in revolt. Whenever this happened the Turks were apt to turn to Britain for help.

By the beginning of the nineteenth century Britain had become keenly interested in the eastern Mediterranean. The reasons for this were partly strategic, partly economic and partly emotional. The British land route to India lay through the Turkish Empire. The British were concerned to prevent Syria and the Persian Gulf from falling under the control of a potential European enemy like France or Russia. They felt insecure because they had no naval base further east than Malta. They began to fear that if the Turkish Empire did break up, Russia would get most of it and would threaten British interests in a particularly vulnerable area. Britain also had important economic interests in the Turkish Empire. Trade with Turkey was valuable because it created a favourable balance for Britain. It was not a large market but in a period when British industrialisation seemed at times to be on the verge of disaster it was impossible to permit Turkey to fall under the control of any power which could impede British trade. British statesmen disliked the reactionary system which Russia had imposed upon Europe, and were determined to prevent any further growth of Russian power. Russia replaced France as Britain's principal enemy. A rational fear of Russian ambitions was reinforced by widespread anti-Russian feeling, verging at times upon hysteria, popularised by the press and deliberately manipulated by certain politicians like Palmerston. 'Russophobia' (as this passion has been aptly named) was spread by pugnacious journalists like David Urquhart (1805–77); after about 1830 it was shared by both political parties, by the court (and especially Queen Victoria) and by the middle-class voters. It played an important – perhaps a decisive – part in drawing Britain into the Crimean War. There was no equivalent anti-British hysteria in Russia. The problems posed by Turkey could have been solved by Britain and Russia acting in sympathetic accord. In fact, when such accord was reached, the problems were solved. But the diplomats were hindered by an increasingly vociferous hostility based upon prejudice and ignorance.

The disintegration of the Turkish Empire in the nineteenth century was started by its Greek subjects. Their struggle began in 1821 and aroused the sympathy of all the European powers except Austria, where Metternich clung to the logical conclusion that if legitimacy was the correct principle for Europe it was also correct for the Turkish Empire. By 1825 the Greeks had not only failed to win their independence; they had also provoked Sultan Mahmud II into summoning the effective military aid of his nominal vassal Mohammed Ali, ruler of Egypt. Egyptian ships and soldiers were more than a match for the disorganised Greek rebels. Britain and Russia regarded each other with suspicion, but neither could afford to permit the extinction of the Christian Greeks by the Moslem forces. The Duke of Wellington was despatched to St Petersburg where his presence flattered the new Tsar, Nicholas I. In 1826 the two governments agreed to act in common to mediate between Greeks and Turks. The latter were invited to save themselves from armed intervention by making peace and granting Greece an autonomous status. The Sultan was surprised by this unexpected alliance. He hastened to divide his enemies by rapidly agreeing to some outstanding Russian demands (Convention of Akkerman, 1826). By this he promised to respect the autonomy of Serbia, to accept the Russian frontier demands in the Caucasus and to permit Russian merchant ships

The Dardanelles and the Bosporus

to pass unhindered through the Bosporus and the Dardanelles. But he refused to make peace with the Greek rebels.

Consequently Britain and Russia (now joined by France) united to present a harsher front towards Turkey. By the Treaty of London (1827) they threatened the Sultan with naval intervention unless he made an immediate armistice. This Mahmud would not do and Codrington, in command of an allied squadron which contained some Russian vessels, consequently sank the Egyptian fleet in Navarino bay.

Far from frightening the Sultan into acquiescence the battle of Navarino merely strengthened his resolve to fight on. His obstinacy was rewarded for the alliance between his enemies rapidly broke up. In Britain the Duke of Wellington (who became Prime Minister in 1828) took the line already adopted by Metternich. He deplored Navarino and feared that an independent Greece would be dominated by Russia. Nicholas I, however, was determined to press the matter to a conclusion with or without his British ally. He wanted not only a settlement of the Greek question but also to force the Turks to carry out the terms of the Convention of Akkerman which after Navarino they had repudiated; consequently he declared war in April 1828. Unexpectedly strong Turkish resistance prolonged the fighting for more than a year. But although both Britain and Austria were hostile to the unilateral action taken by Russia they failed to unite against it. The Russian army eventually crossed the Balkans and approached Constantinople in the summer of 1829. Turkish resistance was at an end; she was without allies and it seemed that there was nothing to prevent Russia from asserting extreme claims, perhaps even a demand for control over the southern exits from the Black Sea.

But at this moment Nicholas drew back. He agreed with a special government committee which reported that partition of Turkey would create a 'labyrinth of difficulties and complications' and 'that the advantages of the maintenance of the Ottoman Empire in Europe are superior to the disadvantages . . . that its fall henceforth would be contrary to the true interests of Russia'. The committee recommended that Turkey should be given lenient terms but that the Russian government should prepare itself for the imminent collapse of Turkey in such a way as 'to ensure that the exit from the Black Sea is not seized by any other power whatever'. This report was the basis of Russian policy for the rest of the reign: in 1829 its spirit shaped the Treaty of Adrianople. Russia advanced her frontier from the northern to the southern mouth of the Danube; the privileges of the Danubian principalities (Moldavia and Wallachia) were increased to the point where they would achieve virtual independence while remaining under the nominal sovereignty of Turkey; the Turks recognised the Russian title to Georgia and Circassia, accepted the idea of Greek autonomy, and promised free passage to Russian merchant ships. To soothe British fears nothing was said about the right of Russia to move warships through the Straits. This treaty was, of course, advantageous to Russia. But it did permit Turkey to survive and it was acceptable to both Austria and Britain. Its terms clearly showed Nicholas's determination not to prejudice his European alliances for the sake of completing Russian ambitions in Turkey.

Although Adrianople allowed the European Great Powers to leave Turkey alone for the time being, Egyptian ambitions were not satisfied. Mohammed Ali's hopes had been checked in Greece. But he had modernised his country, extended his influence into Arabia and the Sudan and hoped to create an Arab Empire out of the Turkish provinces. These activities were rather alarming to Britain but hardly effected either the Sultan or the Great Powers. It was quite different, however, when Mohammed Ali invaded Syria and inflicted a great defeat upon the Turks at Koniah (1832). Not only Syria but also Adana fell under Egyptian control. British interests were dangerously threatened because all the overland routes to India came under Mohammed Ali's influence. The Sultan appealed to Britain and France for help, but both these nations were already fully committed in Western Europe. Meanwhile the victorious Egyptian army was approaching Constantinople and in desperation the Sultan turned to his traditional enemy, Russia. Nicholas was as eager to save Turkey from Egyptian domination as he was to prevent it from falling under British control. Europe was astonished to witness the arrival of Russian troops and warships in Turkish territory not as enemies but as friends. The Russian threat was sufficient to prevent any further Egyptian advance but not enough to drive the Egyptian army out of the Syrian provinces. Turkey and Egypt came to an uneasy agreement whereby Mohammed Ali and his son Ibrahim were invested with a life interest in Syria. Turkish integrity was thus precariously preserved by the first Russian intervention in the Arab portion of the Turkish Empire. Nicholas hastened to exact his price for Russian aid. By the Treaty of Unkiar-Skelessi (1833) Turkey and Russia undertook to support each other if attacked, and Turkey promised to keep the Straits closed against all foreign warships. By these terms Russia hoped to bind Turkey to her as a protectorate. It was wrongly supposed that Turkey would not in future turn to Britain and France for military aid. The agreement about the closure of the Straits was regarded in Britain as a grave defeat. Unkiar-Skelessi encouraged the Russophobes, in spite of reiterated

Russian explanations that it contained no new ruling about the Straits. It became a cardinal aim of British foreign policy to reverse the decision reached in 1833.

The task was not difficult because Unkiar-Skelessi satisfied nobody. Mohammed Ali wanted to get a permanent title to Syria; the Sultan wanted to get Syria back; Britain was worried about her land routes to India; and even Nicholas was dissatisfied by his apparent success. He found himself impeded by British ill-will, not only in the eastern Mediterranean but in Central Asia, Persia and Afghanistan. Britain's alliance with France (1834) alarmed him. It brought support to a French government which he regarded as being dangerously revolutionary; it might be turned against Russian interests in the Turkish Empire. Nicholas also discovered that it was impossible to hold Turkey to the spirit of Unkiar-Skelessi. The Turks continued to look to Britain for commerce and diplomatic support. Both the Sultan Mahmud and Mohammed Ali were old men anxious to settle the Syrian question before they died. Both tried and failed to secure the support of the Great Powers for their Syrian policies. At length in 1839 Mahmud attacked his vassal and was heavily defeated. For the third time within a decade the integrity of the Turkish Empire was threatened.

On this occasion the result of the crisis was strikingly different. Britain and France were able to take an active part in Syria, and Nicholas had already decided that his best interest lay in trying to work with the other Powers towards a multilateral settlement. This offered him two great advantages. First, such a settlement was much more likely to be stable; and secondly, if he could reach agreement with Britain the Anglo-French accord would no longer be necessary. For these gains it was well worth giving up the privileged position achieved at Unkiar-Skelessi – a position which had yielded little and which was anyway due to lapse in 1841. With surprising ease Nicholas united with Britain both against Mohammed Ali and against France which foolishly resurrected its Bonapartist ambitions in Egypt. By 1841 the Powers had forced Mohammed Ali to retreat from Syria. His dream of an Arab Empire was abandoned to his successors. By the Straits Convention the Powers agreed to respect the Turkish declaration that no warship was to pass the Straits while Turkey remained at peace. Both Britain and Russia thus received some security, an illusory security so far as Russia was concerned since Turkey was free to call in British naval aid in case of Russo-Turkish hostilities. But such a war seemed unlikely during the forties. Nicholas was delighted by the establishment of good relations with Britain. To strengthen them further he came to London in 1844, and in conversations with the Conservative ministers then in power he outlined his plans for joint action in case Turkey collapsed. The conversations were embodied in a memorandum which to the British ministers had no binding force but which for Nicholas seemed to have the force of a treaty. In spite of this cordiality the Crimean War was less than a decade away.

THE CRIMEAN WAR AND THE PEACE OF PARIS

What envenomed relations between Russia and Britain had little to do with the Turkish Empire. The Turkish Empire was no nearer to, or further from collapse in 1853 than it had been in 1841. It was the 1848 revolutions and their aftermath which made the Crimean War possible and necessary. These revolutions caused a turmoil in every European country except Britain and Russia. Nicholas was much alarmed by them. They threatened to destroy the 1815 system which suited Russia so well. The Paris revolution which toppled Louis-Philippe was perhaps tolerable. But when Metternich fell, when Austrian Italy claimed its indepence, when Hungary declared itself free from Austrian rule, when the King of Prussia put himself at the head of the German national movement – Nicholas realised that the old landmarks were vanishing and that he must restore them. This he did very effectively. The Russian army put down a national rising in Moldavia and Wallachia. In 1849 it went into Hungary and rapidly defeated the rebels whom the Austrian government were powerless to resist. In 1850, it was Russian pressure which forced Prussia to subject herself once more to the German domination of Austria which had been established in 1815. If, by the end of 1850, it appeared that the 1848 revolutions had been thoroughly repressed, the glory all went to Russia. Nicholas was now the bulwark of the whole conservative system. Only if Russia could be brought down could that system be altered.

France alone had passed through her revolutions without

Russian interference. In 1851 Louis Bonaparte made himself Emperor Napoleon III. That a man of this name should rule France was highly displeasing to Nicholas. But there was little he could do about it and while he remained on good terms with Britain there seemed little danger in it. But Napoleon III was also determined to have friendly relations with Britain. He was anxious to break up the 1815 system. Before he could do this, Russia must be either defeated or humiliated. The Turkish Empire was obviously a good place in which to pursue his ambitions. Britain would be pleased, Russia would react, and the French Catholics, whose goodwill he was anxious to earn, would rally to any policy directed towards sustaining the Christians within the Turkish Empire. It was in this spirit that Napoleon III began to seek special privileges for the Catholics at the Holy Places in Jerusalem. The Sultan, rightly smelling in this the beginning of an anti-Russian combination, eagerly agreed to the French demands.

Nicholas reacted, as had been expected. Perhaps he had been made over-confident by the proof of Russian power during the period 1848–50; perhaps he feared the renewal of the Anglo-French alliance. He continued to think that he had an understanding with Britain. He renewed it by holding conversations with Seymour, the British ambassador in St Petersburg, in 1853. He again outlined his plans for a partition of the Turkish Empire and again received the impression that they had the blessing of the British. Thus fortified, he dispatched Prince Menshikov to Constantinople to demand from the Sultan a fresh definition of Russian rights both in the Holy Places and, in general, as the protector of the Christians in Turkey. Emboldened by the French, probably encouraged by the British ambassador, Stratford Canning, the Sultan refused to grant what Menshikov demanded. Diplomatic relations were broken and Russian troops occupied Moldavia and Wallachia.

But Nicholas continued to negotiate, making use of Austria's good offices. Notes flew back and forth about Europe containing various interpretations of what exactly it was that the Russians were demanding of the Turks and whether they had any historical justification. But while the diplomatists were thus happily engaged, the Turks were determined upon having their war. Never again were they likely to have Britain and France as allies.

The British government, rather nervously headed by Aberdeen, allowed itself to be led by the nose by the Turks. The fleet was ordered through the Dardanelles. To make sure of keeping their allies, the Turks declared war on Russia (October 1853). Within a few weeks they had lost their fleet but won over public opinion in Britain and France. At the battle of Sinope, the Russian navy gained a notable victory. For some reason this was indignantly reported in the Western press as a massacre; it was alleged that the Russians had cleared the wounded from the crippled Turkish vessels with broadsides of grapeshot. This public clamour forced the hand of a weak British ministry. Russophobia had had the last word. Amidst patriotic thunder from Tennyson, the British fleet passed the Bosporus. Britain and France declared war in February 1854. For both powers this was chiefly a war against the 'gendarme of Europe'. It was a popular war, fought by liberal countries against a conservative nation which refused to budge from the 1815 treaties. The defeat of Russia was immediately followed by fundamental changes in the European state system.

The Crimean War is marked by two outstanding features. First, the military incompetence of all the belligerent powers; secondly, the fact that none of them put out its full strength. For Russia it was not a war of national survival; it was a war to retain the European influence which she had exerted since 1815. Nicholas died in 1855 in the bitter knowledge that he had lost the commanding European position which his brother had bequeathed him. But neither he, nor his successor Alexander II, went as far as they could have done in waging the war. They refrained from stirring up Balkan nationalism against the Turks. Similarly, the allies refrained from using Polish nationalism against the Russian government. Both sides were aware of the limited, diplomatic nature of the war. Neither side wanted to take steps which would radically undermine the European state system.

The dominant strategic factor was the British fleet. It attacked the Russian Empire from all sides. In the Baltic, it appeared off Kronstadt, a few miles away from St Petersburg. It bombarded Sveaborg and the Åland islands. Russian civilians standing on Baltic beaches could enjoy the majestic sight of British warships sailing a few hundred yards out to sea. The major objective of British policy in this area – to persuade Sweden to join the allies against Russia – was not attained. But throughout the war large Russian armies had to be kept in the north in order to protect the capital against a British landing. British ships also appeared in the Arctic ocean and in the Far East, where an attempt was made to capture the Russian base at Petropavlovsk in the Kamchatka peninsula. In the Black Sea, British ships and money were made available to Shamil, the anti-Russian Caucasian leader. In fact, the allies achieved very little in the Caucasus. The Russian army achieved its greatest success of the war in this theatre by capturing the town of Kars (1855).

But the most striking effect of British seapower was that it enabled the allies to make their main effort in the Crimea. The decision to attack Sevastopol was made after Austria had interposed her forces between the allies at Varna and the Russians north of the Danube. Nicholas had no wish to provoke war with Austria too. He withdrew his army from the Principalities (Moldavia and Wallachia) and refrained from using the traditional Russian invasion route into Turkey. Allied troops landed in the Crimea in September 1854: Sevastopol was not captured until September 1855. The Russian army fought well in this long siege. The fortifications were rapidly and skilfully completed by Totleben, and Admiral Nakhimov provided excellent leadership for a garrison whose morale never collapsed. In *Tales of Sevastopol* Tolstoy (who took part in the defence) gives a vivid picture of the besieged city. But other aspects of the Russian military effort were less glorious. The field armies which attempted to raise the siege were so stupidly led that they could not even defeat the incompetent allied commanders. The supply system was chaotic; there were no railways into the Crimea. Army supply was possibly even more corruptly mismanaged than it was in Britain.

While the long siege of Sevastopol continued the diplomats tried to find a formula by which the war could be concluded. The great difficulty was the divergence between the British and French war aims. Britain insisted upon terms which would leave Russia powerless in the Black Sea and the Baltic and which would prevent her from advancing further in the Caucasus. France was far less severe in her demands. Napoleon III was already thinking ahead to the next, anti-Austrian stage of his plans for the remaking of Europe. For success against Austria in

Italy and Germany he would require Russian goodwill, consequently he did not want to impose a harsh peace. But the key to the situation lay in Vienna. The Western allies realised that without Austrian intervention in the war it would be impossible to make a really effective military threat against Russia. Their attempts to get Austria to join the war had been fruitless throughout 1855. Buol (Metternich's successor) had broken the Holy Alliance but he hesitated to take the further step of actually attacking the power which had been Austria's main support since 1815. The fall of Sevastopol (September 1855) helped to strengthen his purpose. He was attracted by the prospect of setting up permanent Austrian control of the Principalities, and he feared that if the war dragged on it might encourage national risings throughout Central Europe and the Balkans. Consequently in December 1855 Buol presented Russia with terms agreed between Austria, France and Britain, which Russia had to accept immediately or else find Austria among her enemies. The new Tsar (Alexander II) eventually took the advice of his officials and accepted the Austrian ultimatum. It was argued that Russia was in no state to continue the war and that further military effort would endanger the internal stability of the country. Alexander and his advisers were horrified by the fact that this humiliating decision had been forced upon them by the power which had been Russia's ally for forty years. They went to the Paris peace conference more antagonistic towards Austria than towards the Crimean combatants.

The terms of the Peace of Paris (and associated agreements reached between the powers in 1856) had two main objectives. First to check the influence of Russia, and secondly to strengthen and preserve the Turkish Empire. Russian seapower in the Baltic was diminished by the neutralisation of the Åland islands. In the Black Sea Russia (and Turkey) were forbidden to build naval bases or to keep any warships. Russia lost the special position which she had long enjoyed as the protector of Serbia and the Principalities. Both were placed under the protection of the Powers and remained under nominal Turkish suzerainty. Russia ceded southern Bessarabia to the Principalities and thus lost her ability to control the mouths of the Danube. The river was placed under international control. The integrity of the Turkish Empire was jointly guaranteed by Britain, France and

Austria, while the Turkish government promised to reduce the possibility of future internal revolt by carrying through reforms designed to improve the lot of its Christian subjects.

The most important – and the least wise – portion of these agreements concerned the neutralisation of the Black Sea. This clause was forced through by Britain and Austria. It handicapped Russia far more than Turkey; for whereas it was virtually impossible for any Russian warship to enter the Black Sea it was simple for Turkey (or her British ally) to keep a fleet just south of the Bosporus and to pass it into the Black Sea in case of war against Russia. Thus the results of a century of Russian advance in the Black Sea area were threatened. Even the safety of Russian grain exports was not assured. Britain had not insisted upon such crippling terms even at the end of a twenty-year war against France. Their imposition in 1856 is a fair measure of the anti-Russian sentiment of the period. In other respects the terms of 1856 worked out less badly for Russia than had been feared. Austrian control of the Principalities had been envisaged by Buol. In fact, thanks to a number of developments (which included the goodwill of Napoleon III, the growth of nationalism and the obstinacy of Alexander Cuza) the new Romanian state which emerged during the sixties out of the former Principalities was neither pro-Austrian nor anti-Russian.

ESCAPE FROM THE HERITAGE OF THE CRIMEAN WAR

The Peace of Paris was the greatest check to Russian ambitions since the reign of Peter the Great. For the next fifteen years her foreign policy was dominated by a single motive – to escape from the Black Sea Clauses. From 1856 to 1871 Russia became a 'rogue' or revisionist power – in sharp contrast to the rôle which she had played from 1815 to 1854. Then she had thrown all her weight into the conservative balance – which meant in effect the maintenance of the Austrian Empire. But now Alexander II welcomed any opportunity to further the fortunes of the one European power which did not stand in the way of Russia's southern ambitions – Prussia. Russia's withdrawal from the conservative balance was first seen during the Italian War of 1859. This resulted in the expulsion of Austria from Italy – an event which Russia would not have tolerated during the era of

Nicholas I. The new Prussian minister, Bismarck, rapidly grasped the implications. He hastened to take the opportunity of the Polish revolt of 1863 to draw still closer to Russia. While France and Britain bombarded the Russian government with good advice and polite threats, Bismarck offered concrete help. He guaranteed Russia against the possibility of the revival of the Crimean coalition in Poland. Britain and France had worked together to secure the independence of Italy in 1859; Russia feared that they might do the same in Poland. Alliance with Russia earned Bismarck a free hand in Germany, and enabled Prussia to defeat Austria in 1866. The emergence of a powerful Germany was not altogether to Russia's advantage; but it did involve the weakening of Austria and was consequently a further step towards the abolition of the hated Black Sea Clauses. The final step in the destruction of the Crimean alliance was the Prussian defeat of France in 1870. Russia took the opportunity of French defeat and British isolation to unilaterally reject the Black Sea Clauses in October 1870. This caused great indignation in Britain but without French military aid Britain could not act effectively. Gladstone (who had personally disapproved of the Clauses since 1856) was only concerned to defend the general sanctity of international agreements. He therefore summoned a conference which in 1871 legitimised the Russian action. This confirmed Russia's escape from the most humiliating result of the Crimean War. Although poverty prevented the immediate construction of a Black Sea fleet, Russia was now more fully in command of her destiny. She had regained the status of a Great Power.

PANSLAVISM AND THE BALKAN CRISIS OF 1875–8

After the settlement of the Black Sea question, Alexander II returned to his father's conservative system. The League of the Three Emperors (Germany, Austria, Russia) proclaimed the restoration of the pre-Crimean system. Although the rise of Germany naturally demoted Russia from the leadership of the conservative powers, German supremacy was no threat to Russia. Relations between the ruling families had always been and still were very cordial. Bismarck, the German Chancellor, desired nothing more than the retention of the position which

Germany had gained by the defeat of France in 1871. He was fully aware of the possibility of Austro–Russian conflict in the Balkans, and in the interests of European peace he was willing to exert himself to prevent his allies from provoking a crisis. The advantage for Russia was that Bismarck could be expected to restrain Austrian ambitions in the Balkans and to prevent Austria from restoring the Crimean War coalition. Russia's freedom of action, however, was not restricted to the same degree. Germany had no aims of her own in the Balkans; Bismarck's personal desire was for the dismemberment of Turkey in such a way as to satisfy both his allies. In his view there was little objection to Russian control of the Straits so long as Austria received equivalent compensation on the Balkans. But Bismarck's chief hope was for the maintenance of the stability of the Turkish Empire.

This hope proved vain. Nationalist passions and aspirations in and around the Balkans increased in strength during the post-Crimean period, especially in Serbia and Bulgaria. In Serbia Prince Michael Obrenovich (1860–8) modernised his country's army and administration, eliminated the remnants of Turkish suzerainty and constructed an anti-Turkish alliance of Balkan peoples. In all this he received a little unofficial Russian aid but for the most part he relied upon the nationalist aspirations of his subjects expressed through the elected Skupshchina, or legislative body. After his assassination his Balkan alliance fell into warring fragments, to the relief of both Turkey and Austria. Austria was as much threatened by Serbian nationalism as was Turkey. Many millions of Serbs lived under Habsburg rule and they could not be expected to remain loyal to Vienna when their brothers were throwing off Turkish rule and creating a powerful Balkan State. Indeed, Serbian nationalism presented a greater threat to Austrian integrity than to Turkish. The Austrian reaction against it was consequently determined and prolonged. Bulgarian nationalism was potentially less explosive and more confused than that of Serbia. It was in part a literary and cultural revival encouraged by Russian academics. Hundreds of Bulgarians were educated in Russia and there imbibed, in one form or another, the mystical doctrine of Slav unity. But outside a narrow circle of intellectuals the influence of Russia was not strong. The most explosive issue in Bulgaria during the sixties was concerned with the control of the Orthodox Church hierarchy, traditionally dominated by Greeks. This offended the Bulgar nationalists and turned Bulgarian nationalism in an anti-Greek rather than an anti-Turkish direction.

During the reign of Nicholas I these confused, contradictory and potentially dangerous manifestations of nationalism would have been either ignored or pronounced distasteful. Official Russia continued to be hostile to Balkan nationalism during Alexander II's reign. Little advantage was to be gained from Serbian domination of the Balkans: much was to be feared from the inevitable clash with Austria. But official Russia was not altogether free to shape policy as it liked. Public opinion in Russia – and even an autocrat like Alexander II had to be sensitive to public opinion – was increasingly attracted by the struggles of the Balkan Slavs for freedom. Prevented from taking any active part in the government of their own country, the Russian educated public threw themselves with enthusiasm into the task of encouraging the Balkan peoples to be free.

This meddlesome enthusiasm (the Russian equivalent of late Victorian jingoism and imperialism in Britain) was called Panslavism. Its vigour is attested by the various forms which it adopted. Almost everyone could be a Panslav, whatever his political, religious or social beliefs. The pious conservative Orthodox could equate Slavdom with the adherents of the Orthodox Church. Clearly Russia was the senior Orthodox State; Moscow was the 'third Rome' and the Tsar was the proper head of the Orthodox faithful. Such ideas entitled the Russian government to protect the Orthodox faithful in the Turkish Empire. But after the Crimean War this older form of religious Panslavism was fortified by the addition of a secular variety. This took two forms, depending upon whether the believer was a democrat or a conservative. The former (who naturally, were much distrusted by Alexander II's government) proclaimed the equality of all the Slav peoples – even the Poles and the Czechs. Spokesmen of this viewpoint (like Herzen and Bakunin) demanded the creation of a brotherly federation of Slav republics, the end of exploitation of Slav by Slav and the abolition of Russian autocracy. But a much more influential form of post-Crimean Panslavism was preached by writers like Danilevsky (*Russia and Europe*) and Fadeyev (*Opinion on the Eastern Question*). They adopted a high historical and racist line. The past, they argued, had been dominated by the Latin and the German races; the future belonged to the Slavs. The former world-races had been exhausted and debilitated by their efforts. Europe was now suffering from the results of this exhaustion. It was the duty of the Slavs to redeem this situation by refreshing weary Europe with a new admixture of unsullied blood. Russia must fulfil her destiny by conquering ancient Europe and saving her from herself. The way into Europe lay through Vienna. The treacherous Austrian must be dislodged from the Balkans; this was the policy which Nicholas I had failed to pursue and the result was the Crimean defeat. Failure to follow this forward policy in the Balkans would inevitably lead to Russia being forced out of Europe. These confused but Messianic notions had great influence in Russian intellectual and academic circles, especially in Moscow. The Moscow Slavonic Benevolent Committee (founded in 1858 and copied in other large cities) enthusiastically organised visits and lectures from likely Serb or Bulgar patriots. Other Slavs (Poles and Czechs) were less warmly welcomed. They had been Catholicised, Germanised or Europeanised; anyway, they were not nearly so willing to receive Russian help as were the picturesque and backward Balkan Slavs. Moscow ladies in Panslav knitting circles made garments for the gallant mountaineers. Even Dostoyevsky in his *Diary of a Writer* helped to popularise Panslav emotions. One major figure stood aloof from all this excitement. Tolstoy, the sceptical sage of Yasnaya Polyana, attacked Panslav ideas in the last pages of *Anna Karenina*. Few took much notice of him at the time.

Panslavism was a consoling ideology. It consoled Russians for their defeat in the Crimean War. It directed hostility against Austria, the power whose defection had led to that defeat. Its Messianic elements consoled Russians for the industrial backwardness of their country. The British might have steamships and lots of money but the Russians had a mission to the human race. It consoled Russians for their lack of political liberty. They had no parliament but they could all feel *sobornost* ('togetherness') in the contemplation of Panslav dreams. It appealed to the simple idealism of revolutionaries who merely wanted to do good to Balkan peasants. It filled the gap left by the death of Slavophilism. The Russian peasants had been emancipated (1861) and now Russian energies could be directed towards emancipating the Balkan Slavs. It provided a new way of getting to Constantinople. It provided an excuse for Russian imperialism in Asia. It offered a prospect of restoring Russian domination in central Europe. It was all things to all men.

It was also extremely dangerous. Alexander II and Gorchakov knew very well that if Russia used Balkan nationalism against Austria, Austria could reply in kind against Russia. Those who directed Russian policy during the seventies were not Panslavs but they were not immune to Panslav pressure. Alexander was aware of the growing rift between his government and educated Russia. It was tempting to close the gap by officially endorsing the emotion which was widespread among educated Russians. The alternative was to grant a constitution. Panslav nationalism was, for the autocrat, preferable to the grant of political liberties. It was a very dangerous alternative, certain to lead to conflict with the Germanic powers, and to alliance with Germany's western enemy, France. But these dangers had to be faced once it was granted that a weak and unpopular autocrat should gather public support wheverever it could be found. A symptom of Alexander's drift towards Panslav policy was the appointment of N. P. Ignatiev to the key embassy at Constantinople. During his long tenure of this office (1864–77) Ignatiev made no secret of his Panslav sympathies. He was closely linked with one of the strongest centres of Panslav ideology, the Asiatic department of the Ministry of Foreign Affairs; his presence at Constantinople was a constant encouragement to the activities of the numerous Slavonic Committees in Russia. While St Petersburg spoke with a conservative voice, Ignatyev's official position assured the world that Panslav sentiment was not far from the centre of power. Ignatyev was the hero of the influential Panslav newspapers. His public pronouncements were flamboyantly anti-German and his agents distributed Russian money and support in the Balkan capitals. He was the despair of Gorchakov and the officials of the Ministry of Foreign Affairs. It is doubtful whether he did much to attach Balkan nationalism to Russian interests, yet his continued presence in Constantinople acted as a constant irritant in Austro–Russian relations, a symbol of aspirations which could hardly be contained within the League of the Three Emperors.

The great Balkan crisis of 1875–8 was not prepared in

St Petersburg. Its immediate antecedent was a bad harvest which made it difficult for the peasants of Bosnia-Herzegovina to pay their taxes to Turkey. Prince Milan Obrenovich of Serbia refrained from intervention. His disinclination to go to the aid of his Slav neighbours was encouraged by St Petersburg (although it must be said that the Russian consul in Belgrade, an ardent Panslav, privately urged Milan to succour the Bosnians). Nationalist pressure inside Serbia, which increased in strength after the massacre of Bulgarian rebels against Turkey in May 1876, made it impossible for Milan to remain neutral. In spite of the hopes of the Panslav dreamers, the Serbs were heavily defeated by the Turks. Serbian independence was in danger; a few Russian volunteers helped the Serbs but still the Russian government remained inactive. In conjunction with Austria, Russia forced Turkey to call an armistice (October 1876) and a conference was held at Constantinople. The British delegate, Lord Salisbury, was notably less anti-Russian than his Prime Minister, Disraeli. He managed to reach a compromise agreement with his Russian and Austrian colleagues but the Turks, sensing that Salisbury would not be backed by the British cabinet, refused to implement the agreement. How right they were is revealed by a letter written by Disraeli in December 1876: 'Sal. seems most prejudiced. . . . He is more Russian than Ignatieff'. Having failed to get a European backing for intervention in Turkey, Gorchakov decided to neutralise Austria. He remembered the crushing effect of Austrian hostility in the Crimean War. In January 1877, by the Budapest Convention, Austria declared that she would remain neutral in the event of a Russian invasion of Turkey and agreed to accept Bosnia-Herzegovina as the price of her neutrality. Russia promised not to create a large Slav State in the eastern Balkans and to confine her own territorial demands to the area of southern Bessarabia lost in 1856. After a final attempt to get Turkey to set her own house in order, Alexander declared war in April 1877.

The ensuing campaign reminded the world that Russia as well as Turkey was a declining empire. The Russian army was help up on the Danube by the fortress of Plevna until December 1877. The gallant defence provoked another outbreak of Russophobia in Britain. Not until January 1878 did the Russian army approach Constantinople and even then it was so weakened by disease that any further campaigning was out of the question. During the course of the fighting – so different from Germany's swift military decisions – Austria had remained quiet, satisfied that Russia would stick to the Budapest Convention. But in Britain war fever reached new heights. Queen Victoria chided Disraeli for failing to send the fleet up to Constantinople: 'Oh, if the Queen were a man, she would like to go and give those Russians, whose word one cannot believe, such a beating!' But Disraeli, though just as bellicose as the Queen, was in difficulties for he had no ally and therefore no army. At length, in February 1878 the British fleet was ordered up to Constantinople. It anchored in the Sea of Marmora some fifty miles away; at the same time the Russian army was quartered in San Stephano, ten miles from Constantinople.

The Russian army never covered these last ten miles. The presence of the British fleet was, no doubt, discouraging but what really mattered was the exhaustion of the Russian army. But still, a war had been fought and a victory gained. Russia had to announce these facts in a suitable peace treaty. This was the Treaty of San Stephano (March 1878). In this treaty Serbia was abandoned – the Panslavs had been disgusted by their feeble military performance. It was, besides, essential to leave the western Balkans to Austria. Instead Russia created a big Bulgaria which was to include all Macedonia and a part of Thrace. In fact the Bulgarians had been almost as supine as the Serbs, but they had not yet had the chance to show themselves ungrateful to Russia. The Panslavs consequently showered upon them the praise which had been formerly given to the Serbs. Ivan Aksakov wrote that Bulgaria was 'much more important for us and for the future of Slavdom than Serbia'. The creation of a big Bulgaria – which it was thought would bring Russian influence down to the Aegean – did what the Russian attack on Turkey had not done; it united Europe against Russia. Austria complained that the Budapest Convention had been violated (it had); Bismarck backed Austria; and Britain found herself at last with some allies. Russia's military weakness made war against Europe unthinkable. San Stephano had to be abandoned and a fresh settlement negotiated with Bismarck's aid at Berlin (July 1878).

Although the extreme Panslavs were outraged by the insistence of the European Powers upon an international settlement of the Russo–Turkish conflict, moderate-minded officials like Gorchakov and Milyutin thought that their nation had done quite well out of the Congress of Berlin. The main alteration in the terms of the Treaty of San Stephano concerned Bulgaria. Russia had to agree to a threefold division. Macedonia remained a Turkish province, the central area (named Eastern Roumelia) also remained Turkish but was placed under the supervision of the Powers, and only the northern section remained independent. But even this small Bulgaria seemed at the time to offer Russia a field of increased Balkan influence and thus to be a considerable gain. Taken together with the restoration to Russia of southern Bessarabia and the obvious subordination of Romania to Russian pressure revealed both in the war of 1877–8 and at the Congress of Berlin, the Bulgarian settlement offered both immediate advantage and the hope of future stability. Serbia gained little at the Congress; her claims were not pressed by Russia. Her military performance in the war had disgusted the Panslavs and official opinion was anxious not to offend Austria. Austrian interests were fairly easily satisfied. She gained the right to occupy Bosnia and Herzegovina, and to exercise a dominating influence in the Sanjak of Novibazar, an important strip of territory which separated Serbia from the sea and opened the way for Austrian railway imperialism towards the Aegean. The bitterest disputes of the Congress were between Russia and Britain. Even before he arrived in Berlin Disraeli had persuaded Turkey to permit British occupation of Cyprus. From this base Britain claimed to exert a general protectorship over Turkey-in-Asia, a policy which was directed at the prevention of further Russian encroachment from the Caucasus. Only after a severe struggle in which Bismarck backed Russian claims did the British accept Russia's occupation of the Black Sea port of Batum. To compensate Britain for the alleged gain to Russian power, Disraeli insisted that his country was no longer bound by the international Straits Convention of 1841. In other words, Britain could move warships into the Black Sea at the simple request of the Turkish government and without regard to whether Turkey was at war or not. This was rightly regarded in Russia and Europe as a sign that Britain was still afflicted with the disease of Russophobia.

But the most important result of the crisis of 1875–8 was its

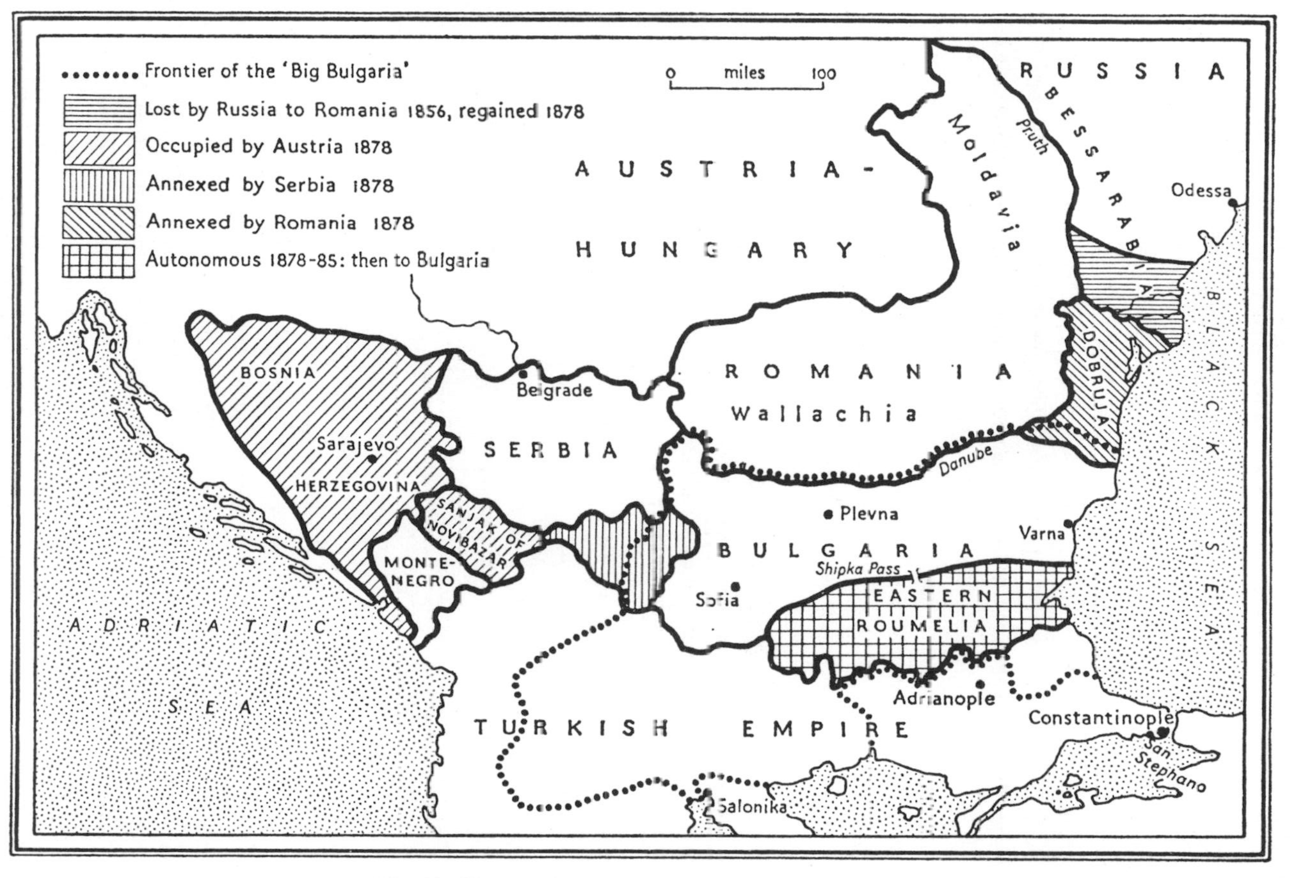

The Balkans after the Congress of Berlin, 1878

effect upon the structure of European relations. Bismarck ceased trying to balance the ambitions of his conservative allies. He decided (1879) to make a firm alliance with Austria which, however ingeniously interpreted by him, was certain in the end to arouse Russian suspicions of a German coalition aimed against her. The worst fears of the Panslavs seemed to be realised. Already in 1878 one of them, General Skobelev, the conqueror of Turkestan, prophesied the shape of the future: 'I . . . entreat you never to forget that the German is the enemy. A struggle is inevitable between the Teuton and the Slav. It cannot be long deferred. It will be long, sanguinary and terrible.' After the Congress of Berlin, in spite of the powerful disinclination of both the Hohenzollern and the Romanov dynasties to end a close association which stretched back into the eighteenth century, Russia and Germany slipped into a hostile attitude. It has lasted without a significant break until the present day.

DRIFTING INTO THE FRENCH ALLIANCE

The move away from Germany was slow and indecisive. On the Russian side there was a strong aversion to the logical step forward, that is, an alliance with France. Powerful emotional reasons made alliance with Republican France unseemly. Such an alliance raised the prospect of a general European war and the end of the Vienna settlement. On the German side Bismarck was eager to maintain the Russian friendship which had served him so well, but developments in the Balkans made his task increasingly difficult. In the end it proved impossible to reconcile the promises made to Austria in 1879 with a friendly attitude towards Russian Balkan ambitions. Even Bismarck was unable to control the repercussions of the fresh Balkan crisis of the eighties

This crisis was caused by the failure of the statesmen at the Congress of Berlin to ascertain whether the settlement which they had made was satisfactory to the Balkan powers themselves. It was not. Serbia was bitterly disappointed by her failure to gain a port on the Adriatic; Bulgaria naturally wanted to regain those territories which had been snatched away from her at the Congress of Berlin. Both Balkan powers eyed each other jealously. Each had the objective of Balkan domination – a significant development from their previous aim of national independence.

The Serbian government, disgusted by Russian desertion in 1878, had moved over into the Austrian orbit. Surprising developments in Bulgaria revealed that this new nation was far less obedient to Russian dictation than the statesmen of 1878 had assumed.

The Panslavs regarded the new Bulgaria with jealous pride. Its existence was the result of Russian effort and sacrifice and they were determined that it should be duly grateful. Russia provided a constitution, Russia dictated the development of Bulgarian railway building, Russians occupied the key positions in the army and the civil administration. Russia also chose a ruler for the new State, Alexander of Battenburg. Although the Prince was a German it was thought that he would be reliable because he had served in the Russian army and seemed, as a man, to be amenable to Russian pressure. Unfortunately for the peace of the Balkans Alexander was not the nonentity his masters desired. He resented the overt Russian control of his country, found it comparatively easy to play off one set of Russian officials against another, and discovered that his independent line secured him the support of his nationalist subjects. He claimed that he would be unable to rule Bulgaria unless the clumsily obvious Russian domination were eliminated or reduced. In 1885 he brought matters to a crisis by announcing his decision to marry a Hohenzollern princess. This was rightly taken by the Russian government to be a symbol of the Prince's intention to align his foreign policy with that of Germany. At this moment the Bulgarian population of Eastern Roumelia rose in revolt against Turkey and offered their allegiance to Prince Alexander. Advised by the great Bulgarian patriot Stambulov, the Prince accepted the offer and relied upon the enthusiasm of his people to protect himself against Russia's anger. Russian officers were withdrawn and Russia stood by while its ungrateful satellite was attacked by Serbia. The Serbs were jealous of the Bulgarian occupation of Eastern Roumelia, they demanded equivalent compensation, failed to get it, attacked Bulgaria and were heavily defeated at the battle of Slivnitza (November 1885). Only an Austrian ultimatum prevented the victorious Bulgarian army from occupying Belgrade.

Prince Alexander was now at the height of his power but his independence was intolerable to Russia. A plot organised by the

Russian government resulted in the abduction of the Prince. In spite of a brief restoration arranged by Stambulov, Alexander thought it safer to retire from a situation in which the next step would probably have been his own assassination. His abdication (1886) permitted the Bulgarian national assembly to elect a successor. It showed no inclination to choose a candidate favoured in Russia. It seemed that Russia would have to take over the direct administration of Bulgaria. This would have been a breach in the Berlin agreement and the prospect caused an international crisis.

Bismarck's task at this moment was made the more difficult by the emergence in France of the hysterical anti-German movement led by General Boulanger. He had to guard himself against the possibility that Russia would do what the Panslavs were already vociferously demanding – ally herself with France. Bismarck extricated himself from his difficulty in two ways. First he made clear to Russia that any unilateral action against Bulgaria would arouse the reaction not only of Austria but also of Germany, Britain and Italy. Secondly, by the Reinsurance Treaty (1887) he offered Russia a German guarantee of the existing situation in the Balkans. He promised to prevent the restoration of Prince Alexander to Bulgaria. He undertook to support the closure of the Straits to foreign warships. He promised German neutrality in case of an Austrian attack upon Russia. He had already published the terms of the Austro–German alliance of 1879 so the Russians knew that Germany was pledged to come to the aid of Austria in case of a Russian attack. By this quibble Bismarck gained a Russian promise of neutrality in case of a German attack upon France.

The Treaty was to run for three years only and was not well received in Russia. The tenuous links with Germany were further weakened by Bismarck's refusal to allow Russia to raise loans in Berlin. But so far as Bulgaria was concerned Bismarck's actions were sufficient to prevent Russia from breaking the Berlin agreement. The Bulgarian assembly elected Prince Ferdinand of Saxe-Coburg to rule the country. Russian hostility was obvious but Russia refrained from action. Bulgaria, whose very existence had seemed such a triumph to the Panslavs, passed outside the Russian orbit into that of Germany. In 1915 the most hopeful creation of the Panslavs declared war upon Russia.

The Bulgarian disaster greatly weakened the influence of the Panslavs. In both the press and official circles it was agreed that Russian national interests – whether or not they benefited the Balkan Slavs – should take priority. The crisis also underlined the fact that Germany was committed to the defence of Austria. The attraction of the French alliance was much strengthened. The Reinsurance Treaty was not renewed in 1890. The Russian government turned to Paris for the loans which Witte's economic policy made necessary. Negotiations between France and Russia opened in 1891 without much enthusiasm on either side, but French investors were already heavily committed to the financing of Russian economic development.

The mutual suspicions, the differences of motive and purpose, the hesitations of both parties can be judged from the fact that the Franco-Russian alliance took three years to reach final signature. France did not want to be drawn into a Balkan war; Russia had no intention of engaging Germany so that France could regain Alsace-Lorraine. Both powers were frightened of the isolation which would follow if Britain joined the German camp – a likely development in the early nineties. The first stage was reached in the autumn of 1891 when a French naval squadron visited Kronstadt; the Marseillaise was played and Alexander III, bareheaded, stood to attention. This was a scene which symbolised the birth of a new relationship between 'revolutionary France' and Holy Russia. But these sensational events were not followed by definite undertakings. Giers, the Russian Foreign Minister, was anxious to prevent the negotiations from going any further. He thought that enough had been done to bring Germany to her senses and persuade her to renew the Reinsurance Treaty. But he was mistaken; there was no German reaction and consequently Giers was unable to protest when General Boisdeffre came to St Petersburg (August 1892) with the draft of a precise military commitment. This was the heart of the alliance as finally ratified (January 1894). France undertook to go to war against Germany if Germany attacked Russia or if Germany assisted Austria in an attack upon Russia. Russia undertook to go to war against Germany if Germany attacked France or if Germany assisted Italy in an attack upon France. If Austria alone attacked Russia, France undertook merely to mobilise. These momentous promises bound Russia to defend the balance of power in Europe: they did not make war with Germany inevitable. In fact both France and Russia hoped to secure the neutralisation of Germany in Europe while they pursued their colonial ambitions. The 1894 alliance was used by Russia to pacify her western frontier while she pushed her ambitions in China. But as soon as the balance of power was threatened the treaty bound both powers either to defend it or to sink into the position of German satellite States – like Austria.

RENEWED BALKAN IMPERIALISM AND THE CLASH WITH THE GERMAN POWERS

The rather surprising result of the Franco–Russian alliance was a decrease in the tension between Russia and the Central Powers. For Russia the French alliance had two great advantages. First it ensured the flow of French investment, and secondly it provided a deterrent against Austro–German initiatives in the Balkans. It protected Russian interests in Europe and freed her to pursue her ambitions in the Far East. This situation suited Germany well and to maintain it she was willing to restrain the Balkan ambitions of her Austrian ally. Tensions within the Turkish Empire were also reduced by the changing attitude of Britain. British interests and resources were fully stretched by imperial adventures, and British statesmen had grasped the implications of the opening of the Suez Canal (1869). This waterway passed under virtual British control in 1882 and its existence helped to moderate the traditional British fear of Russian influence at Constantinople. British statesmen (if not the British public) realised that so long as Britain remained in control at Suez, Russian domination at Constantinople was not much to be feared. Thus by the beginning of the twentieth century one of the fundamental causes of Great Power rivalry in the Turkish Empire had vanished.

The same could not be said for Austro–Russian rivalry in the Balkans. After a brief period of relaxation the two empires lurched into Balkan policies which led directly to the outbreak of a general European war in 1914. The main difficulty was Serbia. In 1903 its pro-Austrian ruler, Alexander Obrenovich, was involved in a rather trivial romantic imbroglio and brutally murdered. The pro-Russian Karageorgovich dynasty was restored to the throne and with it all the Austria fears about the possible results of Serbian nationalism. From this time it became the fixed policy of Austria to destroy the independence of Serbia.

While Austrian intentions were dangerously clear those of Russia were dangerously muddled. Panslavism was much weaker than in the seventies and eighties. Both the general public and the officials were more anxious to secure Russian national interests than the well-being of the Slav peoples. In the early twentieth century the most favoured national interest was control of Constantinople and the Straits. Russians realised that British hostility to this project was far less strong; Russian trade through the Straits was rapidly increasing in volume – indeed the success of Russian industrialisation depended entirely upon the maintenance of the grain trade across the Black Sea. Russia increasingly feared direct German influence at Constantinople. German soldiers and German money had taken up the dominating position previously occupied by the British. Defeat in the Far East (1904–5) had underlined the grave naval disadvantage which Russia would permanently suffer while she was unable to pass warships through the Straits. Russian public opinion, to some extent unleashed by the 1906 Constitution, avidly demanded some compensation for the Far Eastern fiasco. These were all reasons why Russia should take up with fresh energy the traditional drive towards the Straits.

In 1908 Izvolsky, the Russian Foreign Minister, without gaining the assent of either Stolypin or Nicholas II, plunged his country into a major crisis over the Straits question. He met his Austrian opposite number Aerenthal at the Moravian castle of Buchlau and there made with him a rather casual agreement which fundamentally altered the balance of power in the Turkish Empire. Aerenthal wanted to weaken Serbia by annexing Bosnia-Herzegovina. To this Izvolsky gave his consent and thus notably abandoned the Panslav tradition. In return Aerenthal promised Austrian support for a revision of the Straits Convention so that Russia could pass warships through at will. Aerenthal immediately took advantage of his side of the Buchlau bargain, but when Russia attempted to exercise her rights the predictable European crisis was immediately provoked. Neither Britain nor France was willing to back the Russian demand to revise the Straits Convention; Germany was actively hostile and threatened war if Russia attempted to back the Serbian claim to compensation for the Austrian occupation of Bosnia-Herzegovina. Russian strength was not sufficient to permit her to stand alone against the open German threat. With her prestige sadly battered, she was forced to abandon Serbia and to waive her own claims under the Buchlau bargain.

The crisis of 1908–9 envenomed Russo–German relations. For a time after the Japanese War, Nicholas II had hoped to restore the former Hohenzollern–Romanov link and thus to escape from the French alliance which had been of so little use to Russia in her Far Eastern war. The Tsar's personal diplomacy had seemed to reap success when he met the German Emperor at Björkö (1905). But the Russian officials were less enthusiastic, realising that the restoration of Russo–German relations would deprive Russia of French loans at the very moment when they were desperately required. The Tsar's personal diplomacy was quietly shelved – another example of the lack of direction in Russian foreign policy at this period.

Russian responses to the considerable Austrian success in 1908–9 were confused but defensive. Cautious support was given to the Serbian plan for an alliance of Balkan States against Turkey. For Russia such an alliance had the advantage that it tended to minimise Austrian influence in the Balkans. On the other hand a general Balkan war against Turkey might result in Bulgarian rather than Serbian domination of the Balkans. Since Bulgaria was closely linked with Germany such a result would be very damaging to Russian interests at the Straits. An alternative and apparently less risky policy was attempted in 1911 by the Russian ambassador at Constantinople, Charykov. His initiative was not supported by Sazonov, the Foreign Minister, who was ill at the time. Charykov suggested a Russo–Turkish alliance under which Russia would guarantee the integrity of the Turkish Empire against the Balkan States while Turkey permitted the passage of Russian warships through the Straits. But this startling scheme was not backed by Britain – where its effect upon Russophobe public opinion was feared – and was actively opposed by Germany whose strong influence at Constantinople was sufficient to procure the refusal of Turkey. After being restored to health, Sazonov disowned the Charykov proposals.

But whatever the cautious contradictions of Russian Balkan

policy there was no way of controlling the ambitions of the Balkan States. In 1912 they drove the Turks out of Europe and for a time threatened to occupy Constantinople itself. This alarmed Russia; to prevent Bulgarian control of the Straits, Sazonov insisted that the victorious allies should stop short of the Turkish capital. At the conference which concluded this first Balkan war Russia abandoned Serbia. Sazonov realised that if Serbia did particularly well out of the war Bulgaria would demand control of the Straits by way of compensation. He was not displeased when Bulgaria was attacked by her former allies (now joined by Romania) and forced to disgorge some of the territory gained in the first Balkan war. The Treaty of Bucharest (1913) which ended the second Balkan war did something to restore shaken Russian prestige. Serbia gained greatly both in territory and in military reputation. She now seemed strong enough to check any further Austrian penetration into the Balkans. Romania had been drawn firmly into the Russian orbit. She had gained Silistria from Bulgaria and Bulgarian resentment was sure to compel Romania to enlist Russian aid. On the other hand, Russia had not gained her objective at Constantinople. The Straits remained closed, guarded by a much weakened Turkey which more than ever relied upon German support. The Turkish army was virtually under German command and with German financial backing the Turkish fleet was the equal of Russia's.

Russia emerged from the Balkan wars strong in the Balkans but weak at the Straits. The consequence was that she could not afford to admit any erosion of Serbian strength. This explains why Russian policy lacked any flexibility when in 1914 Austria determined to end Serbian independence. With Austria in control of the Balkans and Germany supreme at Constantinople, Russia would lose the freedom of action essential to the being of a Great Power. The Central Powers would be able to close, or to threaten to close, Russia's economic lifeline through the Straits. When Russia went to war in 1914 far more than the fate of the Balkans was at stake. For more than a century Russian statesmen had hoped to extend Russian power by seizing Constantinople. But in 1914 it seemed in St Petersburg that Russian policy was defensive rather than expansionist. For Russia the Straits were a vital interest; for Germany they were merely a diplomatic pressure point. If Germany were not forced to relax the pressure Russia would sink to the status of a German dependency. By a disturbing irony Russia entered the war against Germany in alliance with Britain, the very power which for most of the nineteenth century had worked hardest to exclude Russia from the Straits. When in 1915 Britain tried to force a passage into the Black Sea in order to succour her ally, the German-trained Turkish army was able to resist successfully. With heavy loss of life Britain was repulsed by the power she had so assiduously supported.

RUSSIA AND HER ALLIES DURING THE FIRST WORLD WAR

Russia's relation with her allies during the remaining two-and-a-half years of Tsarist government were confused by suspicion and misunderstanding. They were less cordial than they were to become in the Second World War when a common determination to crush Germany temporarily surmounted deep ideological differences. In 1914 Russia had no desire to crush Germany and after the early campaigns fully realised her inability to do so. Her main enemy was Austria and her principal war aims were to be secured at the expense of Turkey. The total defeat of Germany would merely arouse the Polish problem in an acute form. With the relaxation of German control, the Polish demand for national independence would become irresistible. It was impossible for the Tsarist statesmen to fight a war for the liberty of Poland, a development which Russia had successfully crushed for more than a century. It was desirable to enlist Polish sympathy for the war against Germany but without making any firm commitment to a Russian guarantee of Polish independence at the peace conference. Consequently Russian pronouncements on the Polish question were cloudy and indefinite. They promised the Poles freedom of religion, language and internal administration but strongly resisted French pressure to make a definite offer of independence. At first Germany also refrained from trying to harness the force of Polish nationalism. Only when the German generals got control of German policy in 1916 was such a promise made. Its effect was totally to exclude the prospect of a separate Russo–German peace. The Entente powers promised Polish independence only when Tsarist Russia had collapsed and the United States had entered the war.

Although Russia had no serious territorial objectives in northern or central Europe, the great bulk of her enormous army fought and died in this area. A few divisions could be spared to fight Turkey in the Caucasus; none were available to fight in the Balkans. This paradoxical situation naturally led the Western Allies to fear that Russia would seek a separate peace with Germany. Such a policy was in fact advocated by Witte (who died in 1915) and was sympathetically considered in official circles. Germany tried to open negotiations. But Nicholas II and Sazonov were firmly committed to the principle of allied unity and the German initiatives made no progress. In Western eyes, however, the danger of a separate peace was great and therefore the Western powers were willing to make valuable promises in order to keep Russia in the war. The unexpectedly swift Russian attack on East Prussia in 1914 had helped to save Paris from capture by the Germans. Although the terrible military disasters of 1915 diminished the attraction of the Russian alliance, the Western powers were aware of the likely results of Russian collapse and the consequent German ability to concentrate their army in the West. Russia must be kept in the war by any means that the diplomat or the strategist could suggest.

The strategists suggested that the Western Allies should link up with Russia through the Straits or through the Balkans. The attempt to achieve the former (1915–16) was a failure. The Anglo–French fleet, timidly led, was unable to force the Dardanelles; the Anglo–French expeditionary force was unable to force the Turks out of the Gallipoli peninsula. In the Balkans the Allies were equally unsuccessful. Greece remained neutral until 1917 – and even then only entered the war on the side of the Entente at the point of the gun. Bulgaria waited to see which side was likely to win, balanced up the attractive offers made by both sides and then joined the Central Powers (1915). Romania did likewise but rashly, tempted by the Russian victories in Galicia (1916), decided to join the Entente. Within a few months she had been totally defeated by a German army. Her wheat and oil were of great service to the German war effort and her defeat made the isolation of Russia still more acute. Serbia put up a fierce defence against the Austro–German attack. She remained unconquered until Bulgaria entered the war, and even then the remnant of her army made an epic retreat to the Adriatic and escaped annihilation. Russia could do little for her most reliable Balkan ally either by military action or by diplomacy; indeed Russia had to agree to terms unacceptable to Serbia when Italy joined the Entente in 1915. Part of the Italian price was the promise of Dalmatia after the war. Sazonov protested but he only achieved a mitigation of the Italian claim. By the beginning of 1916 Russia and her allies had been virtually driven out of the Balkans. All that remained was an allied expeditionary force at Salonika whose presence was supposed to threaten Bulgaria. It made no effort to push northwards to join up with the Russian army.

These Balkan disasters helped to keep Russia in the war. It was clear that territorial advantages could come only as the result of a general Allied victory against Germany. Russian determination to keep fighting was also reinforced by the extravagant promises made to her in 1915 at the time of the attempt to force the Dardanelles. Although she stood to gain so much from the success of the Anglo–French venture Russia was nervous about the prospect of her former enemies of the Crimean coalition gaining military control of Constantinople. It was feared that once Constantinople was in Western hands it would not lightly be given up and Russia would have fought the war in vain. Sazonov therefore demanded the agreement of his allies to the acquisition by Russia of Constantinople, the western shore of the Straits, Thrace up to the Enos–Midia line, the Asiatic shore opposite Constantinople and some islands at the entrance to the Dardanelles. Britain agreed without demur to these enormous claims; France put up some resistance but was overborne by the British. By a secret treaty Russia gained the prospect of a post-war settlement which would permanently solve the difficulties of the previous century. To this was later joined a promise that Russia should annex Armenia. These were prizes well worth further sacrifice. To achieve them it seemed acceptable to official and educated Russia to push the nation up to and beyond the brink of revolution.

The overthrow of the Tsar made no difference to the main lines of Russian policy. The Provisional Government continued to fight Germany long after the danger of revolution had become obvious because of the glittering prospect held out by the

Western Allies. There was a close connection between the problem which had plagued Russia and Europe since the beginning of the nineteenth century and the outbreak of the Russian revolution. Had the whole Romanov tradition not been so closely tied to the successful solution of the problem of the Straits, it is possible that more moderate councils would have been followed in 1917 (as they had been in 1856). The Russian peasant masses had no idea that they were being asked to suffer so that their government could get control of the Straits. They were under the impression that they were defending their country against an invader. When Lenin published the secret wartime treaties in 1917 it seemed that he was ending the long Tsarist tradition of conquest by military force.

9

Russia in Asia: A Century of Tsarist Imperialism

THE FAR EAST UP TO 1860

In spite of the denunciation by Moscow of the evils of imperialism, the Soviet Union is the only power in the world today to retain her nineteenth-century empire. India and Indo-China, Java and the Congo are no longer the appanges of the European powers; but the Communist government of Russia continues to rule over the vast spaces of Asia which were conquered under the Tsars. Nowhere is the essential continuity of Russian history more clearly displayed than in the development of Russian Asia.

It should however be noticed that Russian imperialism in Asia was of a very different nature from that of, say, Britain in India. There, a few thousand British ruled and grew rich upon the masses in a land of ancient, settled civilisation. The tribute of the Indus and the Ganges was gathered into fantastic fortunes and flowed into the luxurious living of Mayfair and St James's. But Russian Asia, although greater in area even than British India, was infinitely less populous and far less immediately profitable. In the mid-twentieth century (when the Soviet frontiers in Asia are almost identical to those of the old Russian Empire), only 8 per cent of the Soviet population inhabits three-quarters of the total land mass. The Russian eastward movement cannot be explained by the profit motive alone.

Another explanation which could be offered is the pressure of population, rather like that which carried the Anglo-Saxon peoples across the continent of North America. But this cannot be entertained when we learn that the population of Siberia as a whole only increased from one million in 1800 to ten million in 1917. This was a very slow rate of increase when compared with the size of the human tide advancing across the American prairies in the nineteenth century. A frontier and a frontier life certainly existed in Siberia but never formed the basis of a national myth as in the United States.

It is perhaps best to imagine that Russian Asia has been like a sea, out of which, in the past, the Mongols had come to threaten the destruction of the Russian people. The Russian motive for crossing that sea was partly defensive; its occupation would prevent a similar disaster from occurring again. But with the development of modern communications in the nineteenth century, another motive became important. On the other side of the Asian spaces lay the important interests of Russia's main enemy, the British. A little pressure applied in Turkestan might have desirable results upon the British attitude towards the Near-Eastern question. In other words, the expansion of Russia into Asia might nullify the effects of British seapower. These political motives should be taken into account when Russian imperialism is considered.

Its remoter origins, however, are readily explained by greed. The fur merchants of Novgorod had for long bought their stock from Siberian trappers. In the late sixteenth century it seemed that the source of their supplies might be cut off by one Kuchum, Khan of the Siberian Tatars. The marcher family of Stroganov, with the approval of Ivan the Terrible, decided to defend this valuable supply. They equipped a Cossack, Yermak, to cross the Urals and put down the ambitious Kuchum. This Yermak did and his rapid conquest of Siberia became a legend. The fact that the rivers of Siberia flow from south to north and give little help to the eastbound traveller did not deter those who came after. In 1640, only fifty-nine years after Yermak's first expedition, Ivan Moskvitin, like another Cortez, reached the Pacific. Strong resistance came only when the Cossacks turned south towards China. Their attempts to penetrate the Amur basin brought them face to face with the powerful Manchu dynasty which had just established itself in China. The Chinese troops were disciplined and supplied with firearms, unlike the tribes of Siberia. They were too numerous for the Cossack bands and the Russian government wisely decided to call a halt. By the Treaty of Nerchinsk (1689) the Russians agreed to withdraw from the Amur basin and the frontier was fixed on the line of the Stanovoi mountains, where it remained until 1858. This check proved to be of great importance, for only south of the Amur was there land capable of supporting a large population.

Under tight government control, the fur trade continued to flourish during the eighteenth century. In 1800, the Russian–American Company set up its first station in Alaska and shortly after Nicholas Rezanov, in search of supplies for the Alaskan posts, sailed down to San Francisco where he won the affection of Donã Concepcion Arguello, daughter of the Spanish commandant. Several Russian trading stations were established along the coast of California but they suffered from the fact that communications with European Russia were difficult both because of the distances involved and also because the Russian Pacific ports were icebound for much of the year. It was obvious from the beginning of the nineteenth century that only a move southwards against China could rectify this situation. Yet it was not Russia but Britain which first violated the integrity of China in the Opium War of 1839–42 and forced the Chinese government to open five ports and to cede the island of Hong Kong. Thus in 1842 the Far East ceased to be merely Russia's tradesman's entrance and became a part of the complex diplomatic game.

The British initiative was followed by Nicholas I. In 1847 he appointed N. N. Muravyov Governor of eastern Siberia (capital Irkutsk) with unlimited powers and with instructions to prospect the region of the Amur for a port.

Muravyov was thirty-eight years old at the time of his appointment. He was energetic, ambitious and violently anti-British. In 1853 he wrote to the Tsar: 'We have permitted the English to penetrate into this part of Asia – the same Englishmen who . . . prescribe their laws from their little island to all the continents of the world – except America.' He advocated a Russo–American Pacific alliance, a prospect which was furthered by the sale of Alaska by Russia to the United States in 1867. He took a keen pleasure in the fact that he had repulsed a British expedition to the Kamchatka peninsula during the Crimean War. His appointment caused much dismay among the St Petersburg officials but his forward policy was generally supported by both Nicholas and Alexander II. In 1849 his adventurous subordinate Nevelskoy explored the mouth of the Amur river and established a port there. In 1854 Muravyov, without any serious opposition from the

Manchu government, which was struggling for its life against the Taiping rebellion, annexed the whole Amur basin. The Chinese government was the more willing to concede this region (Treaty of Aigun 1858) because a second and far more dangerous threat had developed from Britain over the matter of the lorcha *Arrow* and her cargo of opium. Still posing as the protector of China, Muravyov seized the whole Manchurian coastline down to the Korean frontier. He eventually got the Chinese to accept this annexation at the Treaty of Peking (1860). At that moment the Manchus were faced by the apparently much more dangerous Anglo–French threat from the south and from the sea. It was worth their while to buy the Russians off with concessions of apparently valueless territory. Muravyov and his equally imperialistic colleague N. P. Ignatyev congratulated themselves upon refraining from taking part in the Anglo–French punitive expedition which burned down the Summer Palace in Peking in 1860. To complete his work, Muravyov founded a new city in the annexed territories – Vladivostok ('Lord of the East'), a fine port but not ice-free throughout the year. This provided a link with European Russia; land communications across Siberia were still very tenuous.

The Treaty of Peking marked the culmination of the first decisive step taken by Russia in the Far East since the end of the seventeenth century. For Russia it was a small compensation for defeat in the Crimea. For Japan it was a portent and a threat. The Far Eastern ambitions of the European powers awoke in Japan reactions very different from those of the Manchu dynasty. The latter tried to appease the Western barbarians; the former copied the techniques which made the barbarians so powerful. Russia wanted to annex the whole of Sakhalin; she also wanted to set up a naval base on the island of Tsushima. The Japanese met these threats skilfully. They negotiated with Russia, using the good offices of the United States. They were much aided by the fact that, after 1860, Russian imperialism was directed more towards central Asia than the Far East. At length, by the Treaty of St Petersburg (1875) Russo–Japanese affairs were amicably settled. Russia kept Sakhalin; the Japanese got full sovereignty of the Kurile islands. This gave Japan time to consolidate the results of her 1868 revolution. She industrialised rapidly and became a formidable rival to Russian Far Eastern ambitions.

THE CAUCASUS AND TRANSCAUCASIA

Expansion in the Caucasus and in Transcaucasia took up much Russian energy during the nineteenth century. The main motives were strategic and political rather than economic. It was necessary to deny Turkey and her allies the north-east coast of the Black Sea. In a south-easterly direction lay the Persian Empire and the Persian Gulf. Control of the Caucasus offered a new means by which Turkey could be threatened, helped Russia in her struggle at the Straits, and offered tempting glimpses of the Indian Ocean.

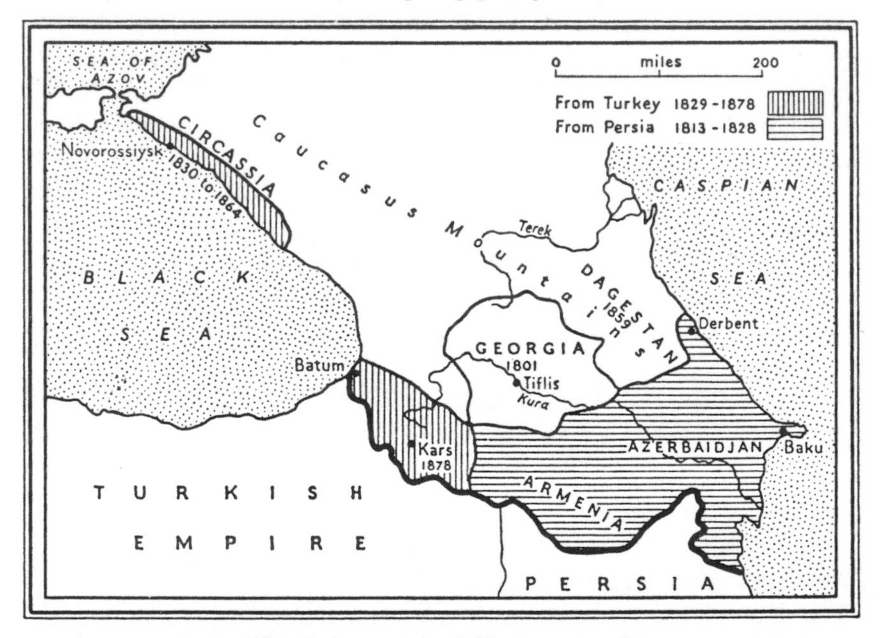

The Caucasus and Transcaucasia

The physical difficulties involved in conquest of the Caucasus were immense. The mountain range stretches for 600 miles between the Black and Caspian seas, with only two easily practicable passages. On the Caspian side a narrow coastal strip was jealously guarded by Persia; in the centre of the range the difficult Daryal pass eventually carried the Russian military road down into Georgia. The Caucasian peoples had all the romantic variety, obstinacy, fanaticism and toughness usually associated with mountain races. Some were Christian – including, fortunately, those tribes who lived in the region of the military road.

Others were Moslems, including the Circassians who lived at the western end and the Checkens who lived at the eastern end. It was from these two peoples that Russia encountered the fiercest resistance. South of the main mountain barrier, but still in high, wild country, lived the two ancient Christian peoples of Transcaucasia, the Georgians and the Armenians. At the beginning of the nineteenth century both looked to Russia for protection against their Turkish and Persian neighbours. Georgia tried unsuccessfully to defend herself during the eighteenth century, but in 1795 a Persian army sacked the capital, Tiflis: in 1801 Alexander annexed the kingdom on the urgent appeal of its last king. After that, Russia was committed to the defence of Georgia. In order to do this, she would have to command the whole Caucasus range.

The first important gains were made at the expense of Persia along the comparatively easy Caspian side of the Caucasus. Here Russia was able to exert her power down the Volga against the Persian-controlled southern coastline. At the Treaty of Gulistan (1813) she acquired Persian Azerbaidjan with the important port of Baku. By the Treaty of Turkmanchay (1828) she got Erivan (Persian Armenia). This enabled her to reach Georgia round the back of the main Caucasus range, but the range itself was still unconquered.

Conquest took far longer than was at first envisaged. During the thirties slow but perceptible progress was made in Circassia at the western end of the range. Here the Russian troops could be supplied from the sea. A chain of fortified posts was built and the tribesmen gradually confined to the most inaccessible regions. The Circassians received some aid from British Russophobes like David Urquhart, who published a Circassian declaration of independence in his periodical, *The Portfolio*. But his attempt to involve the British government in a war on behalf of the Circassians failed. Much more serious for Russia was the powerful resistance which developed on the eastern side of the Caucasus, in Daghestan. Here a guerrilla leader of genius, Shamil, led a holy war against the Russians. For twenty-five years (1834–59) he resisted numerous powerful expeditions. From time to time he threatened the military highway. When he finally surrendered he was treated well. He was taken to meet the Tsar Alexander II, and he was allowed to spend his declining years in Mecca.

After this, the area was rapidly settled. Further territorial gains were made in 1878 when Kars was taken from Turkey. This gave Russia the port of Batum which proved to be a valuable outlet for the oilfields at Baku. But it may be doubted whether the conquest of Caucasia brought all the strategic advantages which had been expected. There was no easy invasion route into Turkey. During the First World War little impression was made upon Turkey by the campaign in the Caucasus. But on the Caspian side, solid advantages followed the defeat of Shamil. The Caspian Sea was now virtually a Russian lake. Its eastern shore could be used as a base for the invasion of Central Asia. The elimination of the Caucasian rebels enabled Russia to turn her attention towards the frontiers of Afghanistan and India. It was in this area that she was to do her most profitable empire-building. The Caucasian campaign may not have led to Constantinople but it did point the way to Tashkent and Samarkand.

THE CONQUEST OF CENTRAL ASIA

The Russian conquest of Central Asia occupied only seventeen years (1864–81). But this had been preceded by a long period of deferred ambition. During the eighteenth century the southern frontier of Siberia had pivoted upon the town of Orenburg. This was the centre of a long line of forts which stretched, east and west, from the northern shore of the Caspian to the Chinese border. It was a bad frontier. It ran through the Kazakh steppe, poor grassland inhabited by nomadic and primitive peoples. To the south lay difficult deserts; and to the south of these the rich and independent Moslem khanates of Kokand, Bukhara and Khiva. Beyond the khanates lay the frontiers of India, Afghanistan and Persia. Obviously, great advantages could be derived from pushing the frontiers southward. A new frontier could be based upon rich and settled regions. It could be used as a base from which to threaten the British in India. For as the Russians moved south the British moved north-west. By the thirties the two imperial powers were beginning to come into contact with one another in Central Asia, and this added yet another conflict to their already poisoned relations. Ambitious Russians hoped to weaken the British grip at the Straits by threatening their Indian empire. Ambitious Britons hoped to defend India by setting up client states in Central Asia. Thus began what Kipling called 'the great game' in this hitherto neglected area. It was a game which the Russians won.

Both the Tsars Paul and Alexander I were interested in the idea of attacking British India. In 1800, Paul dispatched Count Orlov from Orenburg with an army of 35,000 men. His orders show that he had not the slightest idea of the difficulties involved, but he had actually marched 450 miles before Paul's death saved

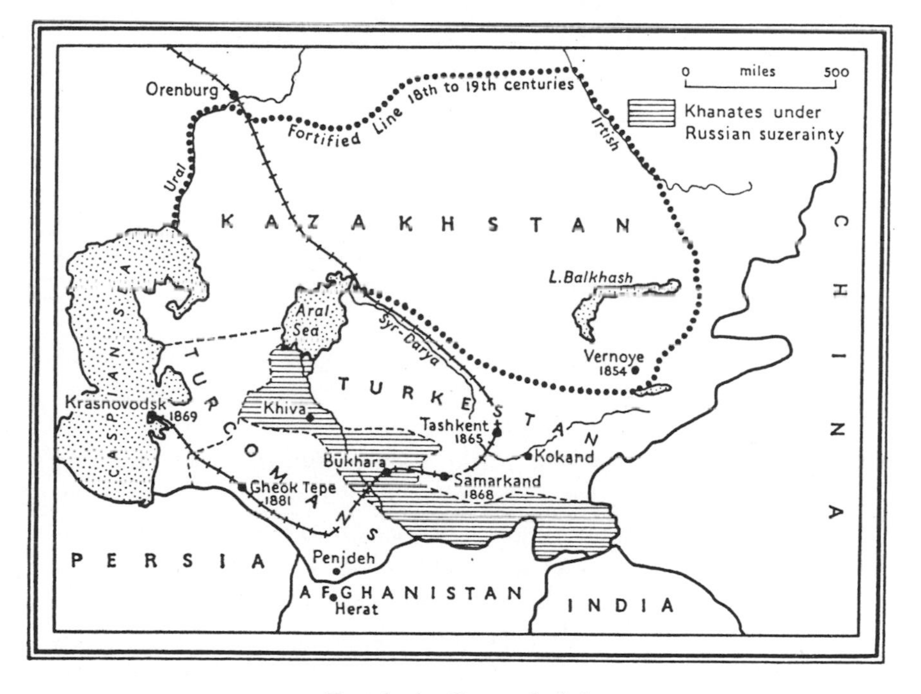

Russia in Central Asia

him from proceeding further with his suicidal expedition. More cautious steps forward were taken during the reign of Nicholas I. Russian peasants were starting to settle in the Kazakh steppes. This provoked an anti-Russian movement among some of the Kazakh tribesmen. They found a good leader in Kenesary Kazymov who (1837–47) resisted the Cossack formations sent against him. Fortified towns were built in the steppes (like Akmolinsk), well forward from the eighteenth-century line. Only one bad mistake was made, in 1839, when an expedition sent from Orenburg against Khiva failed to get anywhere near its destination. Clear signs of British pressure in Afghanistan and the khanates provided still more incentive to penetrate the deserts. The British Russophobes were convinced that a great design was being prepared against India. They were whipped up into a frenzy when the temporarily pro-Russian ruler of Persia laid siege to Herat (1837). They asserted that this was the gateway to India. Palmerston wrote: '[Russia] has opened the first parallels and it would not be wise in us to delay the defensive measures until she has reached the glacis'. The 'defensive measures' involved placing a British candidate upon the throne of Afghanistan. This was attempted (1839–41) and resulted in a dreadful disaster for the British army. After that, the British, like the Russians, concentrated upon consolidating what lay to hand. By the end of the reign of Nicholas I, Russia was prepared for a move against the khanates. She had a fleet on the Aral Sea; she had penetrated 400 miles down the Syr-Darya; she had built a military base at Vernoye in eastern Kazakhstan.

The Crimean War gave Russia an additional motive for pushing towards India. Britain was the main guardian of the hated Black Sea Clauses of the Peace of Paris; pressure on India might help to make Britain change her mind. Another motive which became important at this time was economic. Central Asia produced a lot of cotton. The growing Russian cotton industry was badly hit by the shortage of raw cotton during the American Civil War, and here was a supply ready to hand. But Russia proceeded cautiously. She did not want to be caught by a European crisis with the bulk of her army in Central Asia. Alexander II waited until the Crimean coalition had broken up; he waited until he realised that Bismarck was about to realign the balance of power in Europe. He knew that while this was happening, Bismarck would require the friendly neutrality of Russia. Britain was isolated. She could not prevent the Russian conquest of Central Asia. Nevertheless, British susceptibilities were carefully handled. It was tactfully explained that Russia had no ambitions in Afghanistan or Persia. Forward moves were sometimes excused on the grounds that St Petersburg was unable to control the ardour of the generals on the spot. It was asserted that the Russians were merely doing what the British had long been doing in India – that is, pushing forward to a settled frontier. In fact, there was no British reaction until after the conquest was complete.

The military side was competently handled. Tashkent, the key to Turkestan, fell in 1865; Bukhara in 1867; Khiva in 1873 and Kokand in 1876. Only the south-western corner of Turkestan, inhabited by the fierce Turcoman tribes, remained unconquered. This was accomplished (1879–81) by Skobelev, one of the heroes of the Turkish War of 1877–8.

Skobelev was as able, as energetic and as anti-British as Muravyov. He was already a legend both for his courage and for the eccentricity of his character. It was said that he 'rode to battle clad in white, decked with orders, scented and curled, like a bridegroom to a wedding, his eyes gleaming with wild delight, his voice tremulous with joyous excitement.' The Turcomans called him Bloody Eyes. He seems too to have been a man of great clarity of mind. In 1881 he wrote: 'To my mind the whole Central Asian Question is as clear as daylight. If it does not enable us in a comparatively short time to take seriously in hand the Eastern Question, in other words, to dominate the Bosporus, the hide is not worth the tanning.' It is to be noticed that he does not speak of an occupation of India but merely of the results of a Russian threat. Elsewhere, however, he claimed that not only could Russia break the British hold upon India but that the loss of Indian trade would provoke economic disaster and consequent revolution in Britain. He had equally assured views upon the conduct of colonial wars.

> I hold it as a principle that in Asia the duration of peace is in direct proportion to the slaughter you inflict upon the enemy. The harder you hit them the longer they will be quiet afterwards. My system is this: to strike hard and keep on hitting till resistance is completely over; then at once to form ranks, cease slaughter, and be kind and humane to the prostrate enemy.

It was as well perhaps for Britain that this dedicated imperialist died suddenly in 1882.

In spite of his panache, Skobelev was a thorough organiser and cautious tactician. He built a railway eastward from Krasnovodsk on the Caspian and carefully isolated the Turcoman stronghold at Gheok Tepe. At length, in January 1881, he besieged it and with only 6000 men he captured it from 25,000 Turcomans. He authorised the slaughter of all the male Turcomans and four days of planned destruction followed. The military engineers then pushed on the construction of the broad gauge railway which in December 1881 reached Kizil Arvat, 145 miles from the Caspian. There progress halted; and Russia was still unable to supply a large force on the Afghan frontier when the Penjdeh incident (1885) nearly brought war with Britain. This affair was provoked by the defeat by Russia of an Afghan force at Ak Tepe. This was seen in Britain as a direct violation of frontier agreements and as the first step towards the Russian occupation of Afghanistan. Gladstone could not afford weakness because he already had the reputation of being too 'soft' towards Russia. Yet neither side really wanted war about an Afghan village; the frontier was redrawn and Russia kept Penjdeh.

The crisis encouraged Russia to continue the construction of the Transcaspian railway. In 1886 the Amu Darya was reached (500 miles from the Caspian) and in 1888 the first train steamed into Samarkand. The line was subsequently extended to Tashkent and connected to Orenburg (1905). A branch line which seemed to be aimed at Herat and consequently much alarmed Britain, was constructed from Merv to Kushk. The exact military value of the line was disputed; but it was believed in India that it presented a credible military threat. It seemed that the Tsar Paul's dream had at length been made a practical possibility.

Yet, strangely enough, the completion of the Transcaspian did not lead to a struggle for empire in the east. This was perhaps because both sides now knew that each could deliver the other a nasty blow. A balance of power had been reached. In addition Russia discovered that her Central Asian provinces had a distinct economic value. This lay in the cotton crop which the region produced in increasing abundance. Its availability made Russia less dependent upon the American crop and brought down the price of raw cotton in the manufacturing areas. This happened during the nineties when the Russian textile industry was growing rapidly. The ancient cities of Central Asia sprang to life again – the population of Tashkent, for example, had doubled by 1897. It would obviously be foolish to put all this newly found and badly needed wealth to risk by provoking conflict with Britain over the worthless land of Afghanistan. So, by a curious development, what had started as a Russian threat against Britain turned. out in the end to be a factor in the *détente* between the two nations.

During the century which preceded the battle of Gheok Tepe Russia had annexed a territory about the size of the Indian subcontinent. But only about nine million people lived in it. There was little national consciousness, and an enormous cultural gap between the sophisticated town dwellers and the steppe nomads. The area had been so frequently conquered during the Middle Ages that there were many ethnic and cultural divisions. The only uniting bond was the Moslem faith. Of this the Russians were very frightened, and they did their best not to offend its adherents. Missionaries were not allowed into the new empire. Russian towns were built several miles away from the existing cities so that the inhabitants could be left as undisturbed as possible. In order to soothe the faithful, the khanates of Bukhara and Khiva were left independent client States. The khan of Bukhara was allowed to continue with his barbaric habits of public execution and the operation of the slave trade. This toleration was valuable propaganda. The Russians could point the contrast with British India where the princes were sent to English public schools and made more British than the British. The Russians were not interested in the welfare of their new subjects, and tried to interfere as little as possible. They always kept a very large army in Central Asia, however – they did not pretend to like the people whom they had conquered.

Apart from the independent khanates mentioned above, Central Asia was divided into three administrative areas: Transcaspia, Turkestan and the Steppe region. The administration was, for the most part, directly under the control of the army. The Governors-General were usually serving officers and all their subordinates were seconded from military duties. The standard of administration was not very high. Junior officers frequently gambled heavily; the temptation to recoup their losses from the public purse was irresistible. In 1908–9 Count Pahlen was sent down from St Petersburg to investigate the charges of corruption. He found much to criticise but his conclusion was that the peculations of the administrators did far more damage to the Russian government than to the inhabitants of Central Asia who, after all, were well used to unjust tax-gatherers. They found their new masters preferable to their old. The Russians lightened the general level of taxes. They cleared up many of the old disputes about water rights, they built railways and roads, they encouraged the growth of cotton, they built fine new towns (far more impressive than those of British India), they provided an alternative to the antiquated and biased legal system. The most serious complaint of the natives against the Russians was that they brought in colonists. The influx became particularly great after Stolypin started to 'solve' the peasant problem in European Russia. By 1916 there were about two million Russians settled in Central Asia, most of them peasants, occupying land which the indigenous populations considered as their own.

This was the main cause of the 1916 revolt in Central Asia. It was touched off by a conscription order; hitherto the Russian authorities had not conscripted their new subjects because they did not want to teach them the use of firearms, but at this desperate stage of the war labour was needed behind the front. The revolt was most bitter in those areas where Russian peasants were thickly settled. It was put down by the Tsarist authorities, and the number of deaths inflicted has never been accurately established. One Soviet authority calculated that the population of Turkestan decreased by one-and-a-quarter million in the period 1914–18. This was not a nationalist revolution. Had it been so, the Bolshevik heirs would not have found it easy to hold on to Turkestan. As it happened, they continued to enjoy the latest result of Tsarist imperialism.

THE CLASH WITH JAPAN IN THE FAR EAST

While Russia was conquering Central Asia Japan was making herself a more formidable rival in the Far East. Japanese imperialism was directed towards Korea; by 1890 she had established a commanding economic and political position there. Korea was nominally under the rule of China but the Chinese government was so weak that it could hardly influence the course of events in this outlying dependency. By 1890, China had joined Turkey; a 'sick man' awaiting the physician or the surgeon. As in the case of Turkey, none of the Great Powers could stand idly by and watch China fall under the control of one of their number. The scramble for China opened a new phase in relations between the Great Powers. The prize was enormous in terms both of wealth and power. China was a great market for industrial goods; control of China would give a firm foothold on the Pacific. The

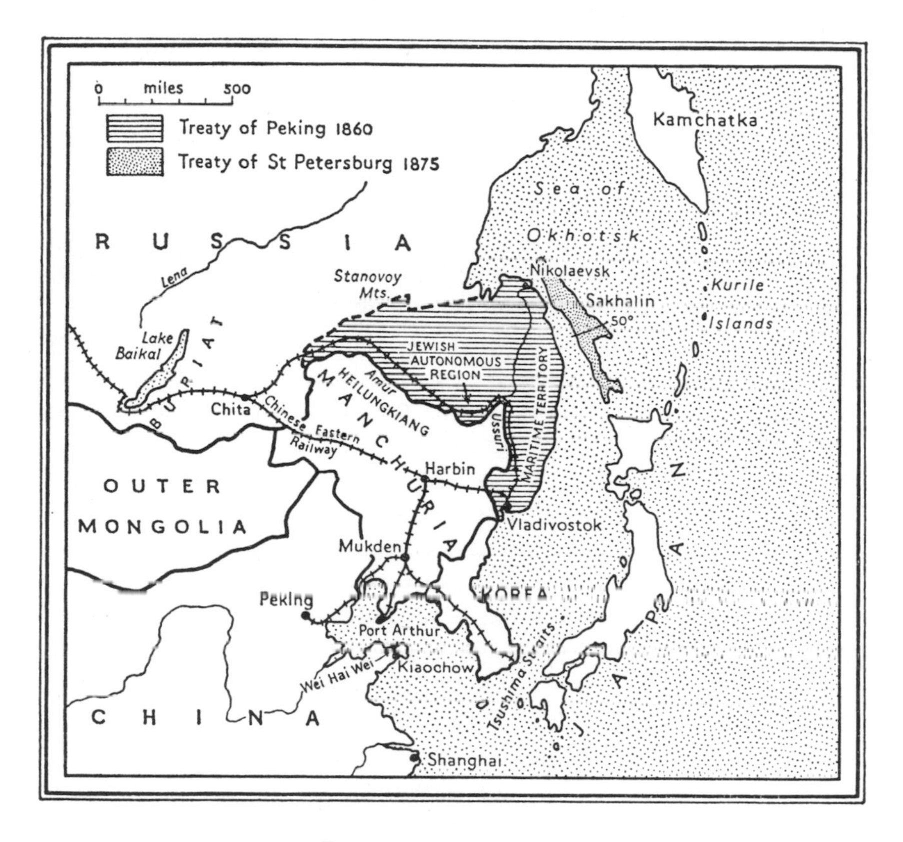

Russia in the Far East

opening of the China question made Russia's earlier Far Eastern conquests all the more valuable. She was the only European power with direct land access. The exploitation of her internal lines of communication would give her an initial advantage over the ubiquitous British fleet.

The possibilities were first clearly seen by Witte. He saw that before Russia could exploit her Far Eastern foothold she would have to link the Amur provinces with European Russia by railway. This was a grandiose project. It would stimulate Russian heavy industry; it would provide a dramatic inducement to the French investors; it would enable the commercial wealth of China to be turned from London to St Petersburg; it would enable Russian political pressure to be put upon Britain; it would please the Germans, who were alarmed by the growing prospect of Franco-Russian alliance; it would open Siberia to peasant colonisation. After skilfully deploying these varied arguments in the right quarters, Witte got his way in 1891. The Siberian railway was to be 3500 miles in length; its eastern terminus was to be Vladivostok; the Tsarevich Nicholas was to head the project, the completion of which was given the highest priority. But it would be several years before it was finished. Until then, Russia was to play a cautious game in the Far East. China was to be preserved until Russia was in a position to enjoy the pickings.

A chance to reveal the temporarily conservative Russian policy towards China soon arose. In 1894 the Japanese decided to force a decision in Korea. They found that their position there was weakened by the shadowy suzerainty still claimed by China. They defeated China in a brief war; at the peace they got not only effective control of Korea but also the Liaotung peninsula with its valuable ice-free port at Port Arthur. Russian ambitions were directed towards this port. Vladivostok was closed by ice for part of the year and it seemed futile to undertake such a vast enterprise as the Siberian railway only to end it at an unsatisfactory port. Russia could have done a deal with Japan in 1895. She could have let Japan keep Port Arthur and helped herself to a warm-water port in Korea. The alternative was to force Japan out of Port Arthur. This was accomplished in 1895 by Russia, Germany and France applying joint diplomatic pressure in Japan. Japan was isolated and gave in without military action being taken against her. The dominant partner in the three-power alliance was Germany. Her main interest in the Far East was diplomatic. In Europe Germany was by now faced by a Franco-Russian alliance; it was therefore in her interests to keep Russia fully occupied in the Far East.

The next initiative in China was taken by Germany. With the intention of giving her Russian ally another push forward in the Far East, Germany seized Kiaochow. This sharpened the China crisis. All the Great Powers had to be compensated – at the expense of China – for the German gain. Russia did best of all. She had already (1896) secured the right to build a railway across Manchuria to Vladivostok. This line (called the Chinese Eastern Railway) gave Russia extra-territorial rights in northern Manchuria; it also shortened by several hundred miles the eastern section of the Siberian railway. Much of the line was open by 1899; only the difficult section round Lake Baikal awaited completion. Improved communications encouraged Russia to drop the policy of defending the integrity of China. In 1898 she demanded the Liaotung peninsula and Port Arthur and the right to build a railway south from Harbin on the Chinese Eastern Railway. She now had a warm-water port with rail communication to European Russia.

These developments alarmed both Japan and Britain. Japan was willing to accept Russian control of Manchuria so long as Russia accepted Japanese control of Korea. Britain was isolated in the Far East and wanted to use Japan as a buffer against further Russian expansion. By the Anglo-Japanese Treaty of 1902, Britain got her way. She did not bind herself to join Japan in a war against Russia; she merely promised to come in if Russia's ally, France, came in on the Russian side. This treaty naturally hardened the Japanese line over Korea. They could afford to fight a war in the virtual certainty that Russia would be isolated.

Since the seizure of Port Arthur, Russian Far Eastern policy had been vacillating between consolidation and further conquest. Witte stood for consolidation. He wanted a period of peace in which Russia could enjoy the economic advantages of her dominant position in northern China. This would mean accepting the Japanese domination of Korea. Other voices, however, encouraged Nicholas II to push forward. There were economic and naval pressure groups with powerful spokesmen in court circles. Such was Bezobrazov, a former Guards officer. Bezobrazov and his friends had obtained some valuable timber concessions in Korea and naturally hoped that Russia would annex this territory. Naval officers argued that the acquisition of Port Arthur was useless while Japan remained in control of Korea. From there, the Japanese fleet could dominate the Tsushima Strait and blockade Port Arthur. Port Arthur would simply become another Sevastopol; useful for defensive purposes but no good as a base for a Pacific strategy. Another powerful voice raised in favour of a forward policy was that of Nicholas II's cousin, the German Kaiser. 'Willy' wrote frequently to 'Nicky'. These two autocrats conducted important business behind the backs of their ministers. 'Willy' sent his cousin a specially designed Christmas card in 1898. He explained its message as follows: 'The two figures symbolise Russia and Germany on the shores of the Yellow Sea, preaching the Gospel, the Truth and the Light in the East'.

'Nicky' was quite willing to be pushed in an anti-British, anti-Japanese direction. He responded eagerly to 'Willy's' warnings about the 'yellow peril'. He was easily flattered when during the Boer War (1899–1902) 'Willy' suggested that it was up to Russia to create a continental league against Britain. At this time, 'Nicky' wrote to his sister: 'You know I am not proud but I do like knowing that it lies solely with me in the last resort to change the course of the war in Africa. The means are very simple – telegraph an order for the whole Turkestan army to mobilise and march to the frontier. . . . The strongest fleets in the world cannot prevent us from settling our scores with England.' It was in this spirit that Germany pushed Russia and Britain pushed Japan. Both European nations had a heavy responsibility for the subsequent war but the ultimate responsibility was with Nicholas II. A stronger man could have resisted the various pressures; he could have concentrated on the Witte policy (the lesser imperialism) and waited for a later opportunity to follow the wider, Pacific imperialism. But even so it might not have been possible to avoid the war. During the final months of peace, the Japanese line was at least as uncompromising as the Russian. Japan was not going to lose the opportunity given her by the British alliance. Both China and the United States were friendly towards her. The dismissal of Witte in 1903 seemed to point to the victory of the extreme imperialists at St Petersburg. Witte's successor at the Ministry of Finance was weak. He was unable to prevent the extremists from influencing the previously cautious policy which the Ministry had followed in Manchuria (where the Russian railway concessions were under its direct control). Rightly or wrongly, the Japanese decided upon a pre-emptive strike. They launched a surprise attack on Port Arthur in February 1904.

The war lasted for eighteen months. Its main feature was Russian naval weakness. By August 1904 their fleet in the Far East had either been sunk, scattered or bottled up in Port Arthur. In 1905 the Baltic fleet, which had laboriously made its way round the world, was ignominiously sunk during a brief battle in the Tsushima Strait. The Black Sea fleet was unable to get out because Russia was at peace with Turkey. This was the price which Russia paid for trying to fight a naval war. The main purpose of building the Siberian railway had been to achieve power in the Far East without decisive naval strength. If Russia had stuck to railway imperialism she might have avoided the disaster at Tsushima. The main feature of the land fighting was the siege of Port Arthur. It lasted 156 days and the final assault cost the Japanese heavy casualties. The Russian soldiers fought bravely but the defence was not as obstinately conducted as was that of Sevastopol. There were still supplies in the town when it surrendered. From the north, along the railway from Harbin and Mukden, the Russian commander-in-chief, Kuropatkin, tried to relieve the Port Arthur garrison. Bigger battles were fought than any in the nineteenth century but after the biggest (February–March 1905) the Russian army evacuated Mukden.

After this Witte was sent to make peace at Portsmouth (United States) in August 1905. He got quite good terms partly because both Britain and the United States sensed danger in permitting Japan to gain too much. Japan had been exhausted by the war and was in no position to press extreme claims. Russia accepted the supremacy of Japan in Korea; she handed over to Japan control of the south Manchurian railway; Japan got Port Arthur and the Liaotung peninsula. But Russia retained the right to keep a fleet in the Pacific; she kept the control of the Chinese Eastern Railway and hence effective rule of northern Manchuria. She was still in a position to profit from the disintegration of China; she was still a Pacific power. Russia continued to straddle the Eurasian land mass, to enjoy access to every sea and ocean. Tsarist imperial expansion in the Far East enabled the Bolshevik heirs to make Russia a world power. This development appeared unlikely when Witte returned from Portsmouth to deal with the revolution which had been provoked by the disasters of the war with Japan. Even more unlikely at that time were the circumstances in which, forty years later, Japan returned to Soviet Russia the conquests made in 1905.

A TEMPORARY IMPROVEMENT IN RELATIONS WITH BRITAIN, 1907

The Russian defeat in the Far East caused the attention of her rulers to return to Europe. The Japanese War had revealed widespread apathy; empire building never had the same popular appeal in Russia as in Britain. Russia reverted to the traditional policy of ousting Turkey from Europe. In the past this policy had been bound to arouse the enmity of Britain but after the Japanese War it became obvious to the British that fear of Russia had been overdone. Her ambitions in the Pacific could be contained by Japan. In the Near East she could be a useful ally against the increasingly dangerous German economic and political penetration of Persia and Turkey. British liberals regarded with favour the halting attempts of the Russian middle classes to set up a parliamentary system. Translations of Tolstoy, Dostoyevsky and Chekhov made Russia less strange, less frightening. The nationalist-minded party of Gladstone could look with favour upon a power willing to support the cause of Balkan nationalism. All these things led to an improvement in Anglo-Russian relations.

Negotiations for an *entente* began in 1906 between the new British ambassador, Sir Arthur Nicholson, and the new foreign minister, Izvolsky. Nicolson (who apparently endeared himself to the Russians by speaking perfect Spanish whenever English would not do) was instructed to get agreements upon Persia, Tibet and Afghanistan. He was to ensure that the two latter countries were to remain independent buffer states and that Persia was to be divided into spheres of influence. The negotiations were slow, chiefly because Izvolsky did not want to affront Germany by uniting with Britain to prevent any German economic activity in Persia. However, the British terms were very attractive, Russia was to have complete control of all northern Persia, including Teheran. This was the richest area of Persia and one in which Russian trade interests were already very strong. Russia had followed a vigorous policy there since the beginning of the reign of Alexander III. She had provided the Shah with a Russian-officered bodyguard and had prevented other Powers from railway building. The ultimate objective of establishing a Russian port in the Gulf was the attraction. In return for the concessions made in northern Persia Britain only wanted to keep control of the area of Persia adjacent to the Indian frontier. At length Izvolsky got a British agreement, permitting German activity in the Gulf area only, and the Convention was signed in August 1907. William II was probably right when he wrote of it: 'Yes, when taken all round, it is aimed at us'.

For Russia, an important result of the Convention was the freedom it gave her to approach Japan, if not as the ally, then no longer as the enemy of Britain. After 1905 it was discovered that Russia and Japan had common interests in China. Both had territorial ambitions in the Chinese Empire and both wanted to prevent the implementation of the doctrine of the 'open door'. This had first been put forward by the United States, and meant that China would remain territorially united but divided into 'spheres of influence' just as Persia had been in 1907. The plan suited the capital-exporters, Britain and the United States.

The Russo-Japanese *rapprochement* bore some useful fruit during the last days of Tsarist Russia. Japan acquiesced in the last little bit of Tsarist imperialism in Mongolia. This large but unpromising desert was held to be important to Russia because the Siberian railway ran so close to its frontier with Siberia. The Chinese revolution of 1911 gave Russia her chance. The Hutukhta, the Buddhist leader of Mongolia, was encouraged to declare his independence of the new republican régime in Peking. Such independence meant Russian protection. Yuan Shi-kai, the President of China, threatened the Hutukhta with war. The latter, quite safe behind the Gobi desert and a Russian alliance, replied: 'Take care lest you be cut up in pieces like a melon'. But in 1913 China recognised Mongolian independence.

In return for a free hand in Mongolia, Russia allowed Japan to annex Korea. The two powers also co-operated amicably in their respective spheres of Manchuria and jointly resisted the attempts of the American railway king, Harriman, to break into the profitable field of Manchurian railway construction and operation. Both powers created useful industrial centres in Manchuria. During the last days of Tsarist Russia, the check imposed by the war of 1904–5 seemed to have been removed.

But the revolution of 1917 and the war of intervention left Russia as weak in the Far East as she was in Europe. Japan had already gained an ascendant position in China and under the guise of intervention against the Bolsheviks she prepared to absorb the Russian Pacific provinces. The White leader Admiral Kolchak stood between the Bolshevik government in Moscow and the Far East, and Lenin could do nothing while he remained. It seemed certain that the Bolsheviks would have to sign an eastern version of the treaty of Brest–Litovsk and that the whole Tsarist empire in the East would have to be sacrificed.

In fact, the threatened disintegration did not take place. In the

Far East, Japan was thwarted by the jealousy of Britain and the United States. In the Caucasus, Tsarist gains were protected by the Turkish revolution. Turkey was too busy struggling against the humiliating terms imposed upon her by the victorious Allies at the end of the First World War to bother about regaining what she had lost to Russia in the Caucasus. In Central Asia Britain alone could have intervened. But Britain was too much occupied by trying to govern an India in which Gandhi had made himself the leader of a powerful nationalist movement to want to add to her imperial troubles by interference in Central Asia. Recovery of the Tsarist colonial territories was also assisted by the mixture of ruthlessness and realism which animated Bolshevik policies. But the most important factor in the Bolshevik recovery of the Asian periphery is connected with a peculiar feature of Russian nineteenth-century expansion. In the Caucasus, in Central Asia and in the Far East Russia had conquered territory, not people. In these occupied territories power was in the hands not of Russian officials but of Russian settlers. Their position was more like that of the French in Algeria than that of the British in India. But, unlike the French in Algeria, these Russian colonists were threatened by foreign domination. For many of them, no doubt, the prospect of a Bolshevik government in Moscow was unattractive; but it was less daunting than the prospect of Japanese or British rule. But why have these Russian colonists not sought independence, like the British in Australia or Rhodesia? Perhaps because they have crossed no sea and consequently have little sense of having left their native land; perhaps because of the centralising tendency of the régimes both before and after 1917; perhaps because of the powerful sense of cultural unity which binds Russians more closely together than any other European people.

10

The Bolshevik Revolution: 1917-21

THE POLITICAL EDUCATION OF THE BOLSHEVIK HEIRS

The party fraction which carried through the revolution of November 1917 had been tried and refined by nearly fifteen years of exile, imprisonment and ferocious controversy. In the course of this period it had so weakened itself and the Social Democratic movement in Russia, it had made so many enemies and had become so small, that it was apparently destined for destruction. But in fact the years in the wilderness had prepared the Bolshevik party for the situation which it found in November. Its doctrine had been clarified, its organisation had been tested and Lenin, its leader, enjoyed an unqualified authority which was based not upon the fear which he inspired nor upon the patronage which he could dispose, but upon the respect in which his unrivalled intellectual and polemical gifts were held. Lenin and his lieutenants were about to enter into their inheritance. What they subsequently did is impossible to understand without some knowledge of the wilderness through which they had passed.

The first Russian Marxists had adopted the doctrine in reaction against the populist and anarchist movements of the sixties and seventies. They hoped to bring a Teutonic order and discipline to the chaotic but futile Russian revolutionary tradition. They hoped to replace the ebb and flow of tyrannicide and reaction with a scientific analysis which, by commanding the assent of all the rational opponents of the autocracy, would lead to political revolution and social transformation. They hoped to break away from the traditional populist worship of the peasantry and its supposedly socialistic inclinations, and to base their revolutionary hopes upon the newly emerging industrial proletariat which during the eighties and the nineties was making its presence felt for the first time in Russia. They were westernisers. They supposed that Russia had no special destiny apart from that of Western Europe and that Russia must embrace the same sort of industrial development which had given rise in the West to a factory system and an industrial proletariat. An unquestioned rejection of Slavophilism and neo-Slavophilism was the first and the greatest of the achievements of the little group of Russian *émigré* intellectuals, of whom Plekhanov was the most influential.

But the application of Marxist ideas to Russian conditions proved to be a task which gave rise to controversies and divisions as fundamental as those which had split the populists. When these theoretical arguments were added to the personal rivalries of the exiled leaders, to the difficulty of maintaining the cohesion of the underground in Russia, to the lack of money and apparent failure, it is hardly surprising that the Russian Social Democrat party broke up into numerous bitterly conflicting elements.

Marxist theory had been designed for the industrially advanced nations of Western Europe. In these, it was claimed, the bourgeois revolution had already taken place. The task of the revolutionary proletariat was consequently fairly simple. It merely had to deepen the already existing class conflict between itself and the bourgeoisie until it had provoked a revolution in which the bourgeoisie would be swept away. In the West, the class enemy was clear enough. He was the factory owner, the boss whom the worker could be expected to hate in the nature of things. All that the revolutionary leader had to do was to make sure that the workers' class feelings were canalised in political channels, that no time was wasted in merely industrial agitation. But in Russia the Marxist was faced by an altogether different situation. The bourgeois revolution had not taken place, the bourgeoisie was as firmly excluded from political power as the proletariat. Further, the proletariat was comparatively small – although growing rapidly – and was swamped in a great sea of peasants. All Marxists agreed that the revolutionary timetable must be observed – first the bourgeois revolution and then that of the proletariat. The dialectic could not be cheated, History could not be hoodwinked.

Some Russian Marxists, including Plekhanov and the enemies of Lenin (Mensheviks) claimed that this dilemma could be met only by patiently waiting for History to catch up in Russia. The task of the proletariat was to co-operate with the bourgeois parties in the struggle against the autocracy. The bourgeoisie must be encouraged to revolt and to seize political power with the customary emphasis upon 'bourgeois' liberties, the rule of law, democracy and so on. The way to socialism passed through bourgeois democracy. This view of Marxist policy had one important tactical implication. It was that the party of the proletariat should be organised for a political struggle, not for secret, underground revolutionary activity. But it also had a great drawback. It implied that the industrial proletariat must be taught to think of the industrial bosses as their political allies – at least until the first revolution had taken place. But at just the time when Plekhanov was first formulating this idea, the Russian workers were striking violently against the bosses. Talk of political co-operation was consequently unrealistic.

Lenin and his disciples (called Bolsheviks after the second Congress of the Russian Social Democratic Party in 1903), took up a line which differed from that described above. Lenin agreed that the proletariat had to find allies, and that it had to kick History through the bourgeois revolution first. But he claimed that this should be done by working with the peasantry. He claimed – and his claim was based upon the facts – that the peasantry was in a revolutionary mood, that it already contained a sort of inner proletariat of poor peasants, and that it would ally with the proletariat in the overthrow of the autocracy. When this had been achieved, he claimed, there would be no need for the proletariat to wait for a long time before it carried out its own socialist revolution. It could rapidly desert its peasant ally by encouraging the class feeling of the poorer peasants against the richer. The failure of the 1905 revolution had convinced Lenin that nothing could be achieved without the consent of the army, and the army represented the class feelings of the peasantry. Lenin, unlike his Menshevik opponents, was impatient. He could not wait for History's wheel to turn through two revolutions. His theory had a quite different tactical implication. He foresaw no long period of co-operation with the liberal groups in a democratic government. On the contrary, in his view it was possible for the party of the proletariat to seize power immediately. But this could only be done by violence and fraud. The Bolshevik party was consequently to be organised as a secret society, which would work underground and which, if it were to be successful, could not afford the weakness of a democratic constitution. Lenin had no intention of forbidding discussion within the party – in fact he was such a formidable controversialist that there was no need for him to forbid it. What he required was that a decision once reached by the Central Committee of the party should be unquestioningly obeyed.

Trotsky belonged to neither of the main Social Democratic groups. He was the only Marxist to play an important part in the 1905 revolutions. As chairman of the St Petersburg Soviet, he showed that it was possible to harness the rather chaotic revolutionary fervour of the city mob – a fervour which the Marxists had done nothing to create and which Lenin regarded with suspicion. After his return from exile, Trotsky showed that he really agreed with the main outline of Lenin's analysis of the peasant situation in Russia. But the brilliant young demogogue could not accept Lenin's ideas about how the party should be governed. The man who had swayed the crowded hall of the St Petersburg Soviet, who had turned his trial into a scathing denunciation of the autocracy, was not likely to accept a type of party organisation which would condemn him to underground activity and to secret committees. At this stage of his career Trotsky required a large audience – the sort of audience which a democratic State could provide. Lenin was content to wield his power in private. The two men remained separated until 1917 when Trotsky (quite rightly) admitted that his opposition to Lenin had been wrong. The type of revolution which both men wanted could be achieved only with the sort of party which Lenin advocated.

In the years after 1905 the position of Lenin and the Bolsheviks seemed to become steadily weaker. There were money difficulties. The Menshevik fraction controlled the party purse and Lenin was driven to some strange means to raise funds. In Tiflis his disciple Stalin organised a large-scale bank-raid as successfully as he managed most other things in his career. Unfortunately the loot was all in 500-rouble notes of which the serial numbers were known to the police. However Litvinov (future Foreign Commissar of the Soviet Union) came to the rescue by changing some of the notes abroad. There were violent quarrels between the Mensheviks and the Bolsheviks about who should have the fortune which had been left to the Social Democrats by the nephew of the wealthy textile manufacturer, Morozov. There were two Morozov nieces and to make quite sure of getting the money Lenin instructed his party that each must be married to a Bolshevik. Unfortunately the party discipline of the older sister's husband was not proof against the fortune which he had netted. He refused to hand the money over to the party. The other match – although probably bigamous – was much more successful financially. The party benefited by some 280,000 roubles. This operation was closely supervised by Lenin. The financial independence which he gained enabled him to resist the blandishments of the Mensheviks. It enabled the party to keep an under ground movement going in Russia, and to make its views known both through its own paper *Pravda* (started in St Petersburg in 1912) and through the six Bolshevik deputies in the Duma. Their position was anomalous in view of Lenin's distaste for bourgeois politics. But he had understood the utility of the Duma for spreading revolutionary propaganda. Unfortunately one of the most trusted Bolshevik deputies proved to be a police spy. This was the sort of penetration against which Lenin claimed that a small party of dedicated revolutionaries would be immune.

But much of the weakness of the Bolsheviks – and indeed of the Russian Social Democrats – in the pre-war years must be attributed chiefly to the ferocious desire of Lenin himself to avoid any compromise which would weaken his own authority. With undiminished relish he turned against a host of heretics. Most of these belonged to the Menshevik clans – for the Mensheviks too had divided into fractions. But some were Bolsheviks of doubtful loyalty. The Machists for example, who attempted to bring Marxist materialism up to date, were subjected to a full-scale philosophical demolition in *Materialism and Empiriocriticism* and their leader Bogdanov expelled from the party. But Lenin's energies were chiefly concentrated upon the struggle with the Mensheviks. With an unscrupulousness matched only by his finesse he got control of the Russian underground for long enough to persuade the secret committees to elect what Lenin called the Prague Conference of the Social Democratic Party. In fact, it was dominated by the Bolsheviks. The Conference (two of the delegates were police spies) obediently elected a central committee packed with Lenin's supporters and backed Lenin in the bitter controversies which were still raging about the funds of the Social Democratic party. The party split was now final. A tiny fraction of the party was claiming to act in the name of the whole. The other European Socialist parties were drawn into these squabbles, and in July 1914 the International Socialist Bureau attempted to mediate. Lenin did not attend but his memorandum – enormously long and vituperative – was presented by a devoted female Bolshevik, Inessa Armand. Lenin proved to be quite unrepentant when his attitude was strongly condemned by the Bureau. He knew he was right; and events were just about to give him some much needed encouragement.

All of Lenin's actions, all his astonishing self-confidence, can be explained on the ground that he distrusted the whole European Socialist movement and his distrust of the Mensheviks derived from the knowledge that they were much closer to their European colleagues than the Bolsheviks were. The events of August 1914 showed that Lenin was right. After years in which the leading European Marxists had extolled the international outlook of their movement, had claimed that their parties would never back a national war, had claimed that a national war was impossible because the working classes would never support it, European socialism collapsed into ignominious nationalism and indeed chauvinism as soon as the war began. The German Socialists fell over each other to vote war credits in the Reichstag; the French Socialists were not tardy either. Only the Russian Marxists – of both brands – refused to support the war; and only the Bolsheviks came out in open condemnation. Only Lenin had the courage to say that the cause of socialism would be best served by the defeat of Russia. He looked to the war to weaken the autocracy, to sharpen the sense of class hatred – in a word to bring the revolution one step nearer. It is not the least of his deeds that he perceived – perhaps luckily guessed – that the mood of patriotism of 1914 would soon turn into just what he wanted. But in the meanwhile his attitude, expressed from the safety of neutral Switzerland, resulted in his party being even further weakened. The leading members of the underground were deported to Siberia; *Pravda* was closed by the government. Lenin maintained only a precarious contact with his followers through junior Bolsheviks like Molotov. Meanwhile the rival Mensheviks, who were not persecuted by the government, made great progress in the trade union and in the War Industry Committees. The latter were dominated by the middle-class parties, with which the Mensheviks had to co-operate. In time this legal activity in support of a 'patriotic' war was to prove fatal to the Mensheviks. In time, Lenin's policy of no contact with the war was to prove strikingly successful – that is to say when the horror and the futility of the war had penetrated even the most unimaginative mind. Then it mattered little that the Bolsheviks were few in number; it mattered only that they alone had been right about the war.

Meanwhile, Lenin continued the work of argument, doctrinal definition and schism from the safety of Switzerland. The attitude which he had adopted towards the war implied that the whole European proletariat should stop fighting against each other and should turn instead against the capitalists. This view was expressed at a socialist conference which met at Zimmerwald in 1915: 'Lay down your weapons. You should turn them only against the common foe, the capitalist governments.' This view fitted in well with one which had been frequently expressed by Lenin (and Trotsky) in the pre-war years – that the Russian proletariat could only succeed in its revolutionary programme if it was assisted by a simultaneous revolution in the West. But it also raised in an acute form the problem of what was to happen to the subject nationalities in the event of a successful revolution in Russia. The Polish Marxists, for example, contemplated without enthusiasm a revolution in Russia which would still leave Poland tied to the chariot wheel of the Russian State. Lenin's tactics on this point were skilful. He did not underestimate the force of nationalism; he had after all witnessed the power of it in 1914. He therefore ruled that each national party could look forward to national independence but that socialist unity would be retained by the international unity of the revolutionary party. There would be many nations but only one party. This formula enabled Lenin to harness national resentments to the task of overthrowing the autocracy. The Mensheviks were more honest and less cunning on this issue. They did not try to hide the fact that the new State would be a unified democracy in which nationalism would play only a limited part. Even some of the Bolsheviks, like Bukharin, protested against Lenin's excessive emphasis upon nationalism. But Lenin crushed him. In one of his most famous pamphlets, *Imperialism – the Highest Stage oj Capitalism*, he showed his Bolshevik critics that twentieth-century nationalism is essentially a struggle against bourgeois imperialism and is consequently a suitable field of endeavour for a revolutionary proletariat. He argued that it must not be confused with the nineteenth-century sort of nationalism which had been a bourgeois movement against feudalism. Lenin always had the best answer – that is, the one which combined a close attention to Marxist niceties with the adoption of the strongest possible bargaining position.

Bukharin also raised the pertinent question: what sort of State will emerge from the victory of socialism? He answered himself, rather naïvely, in the words of Engels, that the State will simply wither away. Lenin, even though at this time he was unable even to pay his hotel bill, threw himself into this controversy with characteristic vigour. He detected in Bukharin's attitude a strong dose of the anarchism which the Marxists had detested in the populists. Such a doctrine gave him far too little room for manoeuvre. So he insisted that after the successful revolution there would be a long, and indefinable, period in which the Marxist party would have to use the State for its own purposes. This period is known as the dictatorship of the proletariat. There would be no withering away of the State until the final overthrow of the bourgeoisie. Lenin intended to promote a social revolution but he did not intend that this should be frustrated by the premature disappearance of state power.

By one of the strange contradictions to which total war gives rise, the Bolshevik cause, almost completely estranged from the European Socialist movement, found an ally in the German General Staff. Ludendorff, who represented an ideological position exactly contrary to that of Lenin, was nevertheless interested in the possibilities of co-operation with the Bolsheviks. He had one thing in common with Lenin: he wanted to create a condition of chaos in Russia. The German government had been in contact with the Bolsheviks since the beginning of the war. The agents of this strange encounter – Helphand, Keskuela and Ganetsky – were strangely motivated. All had been members of various Marxist parties – Helphand, for example, had been closely allied to Trotsky in 1905 – but the war had enabled them to put their knowledge of the European underground to financial advantage. Helphand especially made an enormous fortune out of providing articles in short supply to anyone in any belligerent nation who would pay enough. These socialist entrepreneurs were quite capable of 'fixing' the transport of a few revolutionaries from Switzerland to Petrograd. Ganetsky provided the little party with a considerable sum of money as it passed through Stockholm. Much ink has been spilt in the attempt to determine whether this money came directly from the German government or whether it was part of the profit which Ganetsky derived from his own flourishing trade in contraceptives. Lenin, of course, was quite willing to use the loot of the capitalist world (from whatever source) to further his plans. Those plans were not at all influenced by the wishes of his temporary paymaster.

Ludendorff's final decision to launch Lenin was not taken until March 1917. The first Russian revolution had not done any good to the German cause. Russia was still fighting and indeed might fight still harder under the control of a more efficient nationalist government. The United States was about to declare war upon Germany. Desperate measures were required to bring the war to an end before the effective deployment of American force in Europe. The decision to use Lenin was far more successful than the Germans had hoped; so successful in fact that it not only eliminated Russia from the war but, in the end, helped to eliminate Western Europe from the centre of world affairs. Never did an heir arrive by more devious and crooked ways at the steps of the throne.

FROM MARCH TO NOVEMBER 1917

When Lenin reached Petrograd in April 1917 the Provisional Government had stabilised its position. Its middle-class members held most of the ministries and manipulated the Tsarist bureaucracy without difficulty; it had been recognised by the Allies with enthusiasm; and it had created a working relationship with the Soviet. The gulf between the two bodies had been temporarily bridged; some members of the Soviet had consented to take office in the Provisional Government. Provisional Government and Soviet were linked by their common resolve that the future constitution of Russia could not be determined while the war continued and while the country lived in a popular turmoil. Both groups wanted to postpone the decision about the future until calmer times returned, and both groups agreed that such a decision could only be reached democratically – that is to say, by the voice of a constituent assembly elected by universal suffrage. Both groups agreed to do nothing decisive for the time being – if the people would let them.

There was one issue, however, which could not be postponed; the question of war and peace. The Kadet Foreign Minister, Milyukov, was determined that the Tsarist war aims should be pursued by the Provisional Government. Russia would honour her agreement with the Allies not to make a separate peace; she would continue to insist upon the territorial gains promised in the secret treaties of 1915. In May he informed the Allies of his resolve to continue the Tsar's war. These views, as Milyukov knew, were violently opposed by the Socialist majority in the Soviet. They agreed that the war should be continued but only until such time as all the belligerents could accept a general democratic peace, a peace without annexations, a peace in which there would be neither victors nor vanquished. This policy was hardly less unrealistic than that of Milyukov, but it was at least rather closer to the mood of the masses. The Milyukov note brought the crowds out into the streets of Petrograd and blood was spilt for the first time since March. The Petrograd garrison proved to be quite unreliable: it would not obey the orders of the commanding general, Kornilov, and was not even wholly amenable to the orders of the Soviet. But the most disturbing sign for moderates of all persuasions was the appearance in the streets of crowds of armed working men from the turbulent Vyborg district whose banners carried the Bolshevik slogans demanding peace at any price. These workers were obviously unwilling to accept the assurances of the Soviet leaders that Milyukov's note had been amended. The army was willing to help the Soviet to crush Milyukov's Kadet supporters but was much less enthusiastic about shooting down the Vyborg workers. The result of the May crisis was that Milyukov resigned and that the Provisional Government was reformed upon the basis of closer co-operation with the Soviet. Six moderate Socialists were included in the new cabinet which was now dominated by the Socialist lawyer Kerensky. The move towards closer union with the Provisional Government was fiercely debated in the Soviet.

It was opposed by the Bolsheviks and by many of the left-wing Mensheviks too. In fact this decision turned out to be a fatal one. The result of it was that the moderate Socialists identified themselves with a government which was increasingly out of touch with the demands of the masses for peace at any price, for land and for the socialisation of industry. These May days determined that the moderate Socialists would go over to the anti-popular side of the barricades, that they would be destroyed together with their middle-class allies.

Lenin's tactics immediately after his return had ensured that the Bolsheviks would not share the fate of the other Socialist parties. The original inclination of some of his followers to support the Provisional Government was rudely checked by the declaration of Lenin's policy contained in the 'April Theses'. He demanded an immediate end to the war, the elimination of the police, the army and the bureaucracy, the confiscation of private property in land, and that all state power should be seized by the Soviet alone. He rejected the need for a constituent assembly and claimed that the Soviet already satisfied the democratic aspirations of the masses. By adopting this position he denied his followers the possibility of taking part in any coalition government, but he did ensure that they would identify themselves with the main currents of popular discontent. He relied upon his opponents to dig their own grave, a grave which would contain all the political parties except the Bolsheviks. His tactics were to prod the Provisional Government into unpopular errors and at the same time to restrain his own followers from precipitate action which would enable his enemies to destroy the Bolsheviks before they were ready to seize power. It would be essentially a war of words, of slogans, of ideas. For this Lenin was splendidly equipped, and the fact that the Bolshevik party was so small was actually an advantage to him. Through the newspapers, through the innumerable public meetings which were such a notable feature of the revolutionary months, Lenin could hammer home the one essential point: that the Bolsheviks were the friends of the masses and that all other political parties were their enemies. A lifetime spent in creating political schism had not blinded Lenin to the necessity for mass support; this he would get by avoiding political alliance.

The development of the mass revolution of 1917 is a subject

fitter for the poet or the novelist than for the historian. Pasternak writes:

> I don't know if the people will rise of themselves and advance spontaneously like a tide, or if everything will only be done in their name. Such a huge event cannot be asked for its credentials, it has no need to give dramatic proof of its existence, we'll take it on trust. It would be mean and petty to try to dig for the causes of Titanic happenings. Indeed they haven't any. It's only in a family quarrel that there is a beginning – and after people have pulled each other's hair and smashed the crockery they try to think who it was that started it. What is truly great is without beginning. . . .

In all classes of society people started to ask the most radical question: Why should anyone obey anyone else? The effects of this questioning were nowhere more destructive than in the armed forces. Nearly sixteen million men were called to the colours during the first three years of the war. About seven million were still serving at the front in March and about six million in September 1917. By the latter date this vast army had become a mob. The men would not attack, most of them would not even defend the line. Large crowds wandered about behind the front, terrorising the villages and small towns, shooting their officers, disrupting communications and living off the land. Many more deserted and made their way back to their villages in order to assist in the distribution of the landlord's property. The only authority generally recognised by the soldiers was their own elected committee or Soviet. Years of experience of the village commune had given these peasant soldiers an understanding of such primitive democracy. The Provisional Government had been forced to admit the existence of the committees and had tried, by legalising them, to limit their influence to matters of supply and abuse of authority. It had also been found necessary to abolish the death sentence. But as the summer progressed – and especially after the failure of the July offensive – the government's influence over the committees waned. With increasing frequency, the soldiers shot both their officers and the government commissars sent down to the front. Troops who remained loyal were attacked by the mutineers. This applied especially to the artillery whose greater *esprit de corps* prevented them from falling easy victims to revolutionary propaganda. In August Kornilov, Denikin and other generals attempted to restore discipline by reinstating the death penalty. But as one pessimistic observer wrote: 'What can help? The death sentence? But can you really hang whole divisions? Courts-martial? But then half the army will be in Siberia. You don't frighten the soldier with imprisonment at hard labour. "Hard labour? Well, what of it?", they say. "I'll return in five years. At any rate I'll have a whole skin." ' The Bolsheviks made few converts among the soldiers at the front. Their propaganda (which was chiefly directed at showing that the peasant soldiers had a common cause with the industrial workers) was, however, sufficient to ensure that when the moment came for revolution the army would remain neutral in the struggle for power. The breakdown of army discipline meant that the soldiers would only obey an authority of their own creation. They would have nothing to do with orders which came from far away, orders which demanded that they sacrifice themselves against the Germans or the Bolsheviks. Abstract ideas – patriotism or public order – meant little to soldiers who had grown used to settling their own affairs.

The revolt of the industrial workers was equally spontaneous and equally widespread. The engineering workers of Petrograd, the textile operatives of Ivanovo-Voznesensk and the miners of the Donets basin all seemed to move through the same phases of industrial and revolutionary action – phases which the Bolsheviks used without creating. The first result of the March revolution was vigorous action by the trade unions. The membership doubled between the two revolutions. This movement was accompanied by the usual trade union demands – an eight-hour day (which was gained in March), higher wages, holidays, and payment by time. Such demands were approved by the All-Russian Trade Union Council which, like the Congress of Soviets, was dominated by the moderate Socialist parties. But, as with the army, the more primitive organisations, which were in closer contact with the workers, had more radical ideas than those bodies which officially represented them. These were the factory committees which, especially in Petrograd, fell rapidly under the control of the Bolsheviks. Their policy was syndicalist; they demanded that the factories should be owned and run by the workers. Industrial discipline was undermined by the determination of the workers to expel both the owners and the managers, frequently by force and always against the will of the Provisional Government. Kerensky was bombarded with telegrams from the provinces in which anxious managers requested protection from their own men. They never got it – certainly not the unfortunate engineer Yasinsky who was beaten up by a mob of workers and removed in a wheelbarrow. The revolt of the industrial workers did not make them any richer. Wages failed to keep up with the very rapid inflation of the summer of 1917, and the factory committees proved to be an inefficient mode of organising industrial life. The Bolsheviks did not approve of syndicalism but again they showed wisdom in not trying to halt what could not be halted. The result was that in the ranks of the industrial workers they got a degree of support which no other section of the community gave them. The promise of a Soviet régime was, for the workers, a guarantee that the factories would not fall once more under the control of the bourgeoisie. The armed workmen who lent their support to the November revolution in most Russian towns were not far removed from the peasantry, neither well-educated nor politically experienced. If they had been they would have distinguished between the Bolshevik slogan and the ultimate Bolshevik intention. The dominance of the factory committee ensured that the working men were led by the least experienced.

Over the revolution which swept rural Russia during the summer and autumn of 1917, the Bolsheviks had even less influence than over the army and the factory workers. Their weakness in the countryside was revealed by the Congress of Soviets of Peasant Deputies in which the Socialist Revolutionaries had 537 representatives to the 14 Bolsheviks. Right up to November, the executive committee of this congress was hostile to the Bolshevik policies and loyal to the Socialist Revolutionaries who were playing a leading part in the Provisional Government. But the phenomenon familiar in all social groupings operated in this case too. Power fell into the hands of the local village Soviets who refused to listen to the orders even of those who ostensibly represented their interests, the leaders of the peasant party, the Socialist Revolutionaries. The S.R. leaders were paralysed by the fact that they had to support the general policy of the Provisional Government. If the war was to be carried on the peasants had to be stopped from creating anarchy by seizing the land. Chernov, the S.R. Minister of Agriculture, had to give his support to the following order of his colleague, the Minister of the Interior: 'You are obligated to suppress most decisively any attempts to stir up anarchical confusion. No arbitrary seizures of property and land, no acts of violence, no appeals for civil war and violation of military duty are permissible.' But the peasants took no notice of Chernov, they took no notice of the Provisional Government's land commissars, they took no notice of the warnings of their own deputies that what they did was playing into the hands of the Bolsheviks. They did what they liked, what was close at hand, what they had always wanted to do. The cities seemed far off, the future remote, the present full of freedom. In a peasant land, the S.R.s had seemed at first to have the best chance of dominating the revolution. They had thrown away their chance by linking themselves to a government which they did not control.

Kerensky was not a well-known politician, his government had no popular backing, his manner was frequently bombastic and hysterical and in the end he failed ignominiously. It is therefore easy to condemn him to the rubbish-bin of history (as Trotsky in fact did). Nevertheless Kerensky had an intelligent objective and one which he might have attained had the educated and propertied classes in Russia been less divided and more accustomed to the exercise of political power. Kerensky aimed to preserve the democratic gains of the March revolution from the attacks of both the left and the right. He recognised that both the Bolsheviks and the Tsarists were his enemies; he hoped to play them off against each other. His first move was to secure a military victory. This would help to stem the tide of anarchy and put him into a stronger position to persuade the Allies to accept a negotiated peace. In July the army on the south-western front, where discipline was less eroded than elsewhere, won a convincing battle against the Austrians. But after a few days the Brusilov offensive ran up against German reinforcements, and the apparent victory turned rapidly into an overwhelming defeat. The demoralised attackers fled and behind them a fresh sign of anarchy appeared in the Ukraine: a separatist government in Kiev demanded national independence, a demand which Kerensky was too weak to resist.

The failure of the Brusilov offensive deepened the revolutionary mood of the Petrograd soldiers and workers. Both groups had been strongly influenced by Bolshevik propaganda against the Provisional Government; they were unable to distinguish between a military offensive launched by a Socialist minister and one launched by the Tsarist government. The crowds therefore took up the Bolshevik cry of 'All power to the Soviets' and threatened the Tauride Palace where the executive committee (known as the VTsIK) of the Congress of Soviets held its meetings. This body, dominated by the moderate Socialists, was unable to control the mob. Its members were saved from personal molestation only by the intervention of the Bolshevik leaders. 'Take power, you son of a bitch, when they give it to you', shouted one angry worker to Chernov. But the members of the VTsIK would not; their legalistic, westernised minds could not accept the fact that a body so loosely organised as the Soviet was a proper substitute for formalised, ballot-box democracy. They continued to think that the Provisional Government would remain provisional until the Constituent Assembly met.

Lenin could have seized power in Petrograd during the July days. But he was still cautious; he was not yet sure that the nation would follow Petrograd. He had actually tried to prevent his followers from coming out on the streets but the popular explosion had been too strong even for the Bolshevik leaders. He consequently followed the mob movement until its violence was spent and then persuaded the crowds to go home. As soon as the Kronstadt sailors and the Putilov workers had left the streets, the Provisional Government counter-attacked. On 18 July Kerensky announced that Lenin was a German agent. There was still sufficient dormant patriotism to permit him to seize the Bolshevik paper *Pravda* and to arrest many of the leading Bolsheviks. Lenin himself fled to Finland. His instinct had been correct. The social revolution had not yet gone far enough; at least one of the old landmarks still remained. Kerensky did not follow up his victory. The Bolshevik organisation remained intact and responsive to the stream of orders which reached it from Finland. The apparent defeat of 18 July gave Lenin one great tactical advantage. It enabled him to point out to his followers the revolutionary shortcomings of the Soviet. It had been offered power and it had not dared to grasp it. Lenin continued to pay lip service to the Soviet, but within the party he insisted that the Bolsheviks could not rely upon this institution even if they had a majority in it. There remained only the possibility of seizing power by force, using the Soviet as a screen behind which the Bolshevik party would grasp real power.

The apparent defeat of the Bolsheviks encouraged the forces of the right to attempt a counter-revolution. Its centre was the Stavka (G.H.Q.) and its leader was the new Commander-in-Chief, Kornilov. He was hardly a military conqueror – no Russian general was – but he had won some renown in colonial wars. He reminded the public of his imperial successes by making use of a Turcoman bodyguard which he addressed in their own language. These picturesque savages sat in the corridor with machine guns while their master was interviewed by the Prime Minister. Thus equipped, the old order prepared to do battle with the new.

The generals plotted to restore discipline in the army by the reintroduction of the death penalty and then to march upon Petrograd and extinguish both the Bolsheviks and the Soviet. Kornilov asserted that he would do these deeds in the name of the Provisional Government, but it seems likely that had he got control of Petrograd he would not have rested until he had brought down the Kerensky régime. Kerensky was probably aware of Kornilov's ultimate objective. He foolishly assumed that he would be able to use Kornilov to crush the Bolsheviks and then turn against him. To strengthen his hand in this dangerous game of playing the extremes against one another, Kerensky tried to broaden the democratic appeal of his government. At the end of August he summoned the Moscow Conference, 2414 notabilities from all walks of life, chosen not elected. Kerensky harangued the Conference with such emotion and at such length that the audience had to applaud in order to get him to stop. The Moscow Conference achieved nothing except that Kerensky suddenly realised that rhetoric would not save him from Kornilov once the latter had entered Petrograd. He suddenly dismissed Kornilov from his post (9 September). He need not have bothered. Whatever he did or failed to do was irrelevant to the crisis. As Kornilov's troops approached Petrograd their discipline was undermined and they melted away. They were helpless even against the railwaymen who switched their trains into sidings and branch lines. The Bolshevik-controlled workers, threatened by the right-wing coup, poured into the streets to defend neither

Kerensky nor the Soviet but the revolution. But this mass upsurge was not needed; Kornilov's troops never even reached the capital. The Provisional Government had been saved by its enemies. The weakness of the Kerensky régime had been made clear to all.

The generals' attempted coup benefited the Bolsheviks. They had defended Petrograd against the counter-revolution; Kerensky was suspected of having supported Kornilov; the Provisional Government was fatally compromised. Its policy of continuing the war was already unpopular; its claim to be awaiting the decision of a Constituent Assembly was thought to be merely a device to gain time for a right-wing reaction. All its supporters suffered from the decline of its authority. The moderate Socialist majority in the Soviet was blamed for permitting the Provisional Government to continue in existence. Sinister interpretations were put upon the Soviet refusal to seize power in July. In September it seemed that the revolution was in danger and that its only true friend was the Bolshevik party. The local Soviets in Petrograd and Moscow immediately reflected the swing in the mood of the masses. During September the Mensheviks and the S.R.s were dominated for the first time by Bolshevik majorities. Anarchy spread rapidly in the army and in the villages. The middle ground which the Provisional Government had tried to hold since March was crumbling away. It became clearer that the future lay with either the extreme left or the extreme right.

This was the moment for which Lenin had waited, the moment for which his party had been created. But it needed all his gifts to persuade the Bolshevik Central Committee to follow him. He could only make fleeting visits to Petrograd, equipped with a variety of not very effective disguises. Zinoviev and Kamenev directed an impassioned debate against him. They argued that to provoke a revolution was suicidal; the party merely had to wait for power to be handed to it by a democratic majority. Such a majority might be found in the second All-Russia Congress of Soviets, due to meet in Petrograd on 7 November; it might be found in the Constituent Assembly, for which elections were due at the end of November. In spite of being so close to Lenin for so long, Kamenev and Zinoviev failed completely to understand his mind. He did not want to receive power from a democratic majority. The majority would then become his master. This the Historical Dialectic would not permit. History did not act through majorities but through classes. The Bolshevik party was not the servant of electors but the instrument of the class conflict. It spoke for the 'proletariat' – a mystical entity rather similar to the General Will. At this supreme moment, Lenin acted quite consistently with the ideas which he had developed since the beginning of the century. Without those ideas it is doubtful whether he would have pressed for insurrection at this time.

The final decision to launch the revolution was taken on 23 October. The Central Committee voted 11:2 in favour. Party discipline was such that the minority also prepared to take up position on the barricades. Even now the date for the rising was not fixed. Lenin wanted to proceed immediately but Trotsky persuaded him to wait until the eve of the meeting of the All-Russia Congress of Soviets. Trotsky argued that it would be bad tactics were the Bolsheviks to stage a revolution in their own name. The waverers in the Petrograd garrison would be gathered in if it were shown that the Bolsheviks were acting in the name of the Soviet, that they were protecting it from a counter-revolutionary blow. With considerable skill Trotsky achieved just this. The Bolsheviks in the Petrograd Soviet spread rumours that Kerensky was about to shift the garrison to the Riga front. This would enable him to fill the capital with anti-Soviet troops. In a panic, the Petrograd Soviet authorised the creation of a military revolutionary committee to organise its own defence. It was dominated by Trotsky who, in the name of the committee, not of the Bolshevik party, armed the reliable working men. Meanwhile Kerensky did nothing. He knew that the insurrection was about to take place; he looked forward to crushing it and the Bolshevik party with it.

Trotsky made his dispositions for 7 November so skilfully that there was little bloodshed. The night life of the city was hardly interrupted. By dawn all the key points, with the exception of the Winter Palace, had been captured. Kerensky fled (in a car flying the American flag) to seek out some loyal troops. The Winter Palace was defended for a few hours by a battalion of women and a few officer cadets. Confused fighting went on along the enormous corridors. Somewhere the ministers of the Provisional Government were still conferring. The Bolshevik groups searched everywhere but could not find the cabinet room. At length they were led there by a dignified flunkey with powdered hair. The ministers were not killed: they were escorted across the Neva towards the Peter and Paul fortress. On the bridge prisoners and escort alike had to lie down behind the parapet while some over-enthusiastic revolutionaries loosed off a few volleys at them. Nobody was hurt. The Petrograd garrison remained neutral on 7 November; it was still inactive a few days later when Kerensky returned with some loyal troops and was driven off by the Bolshevik working men. The diarist Sukhanov reckoned that at this moment 500 good troops could have wiped out the Bolsheviks.

By 8 November the Congress of Soviets had been brought to heel. About half its members were Bolsheviks. Most of the other half (Mensheviks and Right S.R.s) walked out in protest against the Bolshevik coup. They thus consigned themselves to the dust-bin of history. The remainder warmly approved of what the party had done. They elected a Council of People's Commissars which was dominated by the Bolshevik Central Committee. Up to the last the Bolsheviks maintained the fiction that they had made the revolution to save the Soviet. In fact, they had shackled it to the party. The fiction that the Soviet became the new governing power has been maintained ever since 7 November 1917.

Outside Petrograd the Bolsheviks were successful in most of the large cities (including Moscow). They had, however, created a situation in which civil war was inevitable. The Stavka at Mogilev served as an anti-Bolshevik rallying point for a few weeks, but soon all the generals fled south to the Cossack country around Rostov-on-Don, and Ensign Krilenko, the new Bolshevik Commander-in-Chief, seized the Stavka with a trainload of revolutionary sailors. At this point Lenin had little faith in the ability of armed men to change events. No sooner had this hunted intellectual climbed upon the throne of the Tsars than he started to address not Russia alone but the whole of Europe. He invited the working men, currently killing each other in north-east France, to overthrow their governments and to follow the Russian example. Both Lenin and Trotsky thought that the Petrograd revolution could only be saved if the revolution was successful in Western Europe. They were as wrong about that as they were right about the possiblity of revolution in Petrograd.

THE TREATY OF BREST-LITOVSK; 1918

The achievement of power did not lead to any slackening in the Bolsheviks' revolutionary momentum. On the contrary, during the next three years the party deliberately intensified the class hatreds which had helped to bring them to power. They intended to rule without allies, domestic or foreign. Those who wanted a moderate democratic revolution and those who were opposed to any revolution at all had to be eliminated. The anarchical tendencies of 1917 had to be intensified and at the same time the foundations of a new discipline had to be created. The Bolsheviks proved at first to be much more successful in their destructive efforts. With the aid of their enemies they destroyed the entire organisation of civilised life in Russia. By 1921 the wave of

The Peace of Brest-Litovsk

destruction had gone too far for even the Bolsheviks. For a time Lenin had to allow society to recover slowly through the primitive incentive of peasant greed.

The most pressing need at the end of 1917 was for a peace with Germany. Lenin had promised it. He realised that the Provisional Government had been fatally weakened by its refusal to adopt a clear peace policy. At first all the Bolshevik leaders were certain that such a peace need not cost defeated Russia too much. They considered that the Russian peace effort would spark off revolutions in all the belligerent nations. They did not believe that the common soldiers would endure another winter in the trenches, another summer offensive, when they got to know that the Bolsheviks were offering a universal and immediate peace. The mutinies in the French army and the German fleet were understood in Petrograd to be the prelude to a general breakdown of military discipline. Time was needed for these manifestations to take effect, and it was time that the Russian peace delegation at Brest-Litovsk sought to gain. For three months Trotsky (the Commissar for Foreign Affairs) and his colleagues held off the German army with words alone. The tactics of Brest-Litovsk were brilliantly executed but the strategy was faulty. The expected European revolution did not occur.

Although Germany's military position appeared to be impregnable at Brest-Litovsk – her army could advance at any moment upon both Petrograd and Moscow through the empty Russian lines – her negotiating position was less strong. The German troops on the eastern front were needed in the west at once for the *Kaiserschlacht* or knock-out blow which, in Ludendorff's view, was to defeat France and Britain before the arrival of the American army in Europe. The German High Command therefore did not want to drive the Bolsheviks into a desperate war which would certainly tie down large numbers of German troops. Secondly, Ludendorff considered that it was absolutely essential to get the whole of the Ukraine under German control so that its economic resources could be used to counterbalance the effects of the British blockade. He was after a frankly annexationist peace which would enable the Reich to tap the riches of central and eastern Europe. But his plans – strikingly similar to those of Hitler in 1941 – were opposed by Kühlmann, the representative of the German civilian government. Kühlmann realised that the signing of such a peace as Ludendorff envisaged would put Germany into a very weak position if by any chance the *Kaiserschlacht* failed; a contingency which Ludendorff refused to consider. In that case, Kühlmann argued, the Allies would be entitled to impose upon Germany just such a peace as Ludendorff intended to impose upon Russia. He consequently wanted to restrict the German annexations to those parts of the Russian Empire already occupied by the German army, and to clothe these annexations in a decent garb of democratic choice.

For much of January and February 1918 Trotsky exploited the rift in the German ranks. He proved to be the equal of the German career diplomats in every twist of the game, although he was embarrassed by the presence in the Russian delegation of a peasant (included for propaganda reasons) whose primitive table manners and frequent intoxication delighted the German staff officers. Trotsky challenged Kühlmann to show that the 'plebiscites' held in the German-occupied territories were genuine expressions of the popular will (they were not); and Kühlmann demanded to khow why the German government should not negotiate directly with the independent government of the Ukraine seeing that the Bolsheviks had constantly demanded the right of national self-determination. Trotsky was able to counter this move by producing another Ukrainian delegation which claimed to represent a Bolshevik government. The inspired quarrels of the two Ukrainian delegations held up the negotiations for several days. Meanwhile in Petrograd the western horizon was anxiously scanned for the coming of the revolution. The Central Committee of the party divided into three groups about the correct policy to follow. Lenin argued that it was pointless and dangerous to prolong the negotiations, for the revolution was not imminent. 'Germany is only pregnant with revolution', he said, 'the second month must not be mistaken for the ninth. But here in Russia we have a healthy, lusty child. We may kill it if we start a war.' Bukharin was for starting the war again; and Trotsky advocated an altogether novel approach summarised in the slogan, 'Neither Peace nor War'. The party at length committed itself to Trotsky's policy; on 10 February he astonished the German delegation by declaring that although Russia refused to accept the peace terms she nevertheless pronounced that she was at peace with Germany. But the gamble failed. After initial confusion and fury, Ludendorff ordered the German army to advance. There was nothing to stop its march through Estonia towards Petrograd. Lenin had been proved right. He brushed aside the renewed demands for a patriotic war and signed the Treaty of Brest–Litovsk on 3 March. It was a heavy price. Russia lost 34 per cent of her population, 54 per cent of her industrial plant and 89 per cent of her mines. The surrender of Germany to the Western Allies a few months later enabled Russia to repossess the Ukraine; but not until the end of the Second World War did she regain control of Poland, Lithuania, Estonia, Latvia and Finland. It is likely that the loss of these non-Russian borderlands simplified the task of the Bolsheviks. But it was difficult to explain this in 1918. Lenin had to marshal all his resources to get the treaty ratified by the Soviet. It provided another reason for hating a régime which had already given many causes for hatred.

THE ESTABLISHMENT OF THE DICTATORSHIP OF THE PROLETARIAT

The Bolshevik control of the new régime was masked by the retention of the Soviet structure. The Congress of Soviets elected an executive committee which in turn chose the Council of People's Commissars. Both the elected bodies were dominated by the Bolsheviks whose Central Committee was the real source of power. But the pretence of democracy was elaborately preserved and all the party's decisions were presented to the world as those of the democratic Soviet. There the other left-wing parties continued to find representation until the middle of 1918: in the Council of Commissars only a few left-wing S.R.s were given minor offices and even these withdrew in protest against the Treaty of Brest-Litovsk. Any other possible centre of authority was deliberately destroyed by the party. The most dangerous was the Constituent Assembly which was already being elected when the revolution occurred. The Bolsheviks held only 25 per cent of the seats in this, the only fully democratic assembly in the history of Russia. Lenin allowed it to meet (January 1918) but argued that since it was hostile to the Soviet régime it could not really represent the wishes of the masses. He contemptuously condemned it to the limbo of 'bourgeois democracy' – that is, a democracy which was manipulated by the middle classes. Its lack of popular support can be estimated from the ease with which the Bolsheviks got rid of it. There was only one session, during which the delegates (mostly S.R.s) could scarcely make themselves heard above the uproar of the armed Lettish guards. At 5 a.m. a sailor named Zheleznyakov asked the delegates to disperse 'since the guard is tired'. The lights were then extinguished and the delegates expelled in spite of the fact that they had armed themselves with candles and sandwiches.

The elimination of the Constituent Assembly cleared the ground for a constitution more to the taste of the Bolshevik party. Protracted and bitter debates on constitutional forms took place in a drafting commission which sat during the first seven months of 1918. Its proposals were ratified by the Congress of Soviets in July. While the drafting commission deliberated, the revolution intensified. The Allies landed troops all round the periphery of the former Russian Empire. The S.R.s virtually declared war on the Bolsheviks. In spite of these distractions, Stalin and his colleagues did their work well. Subsequent constitutional changes have not greatly altered the work done at the crisis of the revolution.

The 1918 constitution created the R.S.F.S.R. (Russian Socialist Federal Soviet Republic). Supreme power was vested in the All-Russia Congress of Soviets. Its executive committee (VTsIK), composed of about 200 persons, could exercise all the legislative powers of the Congress when the latter was not in session. The VTsIK in turn appointed the Council of People's Commissars (Sovnarkom). This last body, which approximated roughly to a Western cabinet, was also empowered to issue decrees, orders and laws. No attempt was made to secure the separation of powers, which was in fact specifically rejected as a bourgeois theory. The members of the Congress were elected by the city and rural Soviets. In the cities, there was one deputy to 25,000 voters: in rural areas, one deputy to 125,000 inhabitants. This gave the cities an unfair weighting in the Congress, a fact which was freely admitted by the authors of the constitution. Although the word' federal' appears in its name, the R.S.F.S.R. was not federal in the Western sense of the word. The Bolsheviks were embarrassed by their 1917 slogan: 'All power to the Soviets'. Many of their supporters believed that they meant what they said and that they were going to create a semi-anarchist constitution in which sovereign powers were going to be exercised by the local Soviets. The party could not afford to refuse outright the demands which it had encouraged; hence the great ingenuity lavished upon the 1918 constitution. Local Soviets were given both electoral and executive functions. They were the primary democratic cell; they were also the instruments through which the central government ruled the provinces. Lenin hoped that they would take the place of the Tsarist bureaucracy, that their revolutionary ardour would compensate for their lack of administrative experience. In this, he was sadly disappointed. The local Soviets were trusted little more than the Tsars had trusted the *zemstvos*. Their power to tax was limited to local needs, and even this limited right was subject to central supervision. The slogan 'All power to the Soviets' was interpreted as: 'Some local power to the Soviets'. There was no provision for national autonomy on a federal basis; it was assumed that when the non-Russian periphery of the Tsarist empire was recovered from those powers which had occupied it in 1918, it would be united with the R.S.F.S.R. on the footing of equal national republics. Perhaps other such republics might be formed in Western Europe after the revolution.

The R.S.F.S.R. was seen by its Bolshevik creators as a merely temporary house for the revolution to inhabit. Some Bolsheviks and most S.R.s had supposed that the revolution would be immediately followed by the disappearance of the State. Lenin had given this point careful consideration in his *State and Revolution*. His view was that the State must survive the revolution for long enough to eliminate the inevitable counter-revolution. Of that there were many signs in 1918. Lenin admitted that while the State continued to exist it would have a class form; this was the 'dictatorship of the proletariat', a temporary condition which would last only long enough to destroy the remaining bastions of feudal and bourgeois class society. According to Lenin the 'dictatorship of the proletariat' would occupy a unique position in the unfolding historical dialectic. Since the proletariat was the majority, little force would be needed for the assertion of its will. Its enemies would rapidly crumble and its task would soon be done. In the view of the party theorists the 'dictatorship of the proletariat' was based upon the Soviet structure. The facts, of course, were quite different. Such 'dictatorship' as existed in Russia was based upon the party, and the party in turn relied upon its inner cadre. Lenin recognised no contradiction between the fiction of Soviet power and the reality of inner-party direction. Like all Marxists he was firmly committed to Rousseau's concept of the General Will: that is, the belief that a single individual may know better than the masses what is best for them and what is most closely attuned to the movement of the historical dialectic. In the Marxist mind the word 'proletariat' came to have abstruse metaphysical overtones. To the eye of commonsense a proletarian is a person exploited in the capitalist economic process. We see him brandishing a spanner and cursing the capitalist. But to Lenin and the Marxist 'theologians' the proletariat was an idea to be put into practice even against the expressed wish of persons who might reasonably be named proletarians. A dictatorships did emerge in Russia in 1918 but it was not the 'dictatorship of the proletariat'. It was, on the contrary, the dictatorship of a tiny group of idealists whose nerve was stiffened by their conviction that they knew the rules of the game and that these rules justified the use of force and fraud.

During the first few weeks of power the Bolsheviks continued to catch the wind of popular approval. A stream of decrees realised the wishes of the majority. A decree on land (which told the peasants to help themselves), on peace, on workers' control of industry, on the nationalisation of banking, on the repudiation of all state debts both internal and external, on the abolition of private ownership of large houses, on the standardisation of all salaries and living conditions, on the reform of the calendar, on the modernisation of the alphabet, on the separation of Church from State and of schools from the Church, on the abolition of the old legal system and its replacement by popularly elected revolutionary tribunals – these all seemed to promise a new heaven on earth, a heaven which could be gained by the destruction of the hated 'bourgeois'. Even the creation of a secret police (December 1918), known first as the Cheka, was invested with popular appeal. Its boss, Dzerzhinsky, announced: 'The Commission appeals to all workers, soldiers and peasants to come to its aid in the struggle with enemies of the Revolution. Send all news and facts . . . to the Commission.' Here was an open invitation to take revenge for all the insults, real and supposed, suffered at the hands of the 'bourgeois'.

But the delights of licensed class warfare could not conceal the fact that the November revolution had not enabled anyone to eat any better or to keep warmer.

The increasingly severe food shortages of 1918 could be met only by extending the range of class warfare. The town worker had to be turned against the peasant as well as the bourgeois. There was some grain in the countryside but there was no means of making the peasant part with it. The currency was hopelessly inflated and besides, there was nothing to buy with it. The peasant ate more and turned his surplus into vodka. The only solution was to force the peasant to give up his grain at the point of the rifle; to turn the poorer peasant against the richer, to terrorise the countryside through the Cheka. During the summer months the campaign against the peasants reached the proportion of a civil war, a war provoked by the new régime, a war which to the peasants seemed very similar to the punitive raids carried out by Tsarist troops. The discontent of the peasants was communicated to some sections of the industrial workers. In June 1918 there was the threat of a general strike on the railways, and in several industrial towns there were pitched battles between the Cheka and the workers. These difficulties naturally encouraged the other left-wing parties which by the middle of 1918 were not only excluded from the government but were also being hounded by the Cheka. The crisis came in July. The left S.R.s assassinated the German ambassador, Mirbach, whom they regarded as a symbol of the hated Treaty of Brest-Litovsk. They intended his murder to be the signal for a resumption of the war against Germany. But the Cheka was too much for the rather amateurish terrorism of the S.R.s, and most of the leaders were rounded up. One woman terrorist, however, Fanya Kaplan, survived until the end of August when she fired three shots at Lenin, two of which struck him. This was the signal for an outbreak of mass terrorism by the Cheka. All the non-Bolshevik groups and parties could now be accused of hostility to the revolution. All could be accused of being the instruments of the foreign powers which by now were beginning to intervene in Russia. The Bolsheviks had concluded a war against Germany in order to fight one against the rest of the world.

It is difficult to estimate the number of victims claimed by the Cheka during the terror of 1918. The former Tsar and his family were among those first eliminated. More than 500 'counter-revolutionaries' were shot in Petrograd alone. In some provincial towns pitched battles went on for several days between the Cheka and the local inhabitants. The terror was specifically directed against the 'bourgeoisie' – a useful word which could be used to include anyone who was not a supporter of the party. The Kadets had already been outlawed. The Mensheviks and the S.R.s were not yet treated in the same way. As the terror slackened off towards the end of 1918, these left-wing groups were once again allowed representation in the Soviets. They were allowed to publish newspapers and they enjoyed considerable influence among the peasants and the trade unions. The Cheka bullied but it did not shoot the Menshevik and S.R. leaders, who while the civil war continued were just as eager as the Bolsheviks to prevent the revolution from being crushed. But in spite of this uneasy toleration, they were not allowed a share in political power. In effect, the R.S.F.S.R. was a single-party state from 1918 onwards. To this achievement Dzerzhinsky and the Cheka had contributed largely. The number of people who entered the Cheka prison in Lubyanka Square, Moscow, may not have been very large. But the Cheka was ruthless, thorough and determined. 'It is time, before it is too late, to carry out the most pitiless strictly organised mass terror, not in words but in deeds. Bringing death to thousands of idle white hands . . . we save millions of workers, we save the socialist revolution'.

THE CIVIL WAR, 1918–21

The attitude of Russia's former allies towards the Bolshevik régime was a mixture of fear, indignation and hopeful self-interest. In the months immediately after the end of the war, it was widely believed in the West that there was a chance that the Bolshevik slogans would inflame the masses of battle-weary soldiers and impoverished workers. Intervention in Russia consequently offered a minimum prospect of keeping the Bolsheviks so busy that they would be unable to help the disaffected Western masses. The Western statesmen assembled at Versailles did not intend to permit the Treaty of Brest-

Litovsk to be reversed – even though it had been imposed upon a former ally by the power which had just been defeated. They were alarmed by the fact that the Soviet government had denounced the peace as soon as the Germans were defeated in November 1918; with considerable ease (at first) the Red Army then reconquered the Baltic States, Poland and the Ukraine. This annoyed some of the Versailles statesmen (like President Wilson) whose main object was to redraw the map of Europe on the basis of national self-determination, and frightened others (like Clemenceau) whose plans for the future of Europe were based upon the idea of a powerful Poland and a weak Russia. The French, anxious for a Poland which would be strong enough to be a real threat to Germany, envisaged a return to the eighteenth-century frontiers, a Poland which included the Ukraine. Intervention in Russia would force the Bolsheviks to accept the Brest-Litovsk terms, would exclude Russia from Europe, would enable France to dominate it once more. The British were more interested in the future of Asiatic Russia. Intervention might hasten the break-up of Russia; out of this Britain might gain security for India by setting up a puppet régime in Central Asia; she might even gain control of the Caucasian oil. Japan wanted eastern Siberia up to Lake Baikal. The Romanians wanted to keep Bessarabia. The Russian corpse seemed large enough to provide all the vultures with a meal. Behind the noises of ideological outrage made by all the Great Powers lurked their expectation of national gain.

But the amount of effort, military and economic, which the Great Powers were willing to put into intervention was limited. The soldiers had no wish to go on fighting. Even worse, there was the danger that Allied troops in Russia would fall victim to communist propaganda. The French army sent to Odessa in 1919 had to be hastily withdrawn because it showed a disturbing tendency to shoot its officers. Even the British troops at Archangel showed something less than enthusiasm for their task. Britain had no wish to help the French to create a great Poland; France was not at all interested in helping Britain to create a buffer state in Central Asia; and America was very hostile to the Japanese ambitions in Siberia. The robbers could not agree to help each other.

No doubt some of these difficulties could have been sorted out if any of the anti-Bolshevik White groups had inflicted a complete defeat upon the Reds. Both the principal White leaders, Kolchak and Denikin, were strongly nationalist. If either had entered Moscow (which became the capital of the R.S.F.S.R. in 1918) he would have demanded a restoration of Russia's pre-war frontiers, perhaps even that the Allies should honour the war-time treaties. The allied leaders gradually realised that a White victory might create even more problems than it solved. They were willing to send enough help to enable the Whites to carry on a civil war which reduced Russia to anarchy and impotence. They were not willing to send enough to ensure that the Whites won. The main objects of intervention were, however, gained. Bolshevism did not permanently infect either Europe or Asia, and Russia was so much weakened that she was unable to effect a major revision of the Treaty of Brest-Litovsk. The allied policy (of which Churchill was a principal advocate) achieved a cynicism worthy of Lenin himself.

The first hope of the Whites was Admiral Kolchak. He had been an Arctic explorer, a quite distinguished naval commander, and he was known to be a courageous and honest man. These qualities gained him the whole-hearted support of the British who encouraged him to seize power by force at Omsk (November 1918). His first enemies on this occasion were not the Bolsheviks but a local Siberian government dominated by the S.R.s. Their policy was for the reconstitution of the Constituent Assembly. This *coup d'état* by Kolchak illustrates an important theme of the civil war: the fact that the moderate Socialist parties were everywhere squeezed out by the extremists. Nowhere could the adherents of liberal democracy collect a fighting force. The civil war would end with a dictator from either the left or the right. Kolchak's government was not based upon the consent of the strongest fighting force in Siberia – the Czech army. These fine troops had been on their way to the Pacific when the revolution had broken out. During the summer of 1918 they had turned back and had nearly captured Moscow. But now that the war against Germany was over they only wanted to return to their new nation. They disapproved of Kolchak's dictatorship but took no action against him. They merely waited along the Siberian railway until it was possible to continue their eastward journey. Meanwhile Kolchak pushed westward with his army of Cossacks and Tsarist officers. He advanced rapidly over the Urals and early in 1919 he threatened the middle Volga. He proposed to unite his army with the British force in Archangel. The northward thrust offered him the hope of supply; but he would probably have done better to concentrate upon joining Denikin's southern army. His advance was checked partly by the stiffening resistance of the Red Army but chiefly by a peasant revolt along his lines of communication. Kolchak's land policy was disliked by the poorer peasants, especially the recent immigrants into Siberia. Behind him was the shadow of the landlord and the factory owner. The mass revolt against him was spontaneous – only a few of the peasant bands were under Bolshevik leadership. Kolchak's retreat turned into a rout. The Siberian towns turned against him, revolted by the barbaric behaviour of the White troops and by the ruthless White terror. Kolchak was eventually caught and shot by the Reds at Irkutsk in February 1920. Siberia remained very disturbed. The peasant bands continued to roam the taiga, little more contented with their new masters than they had been with their old. The Siberian railway was choked with abandoned trains filled with the corpses of those who had died of cold or typhus. But the Bolsheviks held the towns; and the universal anarchy ensured that it would be impossible for the Japanese – who still remained in control to the east of Lake Baikal– to use Siberia as a base against European Russia. The only organisation which could survive in Siberia was that of the Bolshevik party.

It was unlucky for the Whites that their southern front, under Denikin, made no decisive move against the Reds until Kolchak had been defeated. Denikin was a general of peasant origins, a rare product of Tsarist Russia. He had, however, no intention of restoring the Romanovs, unlike many of his officers. Politically he took no clear stand. It was not even certain whether he would, in the case of victory, summon a constituent assembly. Even had he wanted to, it is doubtful whether his officers would have permitted it. His only slogan was: 'Russia shall be great, united, undivided'. This nationalist programme was unlikely to be popular with the Caucasian peoples, the Ukrainians, even with the Cossacks. It could certainly not command the assent of the liberal and educated classes. But although Denikin was not politically resourceful, his strategic position was much stronger than that of Kolchak. In 1918 he had gained control of much of the Caucasus. With his back to the Black Sea he could be easily supplied by the Allies. His Cossack troops, from the Don and the Kuban, were near at hand and were for the most part fanatically opposed to the Bolshevik régime. They were small farmers with horses and cattle, who had much more to lose than their chains by the abolition of private property in land. Just to the north lay the industrial region of the eastern Ukraine. Its capture and retention might persuade a weary Bolshevik government to make peace, perhaps on the basis of the independence of south Russia. Denikin's early victories confirmed the optimism of his followers. In June 1919 he captured Kharkov and Tsaritsyn; in July he advanced on Moscow; in August he captured Kiev and in October reached Orel. Only Tula stood between him and the Bolshevik capital. In the same month the White General Yudenich, operating with British support from Estonia, reached the suburbs of Petrograd. But Denikin's offensive collapsed as rapidly as that of Kolchak and for the same reasons. There was complete confusion in the rear, encouraged by the Reds but chiefly the result of peasant anarchy. The Ukraine was even more turbulent than Siberia. With bewildering rapidity it had been occupied by Germans, a national government, the Reds, and the Whites. The peasant bands turned against the towns and against everything which suggested town influence. The Jews were a favourite target because they were conspicuous and because they were urban. Pogroms occurred on a large scale. Some of the peasant leaders, like the anarchist Makhno, were really skilful. During the autumn of 1919 Makhno's band of about 40,000 men devastated Denikin's communications and threatened to capture his headquarters at Taganrog. Makhno eventually escaped from both the Whites and the Reds and took refuge in Romania. He had permitted no pogroms and imposed a truly anarchist non-government upon the areas under his control. Denikin had nothing to oppose to the spirit of anarchy which convulsed his rear. British supplies remained unused at the Black Sea ports. The White administration was very corrupt, just as the White officers were notorious – even in this age of violence – for their drunken brutality. A British journalist indignantly observed what was happening to his government's money: 'In 1919 we went Denikin 1500 complete nurses' costume outfits. I did not ... ever see a nurse in a British uniform; but I have seen girls, who were emphatically not nurses, walking the streets of Novorossiysk wearing British hospital skirts and stockings.' Harried by the Red Army, weakened by corruption, paralysed by guerrilla ravages along the lines of communication, abandoned by the allied governments, Denikin fell back on the Black Sea ports. With the remnants of his force, he was evacuated by the Royal Navy. For a few months in 1920 Wrangel (with French aid) enjoyed some success in the Crimea and the southern Ukraine. But at length he too had to be rescued from Sevastopol. British troops were evacuated from Murmansk and Archangel in 1919.

But the end of the civil war did not bring immediate peace. In 1919, the victorious Allies had created a new Poland. Russia had not been represented at the peace conference; indeed, during its sessions the Western Allies had been actively fighting against the Bolshevik government. A happy result of the absence of Russia was that the Western statesmen at the Versailles peace conference were able to give the new Poland very generous eastern frontiers. France, in particular, was anxious that the new Poland should be as powerful as possible. The revolution and the civil war had deprived the French of their eastern ally. Who would save Paris the next time the Germans invaded France? The Russian government naturally disliked a peace treaty which gave away Tsarist territory and at which they had not been represented. But Pilsudski (dictator of the new Polish State) was not satisfied even with the generous terms given him. He determined to take advantage of Russia's weakness. He waited until the Reds and the Whites had exhausted one another: although socially conservative (even reactionary) himself, he had no desire to see a nationalist White general like Denikin in power in Moscow. Pilsudski had the most grandiose plans. He was not content with merely restoring his unfortunate nation to the 1772 frontiers – a plan which would have been acceptable to most of his countrymen. He wanted a restoration of the Polish medieval kingdom which stretched from the Baltic to the Black Sea. With this objective, he attacked the R.S.F.S.R. in 1920. At first his Polish colonels carried all before them. But the exhausted heroes of the Red Army rapidly recovered, and by July 1920 the astonished colonels had been chased back to Warsaw by the ragged veterans of Tukhachevsky and Budenny. Now it was the turn of the Bolsheviks to rejoice. They foresaw a Polish revolution; they expected Poland to set Germany alight. But if the Red Army could defeat the Polish colonels, it could not get past the working men of Warsaw who had no intention of being liberated by the Russian proletariat. They threw back the Red Army (August 1920) and once again the Poles advanced into the R.S.F.S.R. By this time both sides had had enough of these bewildering reverses of fortune. By the Treaty of Riga (1921) they came to an agreement which lasted until 1939. Poland got considerably less than the Allies had given her in 1919. Russia had to accept the independence of Finland and the Baltic States, which practically excluded her from the Baltic. In this area she lost nearly all the territory which had enabled her to dominate central Europe during the nineteenth century. But she had regained the Ukraine which the Germans had taken at Brest-Litovsk. For the time being she had ceased to be a European Great Power, but she retained enough territory to enable her to become a world power.

THE SOVIET GOVERNMENT DURING THE CIVIL WAR

In spite of the weaknesses and divisions among the Whites and their allies, the survival of the Soviet régime during the civil war was little short of miraculous. At times, Moscow controlled less territory than had constituted medieval Muscovy. Within this area the harvest of 1920 was only about three-fifths of that of 1916. Industrial production fell still more rapidly. By 1920 it was 13 per cent of the 1913 figure. In mining and in heavy industry the picture was still bleaker. Pig-iron production was 3 per cent of 1913; railway locomotives were not being made at all and it was impossible even to repair the existing ones. The cities were being depopulated. The rich fled abroad (about two million emigrated during these years); millions more went into the country in order to find minimum quantities of food and fuel. The population of Moscow and Petrograd was halved. The régime had certain advantages. It controlled the interior lines of communication; troops could be rushed from one threatened point to another with a minimum of delay. It controlled just enough industrial capacity to provide the Red Army with the primitive weapons required in the civil war; its hold upon the industrial regions of central and northern Russia assured a supply of rifles and machine guns. The Petrograd engineering works even supplied a few tanks when British machines employed by the Whites caused a panic in the front line. This was a war of rapid movement over long distances. The horse came into its own again after its temporary eclipse during the First World War. The Reds employed many former Tsarist officers whose loyalty was uncertain. Such as it was, it was secured either by holding their families as hostages or by placing a politically reliable commissar at their elbows. But at critical times the big decisions were taken by the party. Trotsky was always the man on the spot. His armoured train moved from crisis to crisis. In addition to the ruthless, voluble, imaginative, tireless and invincible Commissar, the famous train also carried printing presses (so that the Commissar's winged words could inspire the weary heroes of the Red Army); machine guns to protect him from the ubiquitous White cavalry; and a motor car so that he could visit and revive distant units. When Petrograd was about to fall to Yudenich, and even the city boss Zinoviev despaired of holding the cradle of the revolution, Trotsky restored morale by summary execution of the local defence committee. But few party members had to be shot. Each one knew that instant death awaited him if he fell into the hands of the Whites; each one knew that he would enjoy considerable power in the event of a Bolshevik victory.

Bolshevik control over the rear areas was imperfect but more efficient than that of their opponents. Lack of such control was fatal to the Whites. The Bolshevik system of 'War Communism' (1918–21) was based upon nationalisation and strict labour control. The system was the product in part of ideology, in part of the desperate war situation. It was tempting to call 'communist' a society in which all were equal because all had nothing. It was tempting to speak of communism when in fact the main development was the growth of state power. This was the main feature of War Communism. As society crumbled away, the power of the Soviet State increased over what little remained. The syndicalist or anarchical tendencies of 1917 were tamed although not finally eliminated.

All industries – even the numerous tiny peasant crafts – were nationalised. Factory management was achieved at first by the direct participation of the workers. When this proved inefficient, single-manager operation was retored. Often the manager was a 'bourgeois'. His powers were resented by the workers but they were as necessary to industry as were those of the Tsarist officers to the Red Army. Industry was controlled from the centre by Vesenkha (Supreme Council of National Economy). Vesenkha had been at first intended to direct the national economy according to a plan. But it never got control either of finance or labour and confined itself to ensuring that such production as there was aided the Red Army. Very rapid inflation made the currency valueless. By 1920 13,000 Soviet roubles represented the value of one rouble in 1913. Workers were paid in food and fuel; sometimes they received a part of the product of their factory which they could then barter for the necessities of life. 'Communist' instincts were satisfied by a government decision to save itself the trouble of collecting worthless roubles by charging nothing for the use of public utilities. Transport, housing, food, fuel, were given without payment to those who held labour cards. The whole distributive process fell under the control of Narkomprod (People's Commissariat of Supply). Producers and consumers were organised into co-operatives. This at least helped to ensure that everybody was equally badly off; the sense of social justice was some consolation for hunger and cold.

Food was strictly rationed according to four grades. Possession of a labour card entitled its owner to a food ration. It was difficult for a 'bourgeois' to get a card at all, and starvation was the fate of those who held the card of the lowest grade or who were not given any card. Unemployed workers were sent out in grain-requisitioning parties which scoured the countryside, terrorised the peasants and generally managed to bring back enough food to maintain the ration. There was a large and flourishing black market in the cities. The party reluctantly permitted it to exist as a safety valve. There the middle classes exchanged their domestic treasures for some grain or a load of wood. The visible poverty and misery of the former ruling class stimulated hatred and contempt. To prevent the formation of any centres of discontent, large numbers of people were arbitrarily moved from place to place. At least five million were conscripted into the Red Army – although it was uncommon for more than 100,000 men to be engaged even at the most critical fronts. Unemployed workers and members of the middle class were called up for forced labour. Even when the fighting stopped in 1920 the armies and the labour gangs were kept at work – on the railways, cutting wood in the forests, putting the mines into production. In economic terms War Communism was very inefficient: production and productivity fell catastrophically. But it achieved its main object. It kept people on the move, it produced just enough food to maintain those groups who were most hostile to the Whites, and it enabled the State to expand into every corner of Russian life. It was not unlike the system which had enabled Britain and Germany to sustain a long war.

War Communism required the support of an enormous bureaucracy. Many of the Tsarist bureaucrats had fled or had been killed. Their places were filled by a recruitment from the ranks of the semi-literate with little experience of administration. Typists who could not type, secretaries unable to read, executives with no education, huddled into freezing offices to provide each other with pieces of paper. Even paper was in short supply. Lenin was sometimes unable to get any ink for his restless and commanding fountain-pen. The new bureaucracy was as inefficient as the nationalised economic structure which it allegedly controlled. But the numerous posts provided a taste of power for those who had never had it: they were given a ration card and they felt that the revolution had really done something for them. The porter who became a house manager, the shopkeeper who controlled ration cards, became unconscious allies of the régime. Their administrative inefficiency was of little importance since there was hardly anything to administer.

Public order was comparatively well maintained behind the Bolshevik lines. The city streets were sometimes rendered unsafe by criminals in search of food or ration cards. Lenin himself was once captured by such a band in the suburbs of Moscow. He was forced at pistol point to hand over his overcoat; he had no money in his pocket. But the low rations, the fear that any further dislocation might lead to complete starvation, the vigilance of the Cheka, and the knowledge that things were even worse behind the White lines, rendered the masses docile. In the country the peasants bitterly resented the grain requisitions. Large mutinous bands roamed the forests but they were never so formidable in Red territory as they were to the Whites. This was in part the result of the policy of general mobilisation which removed the peasant from his home village; in part the fact that the agricultural areas of central Russia were those in which there was the greatest overcrowding and consequently the bitterest hatred by the poor of the rich peasant. The Bolshevik requisitioning of grain was a little less hateful than the White threat to restore the landlord.

THE EVOLUTION OF THE PARTY IN POWER

The most significant development of the civil war years was that of the Bolshevik party itself. Lenin had organised the party to seize power by revolutionary means; now it was called upon to govern a disintegrating society during a period of intense crisis. In some ways this long crisis suited the Bolsheviks. It dissolved all the other groups and institutions and left the party without a rival. Even though the party was weak during these years it was always a little stronger than anything else. It had a virtual monopoly of political power. Behind the elaborate facade of the Soviet system, the R.S.F.S.R. had become a single-party state, the first of the twentieth century. This fact was the most important result of the revolution, the deepest breach with the Tsarist past. It was a political innovation which was to be imitated by many other nations, but none carried it through so consistently as Russia. It is doubtful whether Lenin and his companions realised the significance of what they were doing. They had carried through a revolution in the name of the proletariat, inspired by a faith in the Marxist interpretation of how History works. In several ways they had strained the Marxist ideology to fit their own wishes and impulses. The result of all this was not at all as Marx had predicted. The State did not wither away; it became far stronger in its new, single-party form. The revolution created a new Leviathan, a new type of political structure. By a strange twist of fate, the revolution revealed the answer to Russian backwardness, that problem which had occupied the Tsars since the Crimean War. The single-party State proved in the end to be capable of forcing Russia through the hoop of industrialisation and of making her the equal of the strongest nations of the West. The Russian revolution was supposed by its authors to end the reign of nationalism as a dominant ideal. Class conflicts were held to be more important than national differences. But the main result of the revolution was to enable Russia to become one of the most powerful nations in the world. From the revolution, backward Russia got a modern political system.

The civil war hastened developments which would probably have occurred anyway. The party of revolutionaries became a mass governing party. It had to adopt institutional forms to enable it to carry out its new functions. Party numbers grew rapidly. In January 1918, it had 115,000 members; in January 1919, 251,000; in January 1920, 431,000; in January 1921, 576,000. Numbers multiplied five times in three years in spite of some fairly severe purges. The moral quality of the new recruits was no doubt quite high; it needed courage to join the party during these years. But the level of intelligence and competence sank dramatically. Most of the new recruits were peasants and workers with a very low educational standard. Their enthusiasm and self-sacrifice made them excellent instruments but they lacked the capacity for leadership. The result was inevitable; it has been a general feature of all mass parties in the twentieth century. A division emerged inside the party between the inner-party élite and the rank and file. The forms of democracy were maintained. In theory, the leadership was still elected by the Central Committee, which in turn was elected by the party congress. But in fact power flowed down from the top; the party became the instrument of its leaders. Just as the Soviet system was a mask for the power of the party, so the party itself masked the real power of the inner-party élite.

Significant steps towards an inner-party élite were taken at the party congress of 1919, held during the crisis of the civil war. Such party organisation as there had been was broken by the sudden death of Sverdlov. This old member of the underground had carried the details in his head. He knew all the local secretaries personally and arranged the details by word of mouth. To provide a more permanent structure, the Central Committee of 1919 created three new party organs. A Politburo of five members was to decide on urgent matters which the full Central Committee was too unwieldy to cope with. By creating the Politburo, the Central Committee ensured executive efficiency but it relegated itself to a secondary place in the political structure. An organisational bureau (Orgburo) and a Secretariat were to work in harness to provide what had been previously supplied by Sverdlov's memory. They reviewed party membership, organised purges, kept in touch with the local secretaries, moved key men to important posts. None of the important leaders took much interest in these organisational details. Lenin, Trotsky, Kamenev, Zinoviev, were too busy fighting wars, building a new State and negotiating with foreign powers to bother about party organisation. The job needed an old Bolshevik, an underground man like Sverdlov. Stalin fitted this description. He had no experience of the world outside Russia but he did know the party intimately. In 1919 he was appointed to the Orgburo and from this obscure corner of the party bureaucracy he built up a monopoly of power.

The party drove the nation and the élite drove the party. This inner segment, perhaps 50,000 strong, consisted chiefly of organisation men, administrators, *apparatchiks*. The process of centralisation was by no means complete by the end of the civil war. The need for it was bitterly resented by a large section of the party. But the path towards the domination of the party by the organisational apparatus had been shown. Stalin, Zinoviev and Kamenev rapidly acclimatised themselves to the profound changes introduced by the 1919 Congress, but Trotsky, busy winning the civil war, had little use for organisation. He preferred discipline and industrial planning, fiery words and striking deeds, to the patient collection of dossiers, the careful compilation of a card index, the quiet manipulation and counting of heads. From his armoured train it seemed impossible that Russia should be ruled from behind a desk.

Yet this was just how Lenin did it during the civil war. He never visited a front; he rarely addressed large public gatherings during the crisis of the war. Instead (like Stalin in the Second World War) he remained in his simply furnished Kremlin study reading, annotating and telephoning. He was a splendid committee chairman. He reduced the time available for each speaker to two or three minutes, he was humorous, shrewd, firm, usually right, and nearly always got his way without causing rancour. He never allowed any cult of personality to grow up; the only picture which he allowed was that of Marx. He read everything, remembered everything and even found time to write several long pamphlets on matters of theory. His principal colleagues in the Politburo were Trotsky, Stalin, Zinoviev and Kamenev. He allowed none of them to quarrel deeply with him or with each other. Power seemed to have made him less inclined to seek schism; it seemed to have released in him a geniality which was notably absent in the pre-war years. He talked at length with visitors. They ranged from distinguished foreign intellectuals like Bertrand Russell to the *hodoks* or peasants who continued the tradition of Tsarist times and wandered through Russia to lay their grievances at the feet of the supreme ruler. The flowering of Lenin's personality, the fact that he was an administrator of genius, was an important factor in the victory of the Reds.

While the civil war lasted it was comparatively easy to maintain party discipline but the release from immediate danger brought formidable problems. The party rank and file was bitterly opposed to its exclusion from political power. Since the dissidents could not make their will effective through any other political party, they had to canvass their opinions through the mechanism of the party in power. This raised an acute difficulty. In a democratic society, a political party is a pressure group whose function is to capture public opinion and by this means to get into power, but in Russia the party was much more than a pressure group: it was the nucleus of the state apparatus. Consequently, public disagreements within the party shook the authority of the government to its foundations. Such public disagreements occurred in 1920–1, the most dangerous of them backed by the one organisation which the party had not yet managed to bring to heel – the trade unions.

The union movement had trebled in membership since 1917; by 1920 there were nearly nine million members. The co-operation of the unions was essential for the maintenance of War Communism. The unions were close to the working men and the Mensheviks were still powerful in them. It was consequently dangerous for the party leadership when some of the leading party unionists began to represent the interests of the unions to the party rather than imposing the will of the party upon the unions. Such was the case with the 'workers' opposition', the most formidable party fraction since the revolution. It canalised the feelings of the 'outs' against the 'ins'. It demanded trade union control of industry, it protested against the excessive power of intellectuals in the party, it alleged (quite rightly) that party posts were no longer being filled by democratic election. Its angry denunciations filled the press during the winter of 1920–1. Trotsky made matters worse by joining in the public controversy with his usual gusto, expressing views which were not accepted by Lenin and the other members of the Politburo. Trotsky wanted the virtual elimination of the unions. He wanted labour to be under the direct control of the State, organised in military fashion like his own Red Army. In fact his soldiers were already engaged upon labour tasks. Trotsky wanted a total economic plan into which labour would be slotted in the most efficient way. He contemptuously dismissed the vague syndicalism of the 'workers' opposition': single-man management would be essential to efficient production.

These public controversies alarmed Lenin and the Politburo. They came at a time of acute economic crisis when it seemed that grain deliveries to the towns might cease altogether. If the peasant went his own way – as he threatened to do – then all inner-party controversy became meaningless. The Bolshevik party was essentially urban: without urban life it could not survive. The peasants must be soothed; the town workers must be allowed to have their unions; party unity must be restored. The time for Trotsky's driving revolutionary ardour was over. Everything (except the party) must be sacrificed to ensure primitive social reconstruction. This could only be achieved by the partial restoration of capitalism. Such was the tactical decision taken by Lenin in 1921. Even he could not escape the fact that Russia was still a peasant nation: even in the single-party State the peasant's desire for a free market in grain was supreme.

11

The Party Saved and the Nation Restored: 1921-9

N.E.P.

The nation which emerged from the civil war was the shadow of a shadow. Great tracts of land and valuable means of production had been lost at the Treaty of Brest-Litovsk. What remained was run down beyond hope of rapid recovery. The bare statistics tell part of the truth: by the end of the civil war Russia's national income was only one-third of what it had been in 1913, industry was producing only one-fifth of what was being produced before the war, the coal mines only one-tenth and the iron foundries one-fortieth. The railways had come to a standstill. The mechanism of exchange between town and country had ceased to function. The urban populations, never a significant proportion of the whole, had shrunk still further as famished workers fled to the countryside in order to procure the bare necessities of life. The bread ration was nominally two ounces per day but even this was sometimes not available. A frozen potato was a treasured find. Broken furniture was the chief means of warding off the imperious frosts of a Russian winter. Further and perhaps even final economic disasters were threatened. Locusts and sand blizzards in the Volga region threatened to destroy even the tough basic peasant mode of life, shaped over many centuries to endure any number of disasters. By the end of 1921, thirty-six million peasants were starving and the government had to summon the help of foreign relief agencies. During these terrible months, cannibalism was frequently reported. What remained of Russia was more completely devastated than any other part of Europe in the twentieth century. Even Germany in 1945 rapidly enjoyed the attentions of those whose interest it was speedily to restore to health what they had just destroyed. Russia had no such luck. After the failure of intervention, the Great Powers ignored the plight of a nation which had recently been of their number but was now reduced to chaos.

Between 1917 and 1921 the Communist party had directed a revolution in the name of the proletariat. It proclaimed a 'dictatorship of the proletariat' as an intervening stage between the proletarian revolution and the final withering away of the State. This pause in the march of the dialectic was made necessary by the incomplete nature of the 1917 revolution. External enemies, the powerful capitalist nations, remained to be subdued by the extension of the revolution; and within Russia powerful remnants of the old bourgeois social order still remained unabsorbed in the proletarian state. Consequently, the proletariat must exercise state power until such time as world revolution had removed the external danger, and internal evolution had mitigated the rigours of class conflict. Russia remained in 1921 a class society but a dialectically odd thing had happened. The exploited class – the proletariat – had now become the exploiting class, charged with the task of exploiting the former bourgeois and feudal classes out of existence. The proletariat, of course, was by Leninist interpretation that class of which the Communist party was the mystical representative. There was no need for any vulgar bourgeois voting. The party was placed by history in a representative position. What it did inevitably represented the will of the proletariat, even though a majority of actual flesh-and-blood proletarians showed every sign of wishing not to follow its lead.

Lenin and his friends had to justify themselves before the court of the 'world historical spirit', an entity which they in fact worshipped. Their minds were trained by two decades of metaphysical speculation. They had no difficulty in explaining to themselves the theoretical justification for any action that they might have to take. In 1921 they judged that the preservation of the party was the essential objective even if this meant, as in fact it did, abandoning the cause of the proletarians. In the midst of the surrounding chaos the party alone stood firm. The disintegration of Russian society was not altogether to its disadvantage. The disappearance or demotion of the old governing classes, the reduction of all men to a common level of misery, the destruction of the other possible contenders for supreme power (the other left-wing parties) – all these things helped to emphasise, if only by contrast, the dominating position of the party. Even had Lenin not been prepared by long conviction and struggle for the idea of single-party rule, the circumstances of 1921 would have forced such a choice upon him. The moment was decisive. It was the moment at which the party could have allowed purely national interests to predominate: it could have called a political truce and sought alliance with the other left-wing parties. This would have meant concessions in a democratic direction. The alternative was to retain the political monopoly of the party but to make economic concessions to the strongest class in the Russian State – that is, the peasantry. This was an admission about the events of 1917–21 which no Marxist cared to admit. That is, it had become clear that the essential result of these years had been a successful peasant revolution. The revenge for centuries of serfdom had been taken: a revolution from below had been effected; the peasant mass had become for the first time a force to be reckoned with. This was a strange and contradictory result for a so-called proletarian revolution. It had been in fact a peasant revolution, controlled (inasmuch as anything so elemental could be controlled) by a tiny group of bourgeois intelligentsia. Lenin wisely rejected the rôle of King Canute: he let the waves of peasant revolution break as they would and then proudly proclaimed that he had always intended that the peasants should do what he had no power to stop them from doing. In 1921 Lenin, his party and the peasants started to rule the inheritance left them by the Tsar, his bureaucrats and his landowners.

The decision to conciliate the peasantry, known as the New Economic Policy (N.E.P.), was made public at the 10th Party Congress (March 1921). Just before the congress met, events had occurred which, to the non-Marxist observer, might weaken the party's claim to represent the will of the proletariat. At the military and naval base of Kronstadt the crew of the battleship *Petropavlovsk* touched off a mutiny which was rapidly organised by a naval clerk called Petrichenko. The mutineers were supported by some of the Petrograd workers who were in the midst of a wave of strikes on their own account. The demands of the mutineers were fundamentally anti-Communist. They wanted freedom of speech and of the press for all left-wing parties, the abolition of the specially privileged position of the Communists, full rights for the peasants to do what they liked with land (provided no hired labour was used), and the re-election of the Soviets by free and secret ballot. Lenin, always the realist, commented: 'They do not want the White Guard and they do not want our power either'. More cunning or perhaps more desperate than the Tsars, he did not content himself with repression alone. True, Trotsky and Tukhachevsky enhanced their reputations for rapid action by brutally massacring the Kronstadt mutineers, but while they were doing so Lenin was making concessions which met at least a part of the Kronstadt demands. He would concede neither democracy nor political co-operation but he would allow the peasants to do as they liked with the land.

The essence of N.E.P. was the abandonment of the system of grain requisitioning widely practised during the civil war. Even in those areas not struck down by blight and famine, this system had had disastrous results. The peasants either hid their grain, or they resorted to force to defend it, or they produced only enough for their own needs. In any event, the towns were not fed. Now a graduated tax, payable in kind, was to be substituted. A free market in grain was to be restored – a capitalist device to encourage the peasants to market as much food as they could grow. Perhaps grain exports could be renewed and used to provide the foreign currency so desperately required. The products of peasant industry could also be brought to the free market. As such concessions would obviously give the peasantry a new purchasing power they must have something to buy, so light or consumer industries would have to be encouraged. N.E.P. gave cautious permission to capitalist entrepreneurs, both Russian and foreign, to set up industries whose products would satisfy peasant demand. The State retained full control over heavy industry, transport, foreign trade and banking. Capitalism had been restored to large segments of Russian economic life. It was hoped that its profits could be eventually diverted towards the advance to socialism, an advance which could come about only through the development of heavy industry. In the meantime, the socialist future must be sacrificed in order to conciliate the peasantry.

The results of this decision were profound and far-reaching. The adoption of N.E.P. meant very much more than its strictly economic form. It provided a breathing space – but a breathing space for what? In brief, a pause during which the revolution could be attuned to the traditional facts of Russian life. Order, stability, work-discipline, a sense of continuity, a constitutional and legal framework, all the institutions of civilised life had to be restored. Naturally in this vast task of reconstruction the conservative memories of the people played a great part. Lenin had always proclaimed the need to adapt the teachings of Marx to Russian conditions. The Bolsheviks had always been more 'slavophil' than the Mensheviks. To some extent this tendency had been arrested by the Western exile of the Bolshevik leaders, but after 1921 the specifically Russian nature of the party became more pronounced. The final victory of the 'slavophil' Bolsheviks was proclaimed by the emergence of Stalin as the leader of the party. His rival, Trotsky, was fatally handicapped by his reputation as a 'westerniser', a man who could not be trusted to maintain the balance between the Russian past and the revolutionary present. When Stalin took the next step forward by industrialising Russia under the slogan 'Socialism in One Country', he was acting in a way which would have been intelligible to any of the Tsars: that is, inaugurating a revolution from above with the purpose of enhancing state power. This helps to explain what the N.E.P. breathing space was for: it was to enable the party to recover the traditional powers of the vigorous autocrats, to remould them in the pattern of twentieth-century totalitarianism and to return to the Russian tradition – which had been defied in the period 1917–21 – of revolution from above.

ASPECTS OF RECONCILIATION: MARRIAGE, THE CHURCH, THE INTELLIGENTSIA, LAW AND LITERATURE

The tendency to adapt revolutionary zeal to traditional wisdom can be illustrated from many aspects of Soviet life during the twenties. The evolution of ideas about the basic social unit, the family, may serve as an example. During the civil war, extreme ideas about the equality of the sexes, the right of women to seek abortion, free love and the duty of the State to bring up children, had gained currency. They had been proclaimed by leading women party members, like Kollontai, whose private life luridly illustrated her complete contempt for bourgeois morality. These ideas were embodied in the revolutionary legal codes, and were not altered during the N.E.P. period. In 1918, civil registration of all marriages was made obligatory, the partners were accorded complete equality, divorce was automatically available on the demand of either of the partners, illegitimate children were accorded the same rights as legitimate. For the most part these legal changes consolidated a substantial improvement in the position of women and it was held to be essential to retain the goodwill of female citizens. But during the twenties official ideas about marriage changed significantly. Free love and sexual licence were condemned. Restraint was recommended; hard work encouraged as a substitute for profligacy. The revolution had left behind it an enormous number of orphaned and homeless children. The original solution of large state orphanages for these abandoned children was dropped and replaced by the encouragement of adoption. The stability of the family unit was restored so far as was compatible with the new status of women as equal citizens and workers.

Relations between the party and the Church were another field in which revolutionary theory had to be adjusted to traditional loyalties. In theory, the party was utterly opposed to religious organisations, and in practice the Church had been harried, devastated and accused of being in league with the Whites. Conflict came to a head in 1922 when the Patriarch Tikhon ordered the faithful to resist an order to hand over the Church's silver and gold plate. Tikhon was imprisoned and it seemed that a collision between the party and the Church was inevitable. But to prevent this the party encouraged the emergence of a breakaway group of priests who called themselves the 'Living Church'. This group rejected Tikhon's leadership, accepted the justice of the revolution and in return received official recognition. These developments alarmed Tikhon, who in 1923 confessed his former antagonism to the Soviet régime, promised to accept the revolution, emerged from jail, resumed the patriarchate and denounced the 'Living Church'. The party was now in the happy position of dealing with two centres of ecclesiastical authority. The leadership of the hierarchy was weakened but the peasant could still have his village church.

Not only peasant prejudices were carefully handled during the N.E.P. period. The bourgeois intelligentsia was also invited to accept the Soviet régime. Trotsky had been forced to make extensive use of the professional skill of Tsarist officers in the Red Army. He continued to support the idea that professional skill in all spheres was not to be acquired by the mere acceptance of Marxist ideas. Revolutionary enthusiasm was no substitute for knowledge and training. Until such time as the party could produce experts who were also Marxists it would have to use such experts as were willing to serve the régime. This viewpoint aroused much antagonism against Trotsky; but it was shared by the party leadership during the N.E.P. period. Its application can be seen in the attitude of the party towards law and literature. During the civil war much had been heard of 'revolutionary consciousness' as the basis of law and of revolutionary tribunals as legal institutions. But with the adoption of N.E.P. there was private property to be protected and contracts to be upheld, which meant a return to strictly defined legal rules. Typical of the change was the revised rule about inheritance. In 1918 inheritance had been abolished: in 1922 it was recognised but restricted to 10,000 roubles worth of goods. A professional judiciary was restored and provision made for uniformity of legal decisions. Some concession to 'revolutionary consciousness' was made by the retention of elected judges in the lower courts, but even here minimum professional qualifications were required. Under the 1922 legal codes, the former Tsarist lawyers loyally served the new régime. They filled up the gaps in the codes by reference to their pre-revolutionary experience.

In the field of literature, the party made equally energetic efforts to win the allegiance of the ideologically uncommitted intelligentsia. The N.E.P. years saw much experiment and a good deal of freedom. The party deliberately avoided lending its authority to any literary group which claimed to produce a specifically proletarian culture. Indeed, the possibility of such a culture in the circumstances of N.E.P. was attacked with withering scorn by both Trotsky and Lenin. As highly cultivated and literate men they were horrified by the naïve theories and clumsy practice of the proponents of *proletkult*; they knew that the adoption of a strict literary code would drive out the most gifted and creative writers. Lenin's friendship with Gorky was a useful means of gaining the sympathy of the politically uncommitted writers. Like Gorky he believed that backward Russia could not afford to lose a single useful mind: 'The Russian intelligentsia . . . will long remain the only carthorse that can be harnessed to the heavy load of Russian history'. The 'fellow-travellers' as the non-communist writers were named by Trotsky, were consequently encouraged by the party. They were allowed to express their ideas freely, both in novels and in newly-founded literary journals, so long as they refrained from attacking the régime. Writers like Pilnyak (*Naked Year*), Fedin (*Cities and Years*), Leonov (*The Badgers*), were much concerned with the interpretation of the revolution within a Slavophil rather than a specifically Marxist framework of ideas. They were much influenced by the populists and conscious of the importance of maintaining the rich Russian literary tradition. They tended to see the revolution as a great peasant upsurge, the final victory of Pugachov, and continued to accept Slavophil ideas about the messianic mission of the Russian people. They tended to regard the party as a temporary but inessential result of the revolution. Their books were popular and their ideas fitted in well with the N.E.P. ideology. They also attracted much favourable attention in Russian émigré circles abroad. Writers and other intellectuals who had fled abroad were encouraged to return and helped to reconcile the old revolutionary Russian tradition with the new revolutionary Russian State.

TREATMENT OF THE NATIONALITIES AND THE 1924 CONSTITUTION

One of the most difficult tasks of conciliation facing the party at the end of the civil war was that of the non-Russian nationalities. The war had to some extent simplified the problem: Poland, the Baltic States and Finland had become independent. But as the interventionist powers withdrew their forces, large tracts of the old Tsarist borderlands fell to the Red Army. Belorussia, the Ukraine, Azerbaidjan, Georgia, Armenia, Central Asia and the Far Eastern Province had all by the end of 1922 come under the control of the central government. These areas contained peoples of widely different national cultures and development. It was essential to construct a constitutional house for them all to live in, along with the dominant Great Russian majority. The problem was the more urgent for two reasons:

Soviet Russia, 1922

first because of the chaotic and untidy way in which the original constitution of 1918 (which had created the R.S.F.S.R.) had been stretched to include all these borderlands; and secondly because in the more developed border nations (Georgia and the Ukraine) there were national movements of great strength.

From the theoretical viewpoint the difficulties were considerable. Nationalism was a force which Marx had identified with the domination of the bourgeoisie. Yet after the 1905 revolution Lenin had announced his intention to make use of the anti-Tsarist national forces through the doctrine that nationalism was acceptable in cases where it was not led by the bourgeoisie. This principle had been written into the 1918 constitution, which pronounced the right of national units to secede from the R.S.F.S.R. Clearly some formula would have to be found which could be juggled to justify both the claims of the unitary state and those of the nations. Stalin, the Commissar of Nationalities, who was chiefly responsible for the 1924 constitution, expressed it thus: that national autonomy is a form, but the important thing is the class-content within the form. National autonomy can be allowed where the proletariat is in power. In fact what the party really meant was that it intended to preserve the unitary State at all costs but it also intended to conciliate national sentiment so far as this was compatible with the ideal of the unitary State. This was a subtle policy and one which paid high dividends in the future.

The greatest danger to this policy was that certain local Communists, whose national feeling was at least as strong as their allegiance to the party, would lead the national movement against the central government. This happened in Georgia. There the Menshevik-dominated government was overthrown in 1921 by a Communist coup. Alarmed by the history of co-operation between Georgian Mensheviks and Communists, Stalin decided to neutralise Georgian nationalism by uniting Georgia with its two much hated neighbours, Armenia and Azerbaidjan. The Georgian nationalist Communists naturally resented such shabby treatment at the hands of their fellow Georgian: they protested and they were duly purged. Lenin protested about the allegedly high-handed actions of Stalin. He accused him of 'Great Russian Chauvinism' and claimed that he wielded the party apparatus as if it were still the old imperial bureaucracy 'anointed with a little

Soviet holy oil'. But Lenin must have seen as clearly as Stalin the great potential danger of the national Communist movement in Georgia. Not for the last time in his career Stalin was faced with the fact that the greatest danger to the party was another Communist party.

Such incidents revealed that the 1918 constitution could permit dangerously fissile national aspirations. The party therefore solemnly engineered a treaty of union between the four theoretically separate republics (Russia, Transcaucasia, Ukraine and Belorussia), and proclaimed the right of the All-Union Congress of Soviets to draw up a new constitution. The congress selected a central executive committee (VTsIK) and this in turn selected a drafting commission. Real differences of opinion were expressed on this commission. The unitarians demanded the complete abolition of the national republics: the federalists (among whom the Ukrainians Rakovsky and Skrypnik were prominent) demanded a real share of power for the republican governments. The result was a compromise, but weighted in favour of the unitarians. The Congress of Soviets was established as the sovereign power. Its franchise continued to give more weight to urban than to rural voters. The congress chose a bicameral VTsIK. The Council of the Union was selected on the basis of population: the Council of Nationalities was composed of five delegates from each union republic and one delegate from each autonomous region. This was a concession wrung from the party by Rakovsky and the Ukrainians. The VTsIK (or its presidium) chose the members of the Sovnarkom, the Council of People's Commissars. The Union government reserved the control of foreign affairs, the armed forces and communications. Other departments of Sovnarkom were shared by the Union and the republican governments. These included food, labour, finance and planning. Finally a group of departments were elected by the republican Soviets alone: they had no counterpart in the All-Union Sovnarkom. These included internal affairs, justice, education and health. The constitution established a supreme court and attached to it a procurator who, as a last resort, could appeal against its decisions to the VTsIK. The right of secession from the Union was retained: and this quaint formality accounts for the fact that all the Union republics (originally 4 and now 15) have to possess a frontier contiguous with a foreign power.

Otherwise, it is argued, they could not in fact exercise their legal right of secession. The sentiments of nationalities too small or too backward to enjoy the full status of Union republics were flattered by the creation of ethnically distinct autonomous republics, provinces and districts.

It may be objected that the establishment of the U.S.S.R. as (in Stalin's words) 'not a confederation but a federal republic', was no more than a cynical political manipulation. Behind the elaborate façade of democracy and national rights lurked the power of the party. Such an impatient view gives a false perspective. The democratic side of the 1924 constitution (which permitted people to vote for only one party) seems a travesty to people used to voting for two parties. But for people who were not used to voting at all, this was a concession. For the backward peoples it was a novelty to be accorded the same political rights as the Great Russian majority. The system of interlocking local, regional, provincial, republican and All-Union Soviets gave millions of people who were not members of the party some interest in problems of government and administration. No doubt the Soviet system was in essence little more than a transmission belt for the power of the party. If so, it was one which made use of the energies of many of the most energetic Soviet citizens. No doubt the nationalities policy was at times used along the old imperialist line of 'divide and rule'. This was obviously so in Central Asia where, in 1924, quite new and probably unjustifiable arrangements were made with the sole object of preventing the Uzbeks from dominating the area. But the lack of apparent friction between the peripheral nationalities and the Great Russian majority, the evident effort put into the maintenance of local customs, languages and traditions, the economic growth of large areas which were neglected under the Tsars, the rapid spread of education among the most backward nations: these fact all point to the success of the 1924 constitution. Anglo-Saxon critics who either approve the skill with which the British ruled India or commend Lincoln for maintaining the Union in the American Civil War, should beware of the criticisms they make of the U.S.S.R. With all its blemishes, the 1924 constitution was to some extent a realisation of the old populist dream of a partnership between the Great Russians and the backward peoples.

THE NATIONAL ECONOMY RESTORED

The dispute over the nationalities question had been fierce but it had not divided the party. The same cannot be said about the problems which arose from the economic aspects of N.E.P. These aroused passionate debate which was further exacerbated by the leadership dispute between Stalin and Trotsky. The purely economic aspects were in fact rather clouded by the political implications and by the personal antipathy of these two claimants to Lenin's authority. The basic economic problem raised by N.E.P. was one of timing: for how long could the peasant be conciliated without endangering the revolution and the party? At first some hoped for a windfall from abroad, perhaps a successful revolution in Germany (or even Great Britain) which would help to solve the problem of Russian backwardness; or perhaps it might be possible to revive the policy of Witte and persuade some foreign government to donate the capital which Witte had so skilfully extracted from France. But, as the twenties progressed, such hopes faded. There would not be a European revolution and there would be no foreign aid. The Western powers hastened to reconstruct their former enemy, capitalist Germany, but could spare no crumb for their former ally, Communist Russia. Clear conclusions could be drawn: Russia must go it alone. But how? There were those who argued that the framework of N.E.P. must be retained and that the party must do just what the Tsars had always done – that is, draw away the peasants' profits by taxation, direct and indirect. By means of this 'primitive socialist accumulation' the capital for industrialisation would be collected. Others argued that such a slow process would merely have the effect of strengthening the bourgeois segment of the mixed economy. A dominant peasantry would in time hold the party of the proletariat to ransom and be able to effect a counter-revolution. To prevent this, industrialisation must be forced, even at the expense of the peasantry. The first view, to which the label 'socialism in one country' was attached, was put forward by Stalin during the twenties. The other view was held by Trotsky. The difference between these views was enormously exaggerated by their struggle for power. Yet in spite of the continuous and fine-drawn theoretical arguments, the facts were clear. The 1917 revolution had not created a socialist economy.

The adoption of N.E.P. had been a further retreat from a socialist economy. If N.E.P. was allowed to continue unchecked, it might be impossible ever to build a socialist economy.

The economic developments of the twenties underlined the difficulties and sharpened the political controversy. Agriculture recovered rapidly. Good harvests in 1922 and 1923 restored the production of food to pre-war levels. Peasant prosperity tranquillised a countryside in which the party had little influence. Private trading brought the food to the urban markets and there was even a small surplus for export. Industry recovered more slowly. The towns were unable to produce in sufficient quantity the goods which the peasants, with their new prosperity, wanted to buy. While the price of agricultural products fell, that of industrial goods rose. This imbalance, which Trotsky named 'the scissors crisis', became serious in 1923. It meant that at least a part of N.E.P. would have to be immediately abandoned, and the State would have to restore the balance of a market which was dangerously free. Various means were suggested to achieve this end – means which became involved with the political controversy. Trotsky advocated the deliberate acceleration of the rapid inflation which was occurring anyway, as a means for extracting peasant savings. Others advocated still further reliance upon foreign capitalists who should be encouraged by the grant of economic concessions. In fact the 'scissors crisis' was alleviated by artificially reducing the price of manufactured goods, by exporting surpluses, by encouraging peasant purchasing co-operatives, and by increasing the number of state trading trusts. In other words, the crisis had forced the State to enter the market. An important step had been taken towards a managed economy. Whether this was also a step towards socialism was another matter.

But there were deeper contradictions within the N.E.P. structure which could not be solved so easily. Although industry did eventually recover its pre-war levels of output, there was no guarantee that it would exceed these levels. Energetic attempts were made to restore the basic dynamism of the Witte régime. A gold-based currency (the *chervonets*) was established in order to encourage foreign trade. But although useful, and indeed vital, amounts of foreign credit were earned and used for the purchase of manufactured goods (e.g. Swedish locomotives), the main pre-war reason for investment in Russia had gone. That is, no Great Power was now interested in building up Russia as a potential ally. In the absence of a fresh policy industrial development would merely remain the subsidiary of a dominantly peasant economy, at the mercy of the weather, the rich peasant, and the economic conditions in the capitalist world. To these difficulties the doctrine of 'socialism in one country' seemed to offer a solution. In her efforts to emulate the industrialisation of the West, Russia would do without Western wealth. She would beat the West at its own game but by playing to different rules.

Such objectives were declared by Stalin in 1925. Self-sufficiency was the goal and planning was to be the means. At first Stalin let it be understood that 'planning' meant no more than the rationalisation of a rather cumbrous and inefficient economic system. There had been plenty of evidence of wasteful and contradictory use of resources: industrial prices had risen when they had been intended to fall; the banks had restricted credit to industry at moments when the party had demanded industrial growth. Mass unemployment continued to plague the towns and the gradual transference of rural populations to factory work would entail the most delicate foresight. Such considerations enabled Stalin to popularise the cause of planning – a course of action whose advocacy had earlier cost Trotsky much popularity. What Stalin concealed from the party was that by 'planning' he did not mean mere rationalisation but forced industrialisation and collectivisation.

In the end it was the peasant problem which proved the most intractable. True, the peasant was now feeding the towns, but he was feeding himself still better. In bad years he kept the grain, lived well and let the townsman starve. Peasant agriculture still remained inefficient. Output per acre increased only on the *kulak* lands. These capitalist farmers were more efficient and marketed a higher proportion of their grain. Bukharin and other right-wing party theorists argued (much in the manner of Stolypin) that the towns could be fed by excouraging *kulak* agriculture at the expense of the middle and poor peasants. The events of the winter of 1927–8 showed the danger of this policy. A bad harvest encouraged the *kulaks* to hoard their grain until the prices reached a maximum. Food shortage in the towns forced emergency measures: tax concessions were repealed and grain forcibly requisitioned. The poor peasants were encouraged to turn against their *kulak* neighbours. This was a declaration of war and meant the end of N.E.P.

THE PARTY BUREAUCRATISED

The most important results of N.E.P. were political rather than economic. The restoration of peasant prosperity, the forging of links between the old Russia and the new, were simply means to an end. The main purpose of N.E.P. was to give time for the transformation of the revolutionary party into a governing party. Its refusal to brook any organised rivals was implicit in the destruction of the Constituent Assembly (1918). Thereafter it was obvious that the party would be true to its traditions and refuse to share political power. What was not so evident in 1921 was the growth within the party of a directing class and, within that class, the increasing concentration of power in the hands of a single individual. Such developments were perhaps inevitable: the tradition of autocracy was too powerful to permit of any other. The rise of Stalin should not be attributed to the diabolical cunning with which he manoeuvred his opponents out of the way. The mere command of tactics would not have won him the campaign. The roots of his success lie much deeper. He evolved the sort of personal power which the situation of the party and the traditions of the nation demanded. It is futile to claim that the early sickness (1922) and death of Lenin (January 1924) had a decisive influence upon the course of events. It is true that Lenin's leadership had depended largely upon intellect, personality and success. But these gifts would not have been sufficient to prevent the rapid evolution of a bureaucratic party; the essence of Stalinism was bureaucracy. It is possible that Lenin could have postponed the rise of Stalin; but Stalinism was inevitable in the N.E.P. period.

During the last months of his active life, Lenin had tried to keep the party small, to rid it of the 'opportunist' elements which had joined it during the civil war. But this policy was unrealistic once it was conceded that the party was to have no rivals. It would have to be active in every sphere of political and economic life and for this it would have to be numerous. Between 1924 and 1928 it grew from 472,000 to 1,304,471: the growth is even more sizeable when it is remembered that during the same period thousands were purged and replaced. The new elements far outnumbered the Old Bolsheviks who had joined the party under the most inauspicious conditions, who had been sufficiently idealistic to survive underground work, danger, privation and imprisonment. The new party members were not linked together by a common sense of purpose and a shared idealism. They were careerists, taking their chance in life with the organisation most likely to bring success. Such people were seeking privilege rather than obligation – the privilege of power and the privilege of better living conditions. For such people a career structure had to be created, efficiency encouraged by a system of favours and penalties – in brief, the enlargement of the party made an inner-party bureaucracy essential. This fact becomes the clearer when we examine the human material from which the new cadres were chosen. The horde of new recruits by no means represented the academic or intellectual or professional *élite*, a group of people who might be expected to work together with a minimum of direction. On the contrary, the new recruits were taken from the humblest classes. There was a very low standard of education and an even lower standard of 'political literacy'. It was obvious that the masses in the new mass party were not intended to wield wide power themselves. Their job was to inculcate discipline and enthusiasm in the factories, the unions and the villages. From their place of work they would humbly propagate the party line; they would have no say in the creation of that line. These party masses would require the closest direction and control from the inner-party hierarchy. The pretence of equality and democracy within the party was scrupulously maintained. As in the Soviet system, a series of elections from the smallest party cells upwards culminated in the election of the sovereign organ, the party congress. This in turn elected the Central Committee, and this (from 1919) the supreme executive body, the Politburo.

But with the growth of the party bureaucratic machine, the elective process became further and further removed from the centre of power. A turning point occurred in 1921. Up to the 10th Party Congress, it had been accepted that a wide range of opinion could be expressed within the elected party organs. Individuals and groups could use the platform to criticise the leadership. Such was the case just before this congress. There were at least two formed opposition groups, the Workers' Opposition and the Democratic Centralists. The complaints of both groups were similar; they spoke for the 'outs' against the 'ins'. That is, they complained about the concentration of power at the centre, of the habit of appointing rather than electing to party posts, of the lack of 'democracy' in the party, of high-handed treatment of the trade unions, of the powerlessness of the Soviets. The panic generated by the Kronstadt revolt, together with Lenin's tactical skill, enabled this crisis in party unity to be overcome. But the leadership was determined that it should not happen again and the Congress passed a resolution which condemned any future 'fractionalism'. That is, comrades in future were to debate their criticisms of the leadership only before the full Congress. It was forbidden, upon pain of expulsion from the party, to set up a 'platform' before the meeting of a Congress.

Obviously this move would not have been as decisive as it was, had it not been followed by an effort to ensure majorities obedient to the leadership at each Congress. It was pointless to forbid 'fractionalism' outside the congress hall if, once having got there, a mutinous delegate could find others to share his view. The only way to ensure that the Congress was obedient was to 'pack' it: and this could only be done by a powerful party bureaucracy. Lenin's step in 1921 led inevitably to Stalinism – a fact of which Lenin became aware himself during the last months of his life.

The original party bureaucracy had been very weak. This prompted the Central Committee (1919) to set up a simple machine. An organisational bureau (Orgburo) and a Secretariat were created. Stalin served in the former but it was from the latter (and junior) department, of which he was later appointed general secretary, that he began to develop his power. Perhaps he chose the Secretariat as his base of operations simply because it was a relatively humble niche in the bureaucratic machine. Its great advantages were, first that it was comparatively little supervised by the Central Committee; and secondly, that it was most closely in touch with the provincial hierarchy, with those ambitious men whose feet were set on the ladder of power and who would consequently obey one who could materially advance them. Stalin's power was founded in the middle and lower ranks of the inner-party hierarchy, among the 'outs' who were on the edge of becoming 'ins'. Stalin expected his clientèle to rig the elections to the Central Committee: in return they could expect promotion. In order to increase the patronage at his disposal, Stalin encouraged an increase in the size of the Central Committee: the more places to fill, the more power he could exercise. Essentially, Stalinism was a political system, very similar to such systems in all mass parties. It was the product of ambition, ideology and organisation. It was not, in its origins, the product of terror. The secret or political police (which changed its name from Cheka to GPU during the twenties) was freely used by the party against its external enemies in the nation, but it was not used against party members until Stalin had achieved bureaucratic centralisation. As the manipulator of political patronage he could, of course, wield 'terror' of a sort by the threat to withhold patronage. The threat of purge, the fear of getting a job on the periphery rather than in the centre of power, were important instruments of discipline – but they were no different from the powers employed by the chief of any large organisation in the capitalist world.

The growth of the Stalinist machine tended to strengthen one of the original ingredients of Leninist party doctrine. Lenin had always emphasised the 'voluntary' or dynamic aspect of Marxism. He had trained a party to seize power, not to wait until History had performed its prophesied antics. His rivals inclined to the belief that it was the duty of the Marxist to wait until History had completed its economic evolution. In 1917 political power had been seized but the socialist organisation of economic life was slow to follow. N.E.P. seemed to many to be a declaration by Lenin that the laws of economic development must be given time to catch up with the political revolution. Socialism could be attained only in conditions of plenty, and no mere seizure of political power could ensure wealth. But the Stalinist reorganisation of the party enabled the dynamic Leninist line to be reopened. Socialism (of a sort) need not be awaited: it could be grasped, created, driven forwards, by the adoption of economic planning directed by the obedient party.

STALIN AND HIS RIVALS

Before the onward movement could be taken up again, Stalin had to ensure that his command over the middle and lower reaches of the party gave him complete control at the top as well. He was by no means the 'crown prince' to Lenin. Trotsky, Zinoviev and Kamenev all had more powerful claims to the 'throne'. Stalin was regarded as a dour, dull and industrious bureaucrat. Such claims as he had were weakened by the rumours that Lenin had, during his last days, directed some scathing criticisms at Stalin. But Stalin's trump card was Trotsky himself. This man stood head and shoulders above his rivals in every respect. He was the most brilliant theoretician; he was a demagogue of great power; a writer who could illuminate every subject from history to literary theory; he knew the external world; he was known by foreigners; he had created the Red Army; he had led the 1905 revolution; he had negotiated at Brest-Litovsk; he was Lenin's closest political friend. It seemed inevitable that he should ascend the throne.

But in an age in which revolutionary ardour was giving way to organisation, Trotsky's gifts and the superiority which they gave him over common mortals were a positive disadvantage. He was universally feared in the party as the coming 'Bonaparte', the man who would use his special position in the army to seize power over the party. This fear was absurd. Trotsky had no such plans and military power was far distant from his essentially civilian mind. But he was a Jew; he had not joined the Bolshevik party until 1917; he was brilliant; he was 'not quite like us'. As the party filled up with people who like Stalin had never been abroad, were not intellectuals, had not actually witnessed Trotsky in action in 1917, it became easier for Stalin to point him out as the enemy of party unity, the coming dictator against whom all good party members must close ranks under the leadership of the safe, middle-of-the-road leader, Stalin. These tactics are reminiscent of those by which the British Prime Minister, Baldwin, brought down Lloyd George in 1922.

While Lenin lived Trotsky continued to enjoy his protection, but he did little to prepare himself for the coming struggle. Perhaps he was too confident; perhaps he would not stoop to a struggle for power; perhaps he misjudged the implications of the change from revolution to organisation. The very clarity with which he expressed his views (although these were not notably different from Lenin's) made him singular. Every stand which he made seemed to frighten or annoy the party rank and file. He was an outspoken critic of the trade union group in the party (known as the 'Workers' Opposition'). He called for the abolition or at least the emasculation of the unions. Lenin achieved the same result more tactfully by ensuring that union leadership remained in the hands of party members and that these members put their loyalty to the party before their loyalty to the unions. Trotsky loyally accepted both N.E.P. and the ordinance of 1921 which forbade formed opposition groups within the party. But he was soon alleged to be an enemy of N.E.P. He spoke and wrote enthusiastically in favour of planning and industrialisation at a moment when the party was trying to conciliate the peasantry. 'There may be moments when the government pays you no wages . . . and when you, the workers, have to lend to the State.' He was accused of being the enemy of the workers and peasants and of seeking to build up for himself a new centre of power as the boss of industrial planning. His imagination was caught by the possibilities of industrialisation in a backward country. He wrote (with fatal brilliance): 'Moscow is the capital of the Communist International. You travel a few scores of kilometres and – there is wilderness, snow and fir and frozen mud and wild beasts. . . . Where Shatura [power] station stands, elks roamed a few years ago. Now metal pylons of exquisite construction run the whole way down from Moscow.' He enthused also about the projected power station on the Dnieper rapids – a work which was shortly to become the showpiece of Soviet Russia. Stalin dourly commented that the power station 'would be of no more use to Russia than a gramophone was to a peasant who did not possess even a cow'. In the field of foreign policy Trotsky became associated with rash and dangerous attempts to provoke revolution in foreign countries.

It is hardly surprising, then, that Lenin's illness and death provoked a combination against this apparently rash, ambitious, unreliable and dangerous man. Zinoviev (Leningrad party boss), Kamenev (Moscow party boss) and Stalin (the junior partner) formed an anti-Trotsky triumvirate. Stalin let his partners do all the work while he stood back and, in public, played the part of an honest broker, attempting to moderate the excessive anti-Trotsky hatred of his two partners. When Trotsky attempted to unite all the party dissidents against the triumvirate (Platform of the Forty-six, 1923), the Stalinist-dominated Congress merely accused him of breaking the 1921 ruling about fractionalism. The more he complained about the lack of inner-party democracy and the power of the party bureaucracy, the more it seemed that he was not a good Communist, that his *petit bourgeois* Menshevik past still ruled his mind. The death of Lenin (January 1924) strengthened Stalin's hand. He ostentatiously appeared at the funeral. He ostentatiously associated himself with the public preservation of Lenin's mummified body. He was the guardian of the sacred remains; it was an easy transition to the belief that he was also the anointed heir. This was a brilliant stroke and one which would not have occurred to the rationalist intellectual, Trotsky. The rumours of Lenin's death-bed doubts were effectively scotched.

The triumvirate pressed on rapidly to disarm Trotsky. The 1924 and 1925 Congresses were packed against him so that he could hardly get a hearing. He was reduced to publishing accounts of the 1917 revolution highly unflattering to Kamenev and Zinoviev. These two called for his expulsion from the party but Stalin, with crafty moderation, declared himself content with mere expulsion from the commanding position in the War Commissariat (1925). This effectively disarmed Trotsky and left Stalin free to turn suddenly upon his former allies. As the leaders of the big city parties, Kamenev and Zinoviev were under much pressure to ease the condition of urban workers. Unemployment was widespread; the peasant prospered while the worker starved. The cities wanted a change in the N.E.P. programme. They wanted more investment in industry. Stalin took the opportunity offered by this programme now advocated by his allies. He accused them of being Trotskyites at heart, of being the enemies of the peasant, of N.E.P., of Lenin's memory. He turned for support to the right wing of the party, Bukharin, Rykov and Tomsky. The obedient machine duly threw up a majority. Kamenev and Zinoviev fought hard but against them could be used exactly the same weapon which had been used against Trotsky – fractionalism. The defeat of Zinoviev at the Congress was followed by a thorough purge of the Leningrad apparatus by Molotov. Then the faithful Stalinist, Kirov, was placed on Zinoviev's minor but powerful throne. Only mopping-up operations were now required.

Trotsky provided the initiative which allowed Stalin to put the finishing touches to his tactical masterpiece. Now the reason for allowing Trotsky to remain in circulation appeared. In desperation, Trotsky, Zinoviev and Kamenev united forces against Stalin. This was enough to discredit all three. With considerable but futile ingenuity Trotsky sketched out a programme the very existence of which was a condemnation. He argued that the revolution had succumbed to a *petit bourgeois* reaction. He attacked Bukharin (now Stalin's ally) for being the friend of the *kulak*. Had not Bukharin encouraged the rich peasants to enrich themselves still further? This was to carry N.E.P. far beyond what had originally been intended by Lenin. Lenin had envisaged a temporary retreat, not a complete rout. But the time was long past for ingenious arguments. The delegates to the 15th Conference (1926) never even listened. They shouted down the opposition speakers. Stalin, playing to his right-wing audience, accused Trotsky of wishing to declare war on the peasantry, of stirring up trouble at home and abroad. He won an overwhelming victory. Trotsky was expelled from the Politburo and from the Central Committee (1927). Then, in utter despair, he committed the unforgivable sin. He 'proved' that all the charges against him were true. By means of street meetings and clandestine propaganda he tried to arouse party and popular feeling against the leadership. On the tenth anniversary of the revolution (November 1927), the chief architect of Bolshevik success in 1917 attempted to overthrow the new master of Russia. There was no response. The party was horrified: the city crowds merely gaped at the former popular hero. With great foresight Stalin refrained from the permanent elimination of Trotsky, who was first exiled to the Chinese frontier and then (1929) deported. Trotsky abroad was still to do Stalin great service. The other oppositionists were received back into favour. As repentants they could still be pressed into service against Bukharin and the right.

For, of course, Stalin had merely used Bukharin as a temporary ally. Hardly had Trotsky taken up residence at Alma-Ata than Stalin began to put into effect the economic policy which Trotsky had advocated for five years. But there was an important difference between Stalin's execution of industrialisation and Trotsky's advocacy of it. Stalin's policy of 'socialism in one country' had as its sole objective the strengthening of the Russian State The ideal of socialism was to be subordinated to the growth of state power. Trotsky, on the other hand, had advocated industrialisation as a necessary step forward to socialism. Without the increase of wealth which industrialisation would bring he considered that it would be impossible to move forward from the N.E.P. era into the socialist society which the 1917 revolution made possible. His main objective was not the growth of state power but the development of a new order of economic and social relationships. In his view the creation of a wealthy, just and modern socialist order in Russia was the best way of ensuring that the revolution would extend into Europe and Asia. It was difficult for him to infect his contemporaries with his powerful yet nebulous vision. Stalin's line was safer, it was more in tune with Russian national traditions, it offered a clearer task for the party bureaucracy.

With the elimination of Trotsky it was simple to neutralise the protests of the right-wing leaders. Their defence of the peasantry was depicted as a continuation of the populist mood of the previous century. They seemed to stand in the way of Stalin's bold plan to make Russia strong. Bukharin, Rykov and Tomsky were allowed to recant but they were expelled from the Politburo and thereafter had little influence. This left the Stalinist centre in power. The age of inner-party politics was over.

12

Russia under Stalin: The Second Communist Revolution: 1930-40

'SOCIALISM IN ONE COUNTRY': ITS IMPLICATIONS

The 1917 revolutions had altered the political structure of Russia but a decade later the economic structure remained stagnant. Russia was further from being a world power than she had been in 1914; she was even less equipped to withstand a foreign war and in an even more dangerous position because she lacked allies. The old Bolshevik dream that backward Russia would be saved from the consequences of her backwardness by the provocation of successful revolutions in the more advanced nations had been shown to be futile. The revolutionary word which had seemed to be all-powerful in 1917 was received with indifference in the world of the late twenties. A new course was needed, one that would render Russia less vulnerable, that would give the party a reason for existence, that would enable the people to be disciplined. The era of enthusiasm was over and the era of construction had arrived. To mark this fresh turn a new slogan and a new man appeared: the slogan was 'socialism in one country' and the man was Joseph Stalin.

The slogan had already been used by Stalin in the struggles against both the left and the right of the party. Against Trotsky and the left it had been used to mean that Russia should concentrate upon national survival, should avoid provoking any internal clash between peasant and proletariat, should cautiously follow the guidelines of N.E.P. as laid down by Lenin. Against Bukharin and the right of the party Stalin interpreted the slogan differently. He argued that socialism had not yet been attained, that it could not be reached by continuing along the line of the twenties but only by a determined and dynamic effort of state-building, an effort which would be mainly directed towards the disciplining of the peasantry and the creation of a predominantly industrial, urbanised nation. In brief it may be said that Stalin appropriated Trotsky's domestic programme while rejecting his adventurous foreign policy, while at the same time accepting Bukharin's cautious foreign policy and rejecting his conciliatory attitude towards the peasantry. Stalinism was initially a moderate course: but it was not pursued by moderate means.

Much ink can be and has been spilt in the discussion of Stalin's character and motives. Few men have been so widely attacked and denigrated. Former colleagues like Trotsky used all their considerable literary gifts against the man who beat them at their own game – always a humiliating experience. Since Stalin's death even the Stalinists have hastened to blacken the name of their former master, the release of terror and the relief from sycophancy adding eloquence to their denunciations. Leftish writers throughout the world have reacted with horror to the Stalinist thought-control, a discipline which successfully silenced a whole generation of fellow intellectuals in Russia. Exiles from the various nations of Europe conquered by Stalin's Russia have added their indignant voices to the chorus. Humanitarians have been duly horrified by the purges, the labour camps, the secret police. The Cold War has given a motive for the publication in English of numerous works by Russian defectors describing in vivid – and sometimes, it may be surmised, inaccurate – terms the horrors of Stalinism and the monstrously perverted character of the despot behind the scenes. Above all, what the horde of Stalin's detractors cannot forgive is that he gained his ends: it is highly offensive to the liberal-democratic-radical intelligentsia that a ruler who broke all the rules should nevertheless have been so completely successful. To accept this requires a reorientation of values which the writers of books, as a class, find difficult to accept. Even Hitler is found to be a more congenial character than Stalin because he at least came to a bad end. Stalin died in his bed having achieved more than his allotted span.

To see Stalin clearly consequently requires a more than usually determined effort of sceptical sympathy on the part of the historian. Comparisons with Genghis Khan or Cromwell are not helpful. Even references to Peter the Great or Nicholas I conceal the fact that Stalin adopted a political technique perfectly suited to the conditions in which he lived. He set himself possible goals and with cunning and tenacity he achieved them. Whether he could have achieved them in any other way is no longer worth consideration. Whether he should have set himself other goals is hardly a less futile question: if he had, Soviet Russia could hardly have survived the German attack in 1941. Stalin is best considered as the lineal descendant of the most realistic of the statesmen of the old régime, Witte. To Witte's aim of making Russia a world power Stalin harnessed the political machinery bequeathed by the revolution. This may have been a betrayal of the revolution but it was not a betrayal of Russia. Nothing which has happened since Stalin seized supreme power has challenged the accuracy of his basic assumption, that in the twentieth century nationalism remains a stronger force than class conflict.

The initiation of the Stalinist revolution was the adoption by the 16th Party Congress (April 1929) of the first five-year plan. This was put into operation in 1928 and its completion was announced in December 1932. The predominant motive behind the adoption of the plan is difficult to assess. Later, of course, Stalin was able to argue that he was mainly concerned with making Russia militarily strong: but at the time the danger to Russia from outside was comparatively small. It has been argued that he was mainly concerned to bring to an end the ideologically dangerous situation in which a peasantry, allowed to organise its affairs along the lines of a capitalist free market, permeated the whole of society with an unacceptable set of values. It has also been urged that what Stalin was really after was a means of disciplining his party by means of ceaseless activity – a discipline which was all the more necessary to combat the influence of Trotsky's propaganda. Stalin himself drew attention to the problem of securing sufficient grain. Without more consumer goods the peasants would not market food; more consumer goods could only be produced at the expense of neglecting the heavy industry sector: consequently the peasants, organised as they were under N.E.P., impeded the long-term development of the Soviet Union. Fewer peasants must be made to produce more food; those fewer peasants must be organised in such a way as to make them more amenable to the direction of the party. There had been two victors in 1917: the party and the peasants. Now the party was strong enough to consolidate in the countryside the victory it had already won in the towns. One of the most dramatic results of the Stalinist revolution was to make Russia for the first time a country in which there were more urban than rural dwellers.

The very idea of a national economic plan was a novelty in the twenties. Its relevance to a socialist society had been grasped since the revolution. In 1921 the State Planning Commission (Gosplan) had been set up. Its brief was at first merely to forecast the probable developments of the economy. Its experts, following the lead of the right wing of the party, foretold disaster from any attempt to force the pace of industrial and economic development. In 1929, however, the cautious and rather pessimistic prophecies of Gosplan were reinforced by the fiat of the party. The idea of revolutionary and creative planning was expressed thus: 'Our task is not to study economics but to change it. We are bound by no laws. There are no fortresses which Bolsheviks cannot storm. The question of tempo is subject to decision by human beings.' These brave words recall the 'voluntarist' element in Leninism. Ordinary Marxists said that one must wait for History to deliver the goods: Leninists said that the party was itself historical destiny, and was consequently bound to effect those things which History demanded. The choice of political revolution in 1917 and the decision for economic revolution in 1929 were both dictated by the inner dynamism of the Leninist party. This was not a private wickedness or betrayal of the chaste revolutionary ideal by Stalin.

Elaborate and optimistic goals were set in every field of production. Priorities were declared and the whole weight of the party thrown into the economic battle just as in earlier years it had been thrown into the fight against political enemies. In earlier days the party had assumed that the mere destruction of the old society would lead automatically to the emergence of the new. Stalin showed that economic transformation must be struggled for in just the same way as political revolution. The idea of ruthless direction was not new – both Lenin and Stalin had witnessed its effects in wartime Germany – but for the first time in the modern world a great nation put its economic life upon a war footing in peacetime. A complete break was thus made with the individualist motivation of economic life characteristic of nineteenth-century Europe. The belligerent nations had provided a substitute for individualism in wartime, a substitute which was quickly abandoned as soon as peace returned. Stalin placed Russia upon a permanent war footing in which the future good of the State, the party and the revolution took precedence over the present well-being of the individual. In the planned society, where the plan took the place of the free market, all individual satisfactions had to be postponed until an unspecified future.

COLLECTIVISATION OF AGRICULTURE

The first group to feel the full weight of Stalinist planning was the peasantry. Already, before the implementation of the first five-year plan, great difficulties had been experienced in getting the peasants to bring sufficient grain to market. By 1927 the proportion sold was only about half of what had been put on the market in 1913. This situation had been reached in spite of attempts by the government to increase the payment for grain to a level which would enable the peasant to buy consumer goods. But the terms of exchange had turned progressively against the countryside during the twenties. In consequence the peasant produced more but he also consumed more. This was particularly true of the great mass of farmers – in party jargon, the poor and middle peasants – whose grain deliveries fell off more sharply than the deliveries of the *kulaks* (rich peasants). The latter, who produced only 13 per cent of the total amount of grain, marketed 20 per cent of it: whereas the two classes of poorer peasants, who produced 85 per cent of the total, marketed only 11 per cent of it. One solution to this problem would have been to encourage the *kulaks*, but this would have been ideologically impossible and would in the end have accelerated the rise of a new class of landowners. Another (and ideologically acceptable) solution would have been to organise agriculture in state farms, in which the peasants would have become the paid hands of the State. Such a solution had advantages because it would have enabled the peasant's individual economic motives to be subordinated to the demands of the party. But it was rightly held that any such organisation would entirely alienate the peasant population, and a more moderate course was adopted: the collective farm.

During the twenties some experiments had been made with various kinds of collective. At one extreme was the *toz* or society for joint land cultivation. In this the peasant kept his land and merely joined with his neighbours to share the expenses of machinery. At the other extreme, the commune completely dispensed with all forms of private property: members lived in dormitories and children were looked after in communal nurseries. Between these two forms – neither of which achieved any popularity in Russia – was the *artel*, the organisation which became the basis of the collective farm.

In the *kolkhoz* (collective farm) most production is carried on collectively and the larger means of production are collectively owned, but the members retain some individual private rights. Many of the difficulties of Stalinist and post-Stalinist Russia stemmed from the compromise between private and collective rights embedded in the *kolkhoz*. Each *kolkhoz* has a charter in which its boundaries are carefully defined: the land is held to belong to the State but the *kolkhoz* is given permanent use of it. All the means of production except the very largest and the very smallest belong to the *kolkhoz*: that is to say, garden tools will be the property of individual members, while tractors and other large machinery will be hired from the Machine Tractor Stations (M.T.S.). Members of the *kolkhoz* own their own gardens and houses and a limited number of domestic animals – the number will depend upon whether the area is primarily a grain-producing region or not. An average number per household might be one cow, two sows and an unlimited number of chickens and rabbits. In practice great difficulty is experienced in keeping a balance between the private and the collective activities of the farmers: they tend to concentrate upon their private holdings at the expense of their collective duties.

The collective is managed by an elected committee; this in turn is in theory allowed to elect a chairman but in practice this officer is generally appointed by the party. All the work on the farm is done by brigades of *kholkhozniki*, 50–100 strong. Individual effort is measured in work-days, a unit which is adjusted according to the difficulty of the work done. Additional incentive is supplied by the setting of production norms for each brigade, with penalties for failure to reach the target. Each farm also has an annual production target, set at the beginning of the season at so much grain, cattle, flax, etc. per acre. This target has to be met whatever the weather turns out to be. These first, or state deliveries, are taken at a very low price, much less than the price which could be commanded on the open market. The government can then sell at whatever price it fixes, thus making a handsome profit on the deal. In a good year the State may also take another cut, this time paying a higher price. The Machine Tractor Station must also be paid in kind. Of what remains, the collective keeps some, for seed and reserve; and the remainder of the crop is distributed in kind to the members. The total number of work-days is divided into the amount remaining and each member receives his share of both the grain and money income of the collective. The following table, showing the distribution of grain crops in 1938, a poor but not disastrous year, will illustrate some of the points made above:

	%
A. *Deliveries to the State*	
1. Obligatory deliveries	15·0
2. Payments in kind to M.T.S.	16.0
3. Return of seed loans	2·0
Total	33·0
B. *Sales to the State and on free markets*	5·1
C. *Collective requirements and reserves*	
1. Seed requirements and reserves	18·6
2. Feed requirements and reserves	13·6
3. Reserve for aiding those in need	0·8
4. Other expenditures	2·0
Total	35·0
D. *Distributed to collective farmers on the basis of work-days worked*	26·9

The proportion of grain going to the Machine Tractor Stations should be noted. These played a very important part in the process of collectivisation. Each station served about thirty collectives and provided them with their only source of tractors and harvesting machinery. The whole success of a year's farming operations would depend upon the efficiency with which a station carried out its spring and autumn tasks. There were two reasons for separating the machinery from the farms: first, that it could be more efficiently serviced and employed by specially trained urban personnel, and secondly that these stations could be used as centres of political leadership in the countryside. The party had always found difficulty in maintaining itself at village level. Individual members soon lost their identity in the great mass of peasants. But in the M.T.S., party cells could flourish and bring direction and enthusiasm to a tepid or even hostile countryside. At the height of the drive towards collectivisation the political departments of the M.T.S. were given complete control over the neighbouring collectives, a control which by-passed even the local party organisation. The heads of these political departments were directly appointed by the Central Committee. Most of them were urban communists of Old Bolshevik vintage, who probably knew little about farming but much about the arts of persuasion and control. From 1934, when the main crisis of collectivisation was over, the powers of the political departments of the M.T.S. were merged into the normal party organisation, although they still remained important. But the episode demonstrates the strength and flexibility of party technique: the enthusiasm of the best men could be thrown into a critical phase of economic development just as well as into a battle or revolution.

For although the collective farm had been chosen as the type of organisation least likely to cause resentment, in fact the early years of the plan resembled a civil war more than an economic reorganisation. The chaos may have been caused in part by the over-hasty way in which collectivisation was carried through. At the end of 1927 there were only about 15,000 collectives: by June 1929 there were 57,000: and in November 1929 the party decided to accelerate still further the rate of collectivisation. A political tone was given to the affair by the order that the *kulaks* should be eliminated as a class: by this means the party tried to enlist the sympathy of the poorer peasants. By March 1930 it was claimed that there were 110,000 collectives, containing 14¼ million peasant households: that is, 55 per cent of the peasant population of the Soviet Union. Such dizzy speed, achieved at machine-gun point, threatened to prevent the spring sowing, and Stalin himself intervened to order a slowing down. But collectivisation continued and by 1936 90 per cent of the peasantry were included in about 246,000 collectives. The party had won but had the nation gained?

To strike a balance sheet of collectivisation is impossible because there are ideological factors which defy quantification. Who can define what might have happened had the *kulaks* continued to exist and to flourish? What might have happened to the party had it not had the discipline of carrying through the collectivisation? In human terms, the policy was no doubt very unpleasant: millions of *kulaks* were deported to the N.K.V.D. labour camps and possibly hundreds of thousands were killed in the process of resistance. This was certainly a worse fate than that of the millions of unemployed in Britain and America during the early thirties. The standard of living of the collectivised peasants almost certainly fell; some calculations conclude that most peasants went hungry in 1932 and 1936 and that the average peasant income even at the end of the decade was less than half of what it had been in 1928. By 1940 total grain production had probably risen by about 15 per cent over the 1928 figure; but the number of cattle never recovered from the wholesale slaughter which had taken place during collectivisation. The peasants, however, did not benefit much from the greater production of grain. It was taken from them at very low prices and their labour was consequently one of the main sources of capital accumulation. In one respect it may be said that collectivisation was an unqualified success: it did help to reduce the number of hands required to produce food. The rural population of Russia decreased by about 6 million during the thirties – although it should be noted that nearly 64 per cent of the total was still rural in 1937. Not until 1962 did the proportion of rural population fall below 50 per cent of the total. Grain deliveries were assured and a supply of labour for industrialisation secured.

INDUSTRIALISATION, TRIUMPHS AND ERRORS

Great as was the effort put into collectivisation, however, the main thrust of the five-year plans was towards industrialisation. By 1928 Soviet industrial production had just surpassed the level of 1913: by 1940, according to official Soviet figures, industrial output grew by 650 per cent. This was a spectacular achievement which certainly rivals, even if it does not exceed, the most rapid growth rate of any capitalist economy. It involved the expansion of already existing industries and the creation of entirely new ones – for example, for manufacturing complex machine tools and synthetic rubber. It was backed by rapid scientific and technological progress which led in a very short time to the ability to match and even to surpass the capitalist powers. The present Soviet ability to build the most advanced military weapons and to indulge in the sophisticated technology of space travel is a direct consequence of the decision to industrialise taken in the late twenties.

No foreign capital was available, and some idea of the strain upon Soviet resources may be gained from the following table, showing capital investment in industry:

Year	*Billion Roubles*
1928	1·7
1929	2·6
1930	4·2
1931	7·0
1932	6·5
1933	4·8

An annual average of about 5·0 up to 1938

It will be seen that investment fell off appreciably after the heroic year, 1931. The original investment plan was too optimistic, but a great initial effort was required to overcome inertia.

Investment was carefully directed into the most productive industries. It was decided to starve the consumer industries and to pamper the heavy industry sector. By means of this choice – which could only have been taken by an autocratic government confident of its power of leadership – astonishing progress was made in the selected sectors, as the following table shows:

Commodity	*Unit*	*1913*	*1928*	*1940*
Coal	mil. metric tons	29·1	35·8	166·0
Pig-iron	mil. metric tons	4·2	3·4	15·0
Steel	mil. metric tons	4·2	4·3	18·3
Petroleum	mil. metric tons	9·2	11·5	31·0
Electric power	billion kWH	1·9	5·1	48·3
Tractors	thousands	0	1·3	31·0
Combines	thousands	0	0	10·0

Commodity	*Unit*	*1913*	*1928*	*1940*
Chemical industry	Million roubles 1926–7 value	457	645	6809 (1938)
Machine and metal industry	Million roubles 1926–7 value	1446	3349	48400

In addition to this accelerated growth of heavy industry, military expenditure was kept at a high level. One estimate of the size of the munitions industry is based upon the amount of iron and steel used in it as a percentage of all U.S.S.R. construction. This gives a figure of about 17 per cent in 1932, falls slightly during the mid-thirties and rises to about 30 per cent in 1938. Post-war Soviet figures assert that military production increased by 286 per cent in the period 1933–8 – but it would be natural for the Soviet authorities to exaggerate the degree to which Stalin's foresight had prepared his country for the German attack. The obvious emphasis of the Soviet planners upon economic autarchy also implies a fear of wartime isolation. Although imports of foreign machinery and technicians played an important part during the early thirties, the need for both rapidly fell away as industrialisation got into its stride. Soviet foreign indebtedness (and imports) reached a peak in 1931 – partly because of the disastrous slump in world grain prices. By depriving its already starving people of grain, the Soviet government hoped to be able to purchase essential foreign commodities. But after 1931 the volume of imports dropped very sharply: the 1934 figure was less than a quarter of that of 1931. Economic autarchy had been virtually reached.

Although it is likely that one of the chief motives behind the five-year plans was to create a modern armaments industry, the growth of Soviet industry shows that other and primarily economic motives were uppermost in the minds of the planners. Little effort was made to shift the main centres of industry beyond the Urals, where they would have been safer from invasion from both east and west. The planners concentrated upon developing those areas which were already industrialised: the western regions, which produced over 80 per cent of heavy industrial products in 1913, continued to produce about the same proportion in 1940. Not until the German invasion in 1941 was there a determined effort to move industry eastward. From this it should not be concluded that the party did not envisage a war against Germany: merely that in the event of such a war deep German penetration was not foreseen. During the first years of industrialisation the planners seem to have been agreed upon the need for large-scale production and hence intense specialisation. This decision was copied from American practice and proved to be a mistake in Russia. Products had to be transported all over the Soviet Union, putting an intolerable strain upon the railway system. It was discovered that a single factory produced all the wheels required for railway freight wagons. This error was openly admitted by the party in 1939 (when it was conveniently blamed upon the 'enemies of the people, the Trotskyite Bukharin and bourgeois nationalistic diversionaries and spies'). Molotov called for better regional planning, an idea which implied the creation of autonomous regions each producing all that was necessary for itself. The former specialisation was stigmatised as 'gigantomania'. It is very likely that this shift of policy was caused by the fear that in the event of war the loss of a single factory could cause a complete breakdown of industrial production.

The tug between regional and national planning had been felt earlier when the Ukrainian planners (who based themselves upon the transport-orientated theories of the German Weber) opposed the development of one of the most spectacular aspects of the plan, the Magnitogorsk–Kuznetsk combine, situated in western Siberia. The Ukrainians wanted the capital to go into their own Donbas heavy industry area. They urged that whereas the Ukrainian iron ore and coal were separated by a distance of only 350 miles, almost 1500 miles had to be traversed in order to bring the Magnitogorsk ore to the Kuzbas coal. Whereas a ton of Ukrainian pig-iron took 700 ton-kilometres of haulage to produce, a ton from the projected combine would take 4500; and in addition, even when produced it would still be much further from the engineering plants in which it would be used. The party, however, ruled in favour of the Magnitogorsk–Kuznetsk combine and by 1932, after epic struggles, it had gone into production. The transport crisis which had been prophesied in fact occurred in the mid-thirties. It was met decisively by the party, which had sunk far too much political capital into the project to permit its failure. Freight charges were reduced and energetic attempts made to diminish the transport cost factor by finding coal near Magnitogorsk and iron near the Kuzbas. The original combine, in fact, was divided into two and its component parts separately developed. What is remarkable in this episode is not so much the original error as the rapid recovery from it.

Although electrification played an important part in the plans, and although attention was given to water-power as a source of energy (the great Dniepropetrovsk dam was opened in 1932), the main fuel of industrialisation was coal. Some attempt was made to relieve the pressure on the railways by opening up new fields. Up to 1929 nearly all Russian coal had been produced in the Donbas; but by 1940 significant amounts were being produced in the Urals, at Karaganda and in the Moscow region. It was found preferable to use the low-grade product of local mines rather than better coal hauled long distances. Oil production did not play a large part in the plans. Technology was not advanced and most of the output of the Caucasian wells had to be transported by rail over huge distances. There was little pipeline construction.

Indeed, the transport problem proved to be one of the bottlenecks of industrialisation. The railways got little help from road haulage: there were few roads and little provision in the plans for the development of a road-haulage industry. River and maritime traffic were expanded rapidly but not enough to take the weight off the railways. Under the first five-year plan little capital was available for the railways; yet freight traffic increased by 82 per cent in the period 1928–33. The result was chaos. Huge quantities of produce awaited removal and production was halted for lack of supplies. The party tackled this problem – which might have been foreseen earlier – with characteristic energy. In 1935 a Politburo member, Kaganovich, was put in charge of transport; the railways were given far more money than in the first plan. Electrification, block signalling and the doubling of track enabled them to meet the crisis, although even in 1940 Kaganovich complained about the low standards of maintenance and the frequency of accidents. It was also alleged that extreme winter conditions still impeded railway operation. But however much Kaganovich accomplished, it may be surmised that the transport crisis had much to do with the decision taken in 1939 to develop industry upon a regional basis. Railway tariffs were significantly altered in 1939 in order to penalise very long hauls and cross-hauling. Nevertheless the freight densities remained very high – they are still by far the highest in the world. Lack of attention to rail transport was probably a major weakness in the original plan.

LABOUR DISCIPLINE

The effect of the plans upon the Soviet people was enormous. At the bottom of the scale the masses were assured of employment but had to endure very low living standards – how low is a matter of controversy, but for at least a part of the thirties probably lower than the standards of 1913. Durable consumer goods were not available at all; and supplies of cotton textiles, shoes and house space frequently fell below even the fairly low norms set by the planners. Food rationing was in operation between 1930 and 1935, but a much more effective form of rationing was obtained by the control and diminution of purchasing power. Food prices rose rapidly during the period, much faster than the average wage rates. One calculation shows the following marked decline in the purchasing power of the annual average wage.

Year	*Average annual earnings in terms of 1928 'Food Roubles'*
1928	703
1932	710
1936	344·1
1938	404·1
1940	347·1

It seems that the Soviet masses had to accept a sharp decline in living standards at the very time when extra effort was demanded of them. Absence of the normal material incentives to greater effort raised in an acute form the whole problem of motivation and labour discipline in a socialist society.

The imposition of high demands and low rewards obviously required a reorientation of the relations between the State and the trade unions. During the N.E.P. period it had been accepted that, although the unions should be under the control of members of the party, they should be regarded as the champions of the workers against the employer, even where the employer was the State. Wage bargaining had remained in force and a measure of discipline was provided, as in capitalist countries, by the prevalence of unemployment. The period of the plans, however, was one of acute labour shortage. If the unions were not shorn of their powers, they would be in a very strong position to assert the claims of the workers and, by pushing up wages, to deprive the State of the ability to accumulate savings at the expense of the masses. For ideological reasons the party was loath to leave the unions in a semi-independent position: the monolithic single-party State could brook no rivals, no alternative centre of influence in close contact with the masses. To mark this radical change in policy, Tomsky and other leading unionists of the N.E.P. period were removed from office in 1928–9 and a fresh idea of union function announced by party leaders. All attempts at wage levelling – previously advocated by Tomsky – were abandoned; wage bargaining was no longer to be the chief function of the unions. Instead, the purpose of the unions was to be to 'organise socialist competition of workers and employers for fulfilling and overfulfilling state plans, increasing the productivity of labour, improving the quality and lowering the cost of production'. In other words, the unions were to speak for the State and its economic objectives, to the worker, not for the worker against the employer. Under this new charter, union membership increased rapidly during the thirties – from about 11 million in 1928 to about 25 million in 1940. Although shorn of powers that might embarrass the party, the unions nevertheless played a very important part in maintaining labour discipline. They administered the social services, and by threatening to deprive individuals of benefits they could impose a powerful coercive force; they popularised more efficient means of production; they imposed safety regulations upon managements; they ran clubs, libraries and sporting facilities for workers; they drew women into industrial life. Above all, they helped to avoid strikes. None is known to have occurred in the whole period up to the outbreak of war.

The wage-fixing functions which the unions had lost in 1928–9 were taken over by the State. They proved to be an essential element in the planned economy, by which the whole shape of development could be controlled. The State could thereby take whatever proportion it chose of the worker's wage, without apparently taxing him; differential rates for different industries could wean workers away from consumer industries into coal, iron and transport; above all, free from the pressure of mass unions of the unskilled, the State could create differential wage rates which would reward the strong, the skilled and the willing at the expense of the weak, the ungifted and the recalcitrant. During the thirties average wage rates fell, and the smaller total was more unfairly divided. Piece-work rates, although specifically condemned by Marx, were widely introduced and reinforced with personal and group bonus schemes. A calculation made in 1934 indicates that the highest-paid employees got twenty-eight times as much as the lowest-paid, a rate of difference comparable with capitalist countries. Extraordinary individual effort was also encouraged by the Stakhanovite movement, initiated in 1935 and named after a particularly energetic Donbas miner who both by strength and intelligence increased his production beyond all reasonable expectations. Emulation of this hero was encouraged by a great propaganda campaign, and factory managements were instructed to set their production norms by reference to their strongest workers.

So much for the carrot offered to the minority of the skilled and the strong. But more difficult was the imposition of labour discipline upon the great masses of peasants driven into industry by collectivisation and unlikely to benefit from differential wage schemes or to feel any inclination to emulate Stakhanov. Labour turnover was very high. In 1935, for example, 86 per cent of the workers in heavy industry changed their jobs; absenteeism was serious and labour shortage tempted managements to lure away their neighbours' workforce. Draconian measures were taken to deal with these problems. In 1932, managers were given the right to issue ration cards which could be removed if the employee were dismissed. This condemned him to starvation. Another decree of the same year gave managers the right to dismiss an employee for a single day's absence. In 1938 all employees were ordered to produce employment books when applying for a new job. Another decree of the same year attempted to secure good timekeeping by defining absenteeism as 'arriving for work more than ten or fifteen minutes after work has begun'. On the eve of war, in 1940, a worker was forbidden to leave his place of employment without permission of the employer; skilled workers could be moved anywhere at will; and selected youths and girls were trained for compulsory industrial work. The Soviet work-force was placed under military discipline.

In addition to these measures of labour discipline a degree of terror was supplied by the existence of forced labour in prison camps. Exactly how many people were thus employed and for how long is not exactly known; some estimates speak of 14 per cent of all capital construction being under the control of the N.K.V.D. in prison camps. Some of the largest construction projects in the harshest areas – like the White Sea canal – were wholly under the N.K.V.D. Human beings were harshly used and broken in these remote and cruel camps of whose existence Soviet citizens were well aware. This knowledge encouraged them to endure even the unrewarding toil of ordinary civilian life rather than run the risk of exile to Siberia or the far north. The widespread use of forced labour has since Stalin's death provided one of the most telling charges against his régime. Even Soviet citizens can now read moving personal accounts by some of the survivors of how they lived during the thirties. Could a more humane form of labour discipline have been found? Would a less harsh policy have improved labour productivity? Although Soviet labour productivity doubled during the thirties it still remained very low in comparison with the advanced industrial powers.

Nevertheless the Soviet Union had become the third industrial nation of the world by 1940: Tsarist Russia had occupied the fifth position in 1913. In 1940 only the United States and Germany produced more, and the gap between the Soviet Union and Germany was rapidly closing. This was a gigantic achievement. It altered the destiny of Russia and the shape of world politics. Without the Stalinist revolution the Soviet Union would have become a German colony and its people enslaved by an even harsher tyrant. The Russian war effort of 1941–5 is the best evidence that the Soviet people on the whole approved of what Stalin had done in spite of the appalling harshness of his methods.

STALIN AND THE PARTY: PURGES AND PERSONAL DICTATORSHIP

While Russia was being transformed by the plans, Stalin radically altered the balance of power within the ruling party and further increased the authority of the party within the nation. As Russia grew stronger, the party grew stronger and Stalin became its undisputed leader (*Vozhd*). While few now would deny the desirability of the economic developments of the thirties, the growth of Stalin's personal power is a topic which still causes the keenest controversy. Could the economic transformation have been carried through without this political development? Is it possible to explain Stalin's actions during the purges of 1936–9 on rational grounds? Is it necessary to assume that in becoming a dictator he had also become insane – a paranoiac?

Of the three great leaders of the twenties, Stalin stands out as the one with the most profound knowledge of Russia and the Russians – all the sharper perhaps because he was a Georgian 'outsider'. Neither Lenin nor Trotsky could rival the experience which Stalin had accumulated, first in the party underground and then during the twenties in close contact with the rank and file of the party membership. Although he was not an intellectual of the calibre of either Lenin or Trotsky, Stalin was a voracious reader: he knew the Russian nineteenth-century classics intimately and he knew the history of the Russian nineteenth-century intelligentsia. As a realist (unlike some modern historians) Stalin knew that 1917 was not a magic number: he knew that the traditions of the nineteenth-century revolutionary intelligentsia were preserved in the Russian society (and party) of his own day. It would not be necessary to assume that Stalin was a paranoiac if it was allowed that he had made such an analysis of the recent Russian past. In declaring war against the party – which is in effect what he did – Stalin was merely forestalling an inevitable outbreak of Russian anarchy. Other Russian realists – like Chekhov for example – would have understood very well what he was at. Chekhov's plays have numerous references to the ill-disciplined but potentially heroic rabble of the intelligentsia. 'Masters of culture, which side are you on?' asked Gorky: Stalin made sure that they were on his side; or at least silent if they were not.

During the first five-year plan Stalin dealt gently with his opponents. Some like Bukharin were removed from the Politburo; others, like Tomsky and Rykov, were forced to admit their mistakes publicly. With the fall of the right wing, the left-wingers hastened to make peace with Stalin, all the more easily because he had adopted policies which had been long advocated by Trotsky. Pyatakov, for example, severed his links with the exiled Trotsky and threw himself with energy into the tasks of economic organisation. Thus during the critical years of the plan Stalin was able to command the allegiance of the best minds and the most enthusiastic spirits among the Old Bolsheviks. But this unity was short-lived. By 1934 the passing of the immediate danger of civil war, resentment at the way in which Stalin concentrated power in the Secretariat, realisation that Stalin had made a major error in the field of foreign policy by permitting the rise of Hitler in Germany, fear that Stalin would push industrialisation ahead too fast – all combined to activate an anti-Stalin opposition among the party leadership. It is probable that the oppositionists hoped to use Kirov, Secretary of the Leningrad party, as a lever by which to oust Stalin. Kirov was reported to be in favour of more moderate economic policies; he was not tainted by connection with the old party factions; he was handsome and popular; and he enjoyed the enthusiastic loyalty of the Leningrad organisation. Stalin, who had consistently played the middle against both extremes of the party, was to be beaten at his own game. At the 17th Party Congress (1934) it seemed that the anti-Stalinists had won a victory. Kirov received an ovation: Stalin announced that the second five-year plan would be less ambitious than the first: and the new Politburo (Stalin and nine others) contained only two loyal Stalinists – Molotov and Kaganovich.

But subsequent events showed that at the 17th Congress Stalin was merely conceding a little ground in order to give himself the chance to strike hard against opponents lulled into a sense of security. He had already taken his precautions. Since the beginning of industrialisation the powers of the Secretariat had been greatly increased so that it was able to control the appointment and promotion of all the leading industrial managers. Extensive purging and recruitment (1929 and 1933) enabled the Secretariat to keep up an atmosphere of uncertainty – fear among the less efficient and hope among the ambitious. Total numbers of party members fluctuated widely. In 1928, 1¼ million; in 1933, 3½ million; in 1935, 2¼–2½ million. But so far Stalin had not touched the all-important central cadre of party secretaries, about 150,000 in number, who under the direction of the Secretariat were the central ruling *élite* in the Soviet Union. Much of the work in the Secretariat was done by the faithful Kaganovich. With typical distrust, Stalin moved up another faithful pawn, Malenkov, so that Kaganovich should not be able to build up a little empire. Other men quietly moved up to positions of power within the party were Yezhov and Poskrebyshev, both of whom were to play prominent parts in the purges. Later evidence (that of Khrushchev in 1956) suggests that in making these appointments Stalin acted behind the backs of his colleagues in the Politburo. Possibly the close links between Stalin and the GPU were already complete by 1934. If so, the knowledge would have been another reason why Kirov and his friends were preparing to get rid of Stalin.

As it happened, Stalin struck first. In December 1934 Kirov was murdered in his office in the party headquarters in Leningrad. The murderer was a young Communist, Nikolaev, who was allegedly inspired by the terrorist ideals of the nineteenth-century populists. True or false, the allegation shows the fear which Stalin wanted to inspire – fear of a renewal of that aimless violence which had done so much harm to the revolutionary cause in the nineteenth century. It is likely (but not certain) that the GPU knew of the plot before Nikolaev fired his shot; in any case, Stalin used the murder for his own ends with extraordinary speed. On the very same day he secured a decree from the Central Executive Committee, only later confirmed by the Politburo, depriving those accused of terrorism of any form of trial. This decree remained in force until after Stalin's death and was the means whereby he got rid of his opponents within the party. He did not at once proceed to unleash the Terror. Nikolaev and a few accomplices were shot; and some left-wing leaders including Kamenev and Zinoviev, apparently reconciled with Stalin since 1929, were tried in secret and imprisoned. This trial was a turning-point in the development of the party. For the first time political opposition by Communists inside the Communist party was treated as a criminal offence. But at this point in 1935, Stalin hung back. He promoted a few of his faithful followers – Zhdanov replaced Kirov in Leningrad, Mikoyan was moved into the Politburo, Khrushchev took charge of the Moscow organisation, and Beria sprang into prominence as the author of a revised (Stalinist) account of the activities of the party in Georgia.

But it seemed that the Kirov affair was going to blow over. Both within the party and in the seventh Congress of Soviets the new constitution was the dominant issue. Since this document was chiefly intended for the creation of a popular front of European left-wingers against fascism, it would not have much effect if its passage were accompanied by an outbreak of Terror.

To the gullible (and there were plenty of these in Western academic circles) the 1936 constitution looked like the application of the principles of John Locke to Russian conditions. To many Anglo-Saxons, at least, it appeared that the light had at last dawned in Russia. The constitution remained federal in character. Eleven union republics (each with the right of secession!) were distinguished. Each was based upon a dominant ethnic group. The official reason given for replacing the 1924 constitution was the need to give legal form to the profound social changes introduced by collectivisation and industrialisation. Stalin claimed that there was no longer any class conflict as there had been in the days of N.E.P. Soviet society was a harmonious 'socialist' unity of peasants, workers and intellectuals. Stalin described it as the most democratic in the world. All citizens over the age of eighteen were given the vote: the rule of the 1924 constitution which weighted the urban against the rural vote was dropped. Elections were to be secret and direct. The sovereign power was the bicameral Supreme Soviet. One chamber represented the people at large (one deputy to 300,000 inhabitants); the other chamber represented the nationalities. There were to be 25 deputies from each union republic, 5 from each autonomous region and one from each national area. As in 1924, national feeling was carefully flattered. The Supreme Soviet elected the Council of Ministers, and the Presidium which acted as a joint head of state. The Supreme Soviet had the right to question ministers – a right which has never been exercised. The constitution contains an impressive list of the rights of citizens. They have the right to work, to be educated, to be relieved in sickness or old age, to enjoy sex equality, freedom of conscience, freedom of speech, inviolability of the person and privacy of correspondence. They can own property and enjoy inheritance. An elaborate legal system apparently secures them in these rights. Cases were to be heard in public; judges were declared to be independent; the accused was given the right to be represented by counsel. Nobody could be arrested except with the decision of a court. Voters were given the right to choose any candidate put up by 'Communist party organisations, trade unions, co-operatives, youth organisations and cultural societies'. All these provisions were, of course, an ugly and cynical sham. It was characteristic of Stalin that he proclaimed the rights of the Soviet citizen at the very moment when he was about to unleash the Terror. Then many thousands of people were eliminated without any notice being taken of the constitution. While the Soviet people and the more gullible left-wingers in the world outside were being invited to applaud the 'democracy' of the Stalin constitution, the N.K.V.D. was busy arresting on suspicion, holding prisoners without trial, concocting false evidence, torturing to obtain false confessions, shooting and deporting.

To find Stalin posturing as a Founding Father is bizarre: behind the elaborate facade of a polycentric society, power remained firmly rooted in the inner party. But the importance of the 1936 constitution should not be underestimated. Primitive peoples given some political standing were flattered. Social harmony was promoted by placing the peasantry upon an equal footing with the urban population. Industrious and popular characters who were not necessarily party members (the sort of people who in Britain would be mentioned in the Birthday Honours List) were given public prominence. 'The union of party and non-party' was the slogan at the first election under the new constitution. Without shedding a tittle of his personal power, Stalin had provided a much-needed social cement, a framework for Soviet patriotism. Foreign admirers were dazzled by the clauses which extended to Soviet citizens the classical liberties of the person, liberties which by the end of 1936 Stalin was infringing on a massive scale. The promulgation of the new constitution was turned into an orgy of Stalinist sycophancy: Stalin was described as 'the genius of the new world, the wisest man of the epoch, the great leader of communism'.

Strengthened by popular applause, Stalin turned (August 1936) to the task of eliminating his political opponents, the senior Bolsheviks, anyone who had been for any length of time in any position of authority. The first to go were the leftists, sixteen senior party members including Kamenev and Zinoviev, men who had worked closely with Lenin before 1917. The charge was that of terrorism – including the murder of Kirov and the attempted murder of Stalin, Kaganovich and Zhdanov. A novel feature of the trial was the public admission of guilt made by all the accused. By threats against hostages, by promises of freedom, by appeals to that sense of discipline by which these men had lived their lives, the prosecutor Vyshinsky persuaded these apparently hardened revolutionaries to admit to crimes which they had never committed and to profess fulsome admiration for Stalin which could not have been sincere. No evidence was brought: the court relied upon the confessions of the accused and the bloodthirsty demands of the prosecutor: 'shoot the mad dogs'. This is what happened to them.

Some aspects of the management of the trial of the sixteen must have displeased Stalin. At the end of 1936 Yagoda, the chief of the N.K.V.D. (as the GPU was now called) was 'promoted' to the Ministry of Posts and Telegraphs and from there to prison. His place was taken by Yezhov, the most sinister figure of this period, who became Stalin's chief instrument in the conduct of the purges. The word *Yezhovshchina* entered the Russian language to indicate the arbitrary sadism with which the N.K.V.D. now carried on the blood-letting. Yezhov first purged the N.K.V.D. itself: then with a small group of trustworthy thugs he did his master's will. Khrushchev claimed later (1956) that at this time several members of the Central Committee protested against the *Yezhovshchina*. Postyshev, the party 'boss' of the Ukraine, is alleged to have said: 'I personally do not believe that in 1934 an honest party member who had trod the long road of unrelenting fight against enemies for the party and for socialism would now be in the camp of the enemies'. Postyshev disappeared shortly afterwards and Khrushchev got his job in the Ukraine. But most party members were bewildered and frightened rather than indignant. They inclined to believe the charges made against the highly placed leaders. They heard how these men had confessed their crimes in public, in the full glare of international publicity. Even well-informed foreigners like Sir Bernard Pares believed that the Old Bolsheviks were guilty of the grotesque charges brought against them – espionage, conspiracy to murder, 'terrorist activity of the Trotsky–Zinovievite Counter-Revolutionary Bloc'. Stalin created this myth skilfully. He not only eliminated his opponents: he also persuaded the Soviet people that he alone stood between them and the overthrow of the State and the revolution. Stalin had grasped an important truth about mind control, that it was essential to confuse and disorientate before opinions could be changed. This was the result of the public trials. The Soviet people (like Pavlov's dogs) were harshly jolted out of their accustomed rut of ideas and emotions. Frightened and puzzled, they turned to the god-like figure of Stalin for reassurance.

The second public trial opened in January 1937. It was a more sophisticated affair than the first. The defendants, who included Pyatakov and Radek, admitted to charges of sabotage on behalf of Germany and Japan. They confessed to crimes against the nation as well as against the party. The names of future victims were drawn in. The failings of the industrialisation programme were blamed upon the deliberate sabotage of Pyatakov.

The next blow fell upon the Red Army. Hitherto its loyalty had been unimpeachable. On May Day Tukhachevsky, the Assistant Chief of Staff and a hero of the civil war, appeared in public with Stalin; on 11 June it was announced that he, along with many other senior officers, had been arrested. Tukhachevsky was not brought to trial, perhaps because he could not be persuaded to confess. No legal charges were made against him but he was accused in the press of being engaged in a conspiracy to dismember the Soviet Union in the interests of Germany and Japan. These charges were later (1956) declared to be without foundation. With Tukhachevsky fell the great mass of high-ranking service officers. Thirteen out of 15 army commanders, 57 out of 85 corps commanders, 110 out of 195 divisional commanders, 220 out of 406 brigade commanders, were executed. The casualties among the ranks from colonel downwards were also very heavy. Thus Stalin reduced to political impotence the one organisation in the Soviet Union with the power to oppose the party. It remained to be seen whether in so doing he had destroyed the army's military efficiency. Western estimates of the utility of the Soviet Union as an ally were diminished and consequently the army purge seriously undermined Stalin's avowed policy of seeking alliance with France and Britain.

The final act in the public drama of the purges occurred in March 1938. Twenty-one defendants were tried; among them were some of the most celebrated Old Bolsheviks – Bukharin,

Rykov, Yagoda and Rakovsky. The charges were familiar although a few novel and melodramatic twists were added. Yagoda was accused of poisoning the writer Maxim Gorky – a deed which was certainly in keeping with his reputation; and Bukharin was alleged to have plotted to murder Lenin and Stalin in 1918. This charge he denied and one of the witnesses called by Vyshinsky, an S.R. called Kamkov, turned awkward in the witness box. Nevertheless, most of the defendants were shot. The sign of the end of the purges was the removal of Yezhov, who in December 1938 was replaced as the chief of N.K.V.D. by the Georgian Beria. For a time Yezhov was Commissar of Water Transport and then he too disappeared. It was rumoured that he died insane; he certainly knew too much to be allowed to live.

The purge fell most heavily upon the Old Bolsheviks. As a class they were totally eliminated. Most of those reckoned to be Stalinists also suffered. Of the members of the Central Committee elected at the 17th Congress (1934), 70 per cent were killed by 1938. Of the 2000 delegates at this Congress, more than 50 per cent disappeared. Below this top level in the party, the casualty rate was equally severe. The number of party members purged between 1934 and 1937 was one-and-a-half million; there were three-and-a-half million members in 1934, two million in 1937. Proved loyalty to Stalin was no guarantee of survival. For example, Rumyantsev, a loyal Stalinist who for years had been the 'boss' of the Smolensk party organisation, was purged along with all his relatives and friends. His 'crime' was really that he had been in office for a long time and had collected around him a loyal body of subordinates. Thousands of party members were denounced by others who hoped to save their own skins. This was afterwards remembered as the most humiliating result of the *Yezhovshchina.* Private loyalties and the restraints of civilised life were cast aside in a frantic attempt to avoid death or the camps. The loss of self-respect among the most gifted of his subjects helped Stalin to impose his personal dictatorship. By 1939, Stalin had cut all the links with the past. For the new industrial society he had created a new party, dominated by himself and a few loyal subordinates. The rank and file rejoiced in the mere fact of survival. The average age of the party had dropped. The new men had for the most part reached manhood since 1917. They were able, ambitious and uncritical. They were

Stalin's dogs. The Terror halted as suddenly as it had started. In the labour camps prisoners were surprised to find themselves working alongside the N.K.V.D. agents who a few months before had threatened and tortured them. Stalin carefully swept away his own broom. How many persons were detained in the GULAG camps? Estimates vary between seven and fourteen million. Camp life became a widespread Soviet occupation. The prisoners made a significant contribution to the national economy. For a few hundred grams of bread per day they chopped down forests and built power stations. Their memoirs reveal the inextricable confusion of good and evil in Stalinist Russia.

It is evident that Stalin went much further than many of those who have wielded Terror. He wanted people to obey: he also wanted them to think Soviet thoughts. Stalin never underestimated the influence of writers and artists. He was himself an omnivorous reader of the Russian classics and of Russian history. He took much trouble to ensure that creative work fitted in with the needs and aspirations of Soviet man. Unlike his predecessor Nicholas I he did not aim merely to silence the writers. He wanted them to speak eloquently and movingly but to say only such things as were socially useful.

STALIN AND THE INTELLIGENTSIA

During the twenties there had been much controversy about the function of writers in a socialist society, but it was carried on in an atmosphere of relative freedom. It was a period in which Russia contained a great deal of literary talent. The revolution released the urge to experiment with new forms. The most talented writers (the 'Fellow Travellers') on the whole avoided political commentary in their works. But a group of younger writers argued that since Russia was still a class society it was necessary for writers to be committed to the class struggle and that the supporters of the proletariat should champion its cause in their literary works. This movement was known as *proletkult* and its views were expressed with the force common in ideological debate between Russians. The debate was joined by the leading intellectuals in the party – notably Trotsky, Lunacharsky and Bukharin. These party leaders disapproved of *proletkult*: their objections were tellingly expressed by Trotsky:

> There can be no question of the creation of a new culture, that is of construction on a large historical scale during the period of dictatorship [of the proletariat]. The cultural reconstruction, which will begin when the need of the iron clutch of a dictatorship unparallelled in history will have disappeared, will not have a class character. This seems to lead to the conclusion that there is no proletarian culture, and that there never will be any. . . . Such terms as 'proletarian literature' and 'proletarian culture' are dangerous, because they erroneously compress the culture of the future into the narrow limits of the present day. They falsify perspectives, they violate proportions, they distort standards and they cultivate the arrogance of small circles which is most dangerous.

Trotsky's views were proclaimed by the Politburo in 1925 and consequently freedom was retained for all forms of literary expression. An association of writers (RAPP) was formed but it remained out of direct party control. At the beginning of the five-year plan, RAPP encouraged its members to throw their talents behind the drive for industrialisation. Two writers, Leonov and Pilnyak, produced novels of some power in spite of the unpromising brief. Leonov's *Sot* (1931) has a conventional setting – the construction of a large paper-mill in north-east Russia. But Leonov is not much concerned with the technical aspects of industrial construction: his theme is a human one worthy of a good novelist, that is, the transition from the old into the new. The most interesting characters are those for whom this transition is painful and complicated, like the engineer Burago who serves the plan faithfully but is far from accepting the party orthodoxies. Another good novelist, Sholokhov, depicted a different side of the plan in *Virgin Soil Upturned*, set in the Don Cossack area and depicting, with much sympathy towards the vanquished, the frustrated ambitions of the small independent landowners when challenged by collectivisation. It is easy to find some absurdly naïve novels of this period. Gladkov (*Power* and *Cement*) and Kataev (*Forward, Oh Time*) were two of the most uncritical 'plan' authors. Kataev wrote about the great chemical combine at Magnitogorsk and included in his novel long descriptions of the technique of producing concrete. Such books bored the insatiable reading public; a typical complaint came from the members of the municipal library at Rostov-on-Don. 'Write more about love, about marriage, paint the way of living. . . . Give striking unforgettable types of heroes of our days. . . . Besides we want to laugh . . . and not only to smile gingerly. . . . Write in a simple correct language. Learn that from the classics.' The last remark recalls that in one important respect Stalin and the party did not cut the links with the past: nineteenth-century Russian literature was in constant demand and full supply during the twenties and thirties.

Stalin was keenly interested in literature and maintained a close contact with the leading writers. Maxim Gorky, the patriarch of Soviet letters, was a member of the Kremlin circle. Promising writers, like Leonov, would receive peremptory and terrifying invitations to Kremlin dinner parties. It is possible that Stalin's vanity had been injured during the twenties when his views on literary and ideological matters had never been sought. From 1929 he started to put this right. The first group of intellectuals to feel his authority were the philosophers. They were summoned to his office (1929), accused of 'rotten liberalism' and collapsed ignominiously into sycophantic flattery. The historians came next. They were ordered to throw out their 'Trotskyist contraband' and to busy themselves with the depiction of the past as the past appeared to the General Secretary. The 'archive rats' were to be disciplined: the mere facts were not to be allowed to stand in the way of Stalin's interpretation of the past. The historians had a bad time for, as the purges progressed, more and more names had to be expunged from recent history until at last only one name was left. After changing from year to year, history at last settled down in 1938 when the *Short Course on the History of the All-Union Communist Party* appeared, probably written in part by Stalin himself. This became the central item of the instruction offered in the party schools. The lawyers also suffered; the most eminent, Pashukanis, was shot in 1937. His 'crime', apart from the mere fact that he was very influential, was that he stood for an interpretation of law which was inconvenient in the thirties. He had argued that under socialism the law would wither away, since law was an expression of class domination and under socialism there was neither a dominating nor a dominated class. In the N.E.P. era this theory had been acceptable since theorists had agreed that a class structure still existed; but by the mid-thirties Stalin proclaimed that the transition to socialism had been made. Nevertheless he had no intention of dispensing with the laws. Vyshinsky was set to work to produce a new theory which would explain why law was necessary in a classless society.

Stalin's attitude towards the scientists was in marked contrast with his treatment of lawyers, philosophers and historians. Scientists were useful and it was unlikely that their views had any political or social implications. The search for scientific truth could be allowed to proceed unhindered; a dynamic scientific culture was quite compatible with thought-control. Consequently Kapitsa, the greatest of Soviet physicists, was permitted unheard-of freedom. He worked at Cambridge with Rutherford and was accorded the most lavish facilities in Leningrad. By 1941 nuclear research was at least as far advanced in Russia as in any other country. It was slowed down during the war, possibly because Stalin thought that its military possibilities could not be harnessed in time to be militarily useful.

But Stalin's most subtle attentions were reserved for the creative writers. In 1932 RAPP was abolished and replaced by the Union of Soviet Writers. This body was under direct party control; for a writer not to belong to it was dangerous. Membership was lucrative and prestigious. The new organisation was equipped with a new literary theory – Socialist Realism – which Stalin himself had a hand in defining. This doctrine was much less crude than the 'cement and enthusiasm' theory which preceded it. It gave scope to good writers and it enabled them to provide readers with interesting literature. At the same time, it enabled the party literary critics to keep a tight control over writers and to persuade them to produce the sort of emotional climate which was politically desirable. This was a censorship quite different from that of Tsarist days: then its purpose had been chiefly negative. The doctrine of Socialist Realism was used to dragoon the writers into a positive affirmation of certain values, an affirmation less obvious but more telling than had been achieved under the direction of RAPP.

The theory of Socialist Realism (officially proclaimed in 1934 by Zhdanov and Gorky) was both liberating and constricting. It liberated the writer from the need to earn his living by the production of sheer propaganda designed to stimulate economic effort. Zhdanov assured the novelists that they would be deprived of only one right, 'the right to write badly'. On the other hand, the range of experimentation was limited. The writer was to affirm positively the values of Soviet society, as those values were currently defined. Novels should contain due proportions of three ingredients; *narodnost*, *partiynost* and *ideynost*. *Narodnost* means (in Stalin's own definition) that art should be national in form but socialist in content; *partiynost* means that work should contain identification with the aims of the party; and *ideynost* means that work should have ideological content and should not be merely entertaining. The work of the Socialist Realist should be mainly didactic. This was a trend in Russian critical thought which went back to the radicals of the sixties like Chernyshevsky. Soviet types were to be portrayed, men and women inspired by the ideals of socialism, with clear powers of leadership. They were to be shown in situations in which their personal decisions and heroism decisively altered the course of events. The future was to be depicted in glowing colours. The past was not to be ignored, especially such episodes as could be put to good account in a patriotic manner. Language was to be simple and direct, to avoid experiment and to be easily intelligible to the mass of readers. Pornography was to be avoided, and this was taken to mean the avoidance of any frank treatment of sexual matters. Socialist Realism was very prudish. Any attempt to see the world and life from a sharply individualistic viewpoint was deplored. Literature was to be utilitarian, to serve the greatest good of the greatest number.

Some of the most gifted writers were unable to tolerate this régime. Babel wrote little and disappeared in 1938; Pasternak was silent under the attacks of the official critics. He occupied himself by translating Shakespeare. Olesha (perhaps the most original novelist of the twenties) promised to re-educate himself, and remained silent. Mandelstam, a poet who composed a much-enjoyed epigram about Stalin, disappeared. But most writers submitted to the official doctrine and some produced some very good work. Sholokhov, author of *And Quiet Flows the Don*, annoyed the literary bosses by writing tragedy but wrote the most praised novel since the revolution. Leonov and A. Tolstoy also wrote interesting books, although the former was also in trouble with the critics. His hero in *Road to the Ocean* (1935) is a

Communist who is approaching death. He tries to balance up the meaning of his life and decides that although his dedication to duty has deprived him of many of the common joys, his involvement with the victory of the party has made all worth while. But this affirmation was hesitant enough to make it distasteful to the official guardians.

Patriotic literature was also encouraged by the Stalin literary code. Pavlenko wrote an account of a war against Japan (*In the East*), but his conclusion shows that the Soviet victory has benefited the whole of mankind. Shpanov (*The First Blow*, 1939) was more narrowly nationalistic. His story concerned a war against Germany, but a few weeks after its publication the Nazi–Soviet pact was signed and the unlucky Shpanov's book was caused to disappear.

Stalin was probably not altogether displeased with his writers. It is true that official complaint was made sometimes about the absence of a Tolstoy or Dostoyevsky. Stalin began a prize system to encourage a higher quality of work with considerable material incentive. But it may be surmised that any writer who brought Tolstoy's sceptical and anarchical spirit to the literary scene would have been rapidly silenced. Tolstoy's patriotism was, however, proclaimed as a model: and one of the more curious publishing events of the thirties was Sergeyev-Tsenski's *The Ordeal of Sevastopol* (1937–8), a novel of enormous length which described in glowing terms the heroism of the Russian people – of all classes – in the face of the Anglo-French invasion of the Crimea. Dmitri Donskoy and Ivan the Terrible were also used by writers and film-makers as symbols of Russian patriotic fervour, particularly in defence of their native land.

A PROVISIONAL BALANCE SHEET

Any attempt to evaluate the Stalin revolution must take into account both the elements of destruction and those of creation. Never before in the history of Russia had so complete a change been wrought so quickly in so many aspects of life. In some ways Stalin tried to restore the link between the new Russia and the pre-1917 past; in others, for example by the destruction of the Old Bolsheviks in the purges, he tried to cut loose from the immediate past. The most striking fact is the complete obedience which Stalin obtained from a people not obviously given to this virtue. Even Hitler never managed to cow the German generals as Stalin managed to silence the Russian intelligentsia. He not only bullied his people and his party, but also made them accept a myth in which Stalin appeared as the benevolent father-figure leading the happy masses to an even happier future. By the late thirties time seemed to be on his side. A new generation would not remember the truth about the past: there were no authoritative voices left to correct the official versions. Only Lenin's name was now linked with Stalin's, and in the eyes of the simple, Lenin was a saint whose bones were to be revered. In the eyes of the more sophisticated he was the omniscient wise man whose mantle had fallen upon Stalin. The worst stages of industrialisation were over and it seemed possible that shortly the masses would be allowed to consume more of what they produced. No centre of resistance to Stalin was left. Society had been stirred up to its very depths and all the cohering lumps smoothed out. With the murder of Trotsky in 1940 the last voice from outside was silenced. Stalin was now not only the task-master but also the only, unrivalled myth-maker of his people. He set the daily work, provided the daily bread and, to an increasing extent, guided the thoughts and emotions of the toiling masses. But this unrivalled degree of control within Russia was in sharp contrast to Russia's position in the world at large. There other men were the masters of rival and hostile systems of thought-control. It remained to be seen whether Stalin retained a sufficient sense of reality to survive in Hitler's Europe.

Stalin remains one of the most mysterious figures of modern history. He lacked the gift of communication with the masses. He left no table-talk, there are few captured documents, there are few reliable memoirs. Little is known of his motives: only his deeds speak for him. He retained a god-like detachment from the society which he dominated. He remembered that his people were still simple and knew that simple people wanted something to worship. Party and the revolution were too abstract for them. Stalin revived the simple patriotic emotions. He replaced the worship of the Christian God with the adoration of Stalin. Everywhere his portrait, like a twentieth-century icon, challenged the adoration of the masses. His lightest word on any subject was taken to be the truth. Like a god he both created and destroyed;

like a god he did these things for no reason apparent to human eyes. His arbitrariness proved his divinity. He was infallible, omniscient and omnipotent. He understood the Russian people (which is to say the Russian peasants) perfectly. The more he drove them out of the villages into the unfamiliar life of the towns, the more they would yearn for the retention of some of the old landmarks. For ideological and political reasons he could not allow them to worship freely in the Church, but they could worship Stalin, the god-like ruler. Stalinism was a new 'opium of the people'. He built a temple to materialism and installed his own image in every corner of it.

13
The Foreign Policy of the Soviet Union: 1919-41

WORLD REVOLUTION VERSUS NATIONAL INTEREST

During the two or three years after it seized power the Bolshevik party was in a dilemma about the basic aims of its foreign policy. It could not decide whether its first duty was to Russia or to the revolution. To confuse the issue still further, most Bolsheviks felt (with varying degrees of conviction) that the fate of the Russian revolution depended upon further revolutionary success in Western Europe. But by encouraging revolutionary elements in Western countries the Soviet government made it less likely that foreign capitalist States would enter into friendly relations with it. As a nation, Russia desperately needed the friendship of her neighbours. She needed peace, trade and economic help. The dilemma was at root a matter of timing. If world revolution could be secured quickly it would solve all problems, but if it were too long delayed, Soviet Russia would have to look to her own survival. This would mean a return to traditional international diplomacy. During the civil war period the Bolsheviks tended to place their strongest hopes upon international revolution. This was in part for ideological reasons; in part because the hostility of the Great Powers left them with little choice; in part because they were cut off from the outside world by the White armies and knew little of what was going on there.

Although the Bolsheviks were cheated of the expected European revolution, many events of the months immediately following the end of the war encouraged them to believe in its imminence. European society had been shaken to its foundations. Poverty, hunger, disease and national aspirations created a revolutionary situation. Lloyd George (not Lenin) wrote: 'The whole of Europe is filled with the spirit of revolution. There is a deep sense not only of discontent but of anger and revolt amongst the workmen against pre-war conditions. The whole existing order in its political, social and economic aspects is questioned by the masses of the population from one end of Europe to the other.' Germany was angered by the peace terms imposed upon her. There were attempted Communist risings in Germany in 1919 and again in 1921. Bela Kun led a briefly successful Communist government in Hungary (1919). Even in Britain, non-communist working men showed sympathy towards the new régime in Russia. They refused to load cargoes of arms destined for Poland. They showed a strong disinclination to go to Russia to fight against the Bolsheviks. The Russian revolution had aroused much sympathy among the Western industrial masses. The Russian leaders made the mistake of thinking that this sympathy could be directed into a specifically pro-communist channel.

The existence of so much revolutionary ardour in the Europe of 1919 encouraged Lenin to define and dominate it. He hoped that the successful Socialist party in Russia would enjoy the leadership of the powerful European socialist movement. If Moscow did not make a bid for the control of the European left, it would again fall under the influence of the bourgeois socialists. These men, Lenin claimed, had betrayed the cause of the European proletariat in 1914. They had encouraged their parties to support a capitalist and nationalist war. Now that the war had ended in disaster for vanquished and victors alike, it seemed likely that the working men of Europe would turn to the one socialist party which had condemned it from the beginning – the Russian Bolshevik Party. To achieve Russian control of international socialism, Lenin established the Communist International (Comintern) in 1919.

The results were rather disappointing. The Comintern failed to unite the left-wing parties. In fact, it divided them. In spite of elaborate efforts to prove the contrary, it was clear that the Central Executive Committee of the Comintern (IKKI) was dominated by its five Bolshevik members. Instead of leaving the question of doctrinal unity flexible, Lenin applied to the Comintern the same passion for exact definition which had previously governed his conflict with the Mensheviks. He hammered home his ideas in one of his last writings, *The Infantile Disease of 'Leftism' in Communism.* This shows that he assumed that every foreign revolution would pass through exactly the same stages as the Russian revolution. The tactics of the Bolshevik party were to be followed down to the last detail. His authoritarianism was best revealed by the '21 Conditions' (1920) which had to be accepted by any foreign party which wanted to join the Comintern. They included an undertaking to accept all the decisions of IKKI and to expel any member who voted against their acceptance. As a result, the foreign parties were reduced to obedience but they were also rendered impotent. In several nations (particularly in Germany) much might have been gained by friendly assistance to a coalition of left-wing parties. Such a procedure would have enabled the left to make progress in a democratic manner. But Lenin did not want this. He insisted that the foreign Communist parties should seize power by revolutionary means, just as he had in Russia. It was possible for foreign Communist parties to co-operate with other left-wing groups. Frequently during the history of the Comintern (it was dissolved in 1943) the order went forth from Moscow for a 'popular front'. But for a national Communist party merely to join such a 'front' was almost a guarantee of its failure, for it was known that should the 'front' succeed, it would, on orders from Moscow, kick aside its former allies and seize power illegally. It must be admitted, however, that the impotence of the Comintern was in one way an advantage. Its existence made normal diplomatic relations with foreign powers rather difficult. Foreign governments complained that Moscow spoke with two voices – the voice of diplomacy and the voice of subversion. But soon the extreme weakness of the Comintern became apparent to all. The German government was the first to grasp this fact and the German Communist party (K.P.D.) was by far the biggest outside Russia. Long before its disappearance the Comintern was seen to be as futile as that other international organisation of the inter-war years, the League of Nations.

By the end of 1919 the crisis of the civil war was over. The Soviet government turned with more emphasis to more regular diplomatic contacts. It was essential for the recovery of Russia to establish trade links. For their part, the capitalist powers, faced by the problem of unemployment, could not afford to neglect any commercial outlets. By 1921, Russia seemed to be well buried away behind the ring of new States in eastern Europe. Her military weakness had been shown by the precipitate retreat of the Red Army from Warsaw. Lenin's hopes were mainly fixed upon Britain. Britain had to trade, and the British government was frightened of communist agitation in India and China. In the peasant colonies of Asia the Comintern was much more of a threat than in Europe. Full diplomatic recognition was out of the question, thanks to the ideological horror which Bolshevism inspired in the British Conservative party, but a trade agreement was finally signed in March 1921. It bound Russia not to meddle in the British Asian colonies: it passed over for the time being the vexed question of the public and private debts repudiated by the Bolsheviks when they seized power.

The trade agreement of 1921 was a turning point in Soviet foreign policy. It was rapidly followed by similar agreements with Germany and Sweden. It aroused the interest of American capitalists, who feared that they might be excluded from a lucrative market and source of raw materials. Lenin was able to tempt them by offering mineral concessions. It also led to the resumption of full diplomatic relations with three bordering nations, Afghanistan, Persia and Turkey. Friendly relations with Afghanistan and Persia were a part of the price paid for the British trade treaty. As regards Turkey, the 1921 treaty marks an important break with traditional Tsarist policy. Turkey, like Russia, had suffered badly from the war. She had lost her empire to the British and much of Asia Minor to the Greeks. Her leader, Kemal, was anxious to do a deal with Russia which would leave his hands free to oppose Greece. In a treaty of March 1921 he admitted Soviet control of the Caucasus and Transcaucasia and promised to co-operate with Russia in an international agreement about the Straits. In return, the Soviet government pledged its support for Kemal against the humiliating terms imposed upon Turkey by the Western powers. Most significant of all, the Turkish Communist party was abandoned to its fate. This was a small price to pay for a treaty which gave the Soviet Union security in Transcaucasia, but this betrayal of international communism marks the end of a brief period of revolutionary enthusiasm in the conduct of foreign affairs. With the beginning of N.E.P. (which coincided exactly with the Turkish and British treaties) foreign policy became distinctly more nationalist in tone. World revolution was not dismissed as an ideal but it was postponed, and its arrival predicted for Asia rather than for Europe.

THE FOREIGN POLICY OF N.E.P.

The foreign policy of the Soviet Union during the N.E.P. period was closely linked to the ebb and flow of the domestic crisis. Either of two developments in foreign policy could have prevented the rise of Stalinism. If the Bolshevik party could provoke a revolution in one or other of the highly industrialised countries (Germany was the most likely candidate), isolation would be ended and the economic means available for the reconstruction of Russia. If one or more of the rich nations could be persuaded to give economic aid (and here Great Britain seemed the most likely partner) then the rigours of Stalinist industrialisation could be avoided. In other words, it was necessary to find a substitute for Witte's French alliance. Up to 1914 Russia had been essential to France, and France had paid for Russia's industrialisation. The purchase had been of great value to France: it had saved Paris in 1914. But had the Russian alliance any market value in 1921?

It would seem not, Russia had been excluded from the Versailles peace conference. Russia was ruined; her new frontiers left her vulnerable to attack; the new Poland and Czechoslovakia cut her off from Germany. Germany was weakened, disarmed, humiliated by reparations. France no longer needed Russia. Her needs could be met by threatening Germany on the east by means of an alliance with the new east European powers. The end of the war of intervention did not end the Anglo-French hostility to the Bolshevik government. Bolshevism was an unpleasant political doctrine which might have dangerous repercussions upon populations excited by war and angered by unemployment. The new rulers of Russia refused to pay the money owed by the old régime. Besides, it seemed impossible that Bolshevism could survive. If it were deprived of foreign aid it would surely collapse. Even if the Bolsheviks survived Russia would always be weak, and it might prove useful to preserve a weak Russia for a few years. There might come a time when both Germany and Japan could be diverted by encouraging them to help themselves to some portions of the Soviet Union. Russia was no longer required in the European or the world balance of power. She could be ignored, reviled, used as a bogey to frighten timid electors into the acceptance of conservative policies, offered as a bait to divert the attentions of revisionist powers. Such views were widespread in governing circles in Britain and France but they were not universal. Some (like Lloyd George) saw that the withdrawal of the United States from Europe made it essential to procure the alliance of Russia. Germany would not always be weak; Poland could not provide a containing alliance. At the Genoa conference (1922) he tried to bring Russia back into Europe. He failed and his Conservative colleagues did not forgive him for even trying. They forced him out of power and started to prepare a European system which would contain Germany without Russian aid. Others in Britain rather admired the Soviet Union. The working masses who had refused to load arms for Poland had applauded the brief Soviet victory in the Polish war (1920). When the Labour party came briefly to power (1923–4), its leader Ramsay MacDonald attempted to foster trade relations with the Soviet Union. At the moment of success he was thrown out of power.

In the face of so much Western, and especially British, hostility it might have been wise for the leaders of the Soviet Union to play down the ideological content of their creed. But it was impossible for them to do so. They still believed in the proximity of world revolution. The books said that it must happen: it was difficult to imagine the future if world revolution turned out to be an illusion. Even if the Comintern failed to provoke world revolution, it could be used as an additional weapon in the Soviet diplomatic armoury. It could foment trouble in the Asian colonies of the European Great Powers. Nationalism was already stirring in Asai and nationalism was the friend of the Comintern wherever it was directed against imperialism. The British had already shown their sensitivity to the Comintern in the 1921 trade treaty. The weakness of this policy was that it tended to make foreign governments still more hostile. They objected to the thinly veiled Soviet interference in their own domestic affairs. They felt threatened by a power which they were determined to ignore, and wondered what would happen if there should be mass unemployment or a financial crisis. The Comintern was like a spectre at the capitalist feast. It encouraged the capitalist governments to ensure that the poor were not too discontented, but it did not frighten them into friendship with the Soviet Union.

Although the Comintern made threatening noises in most parts of the world, it was chiefly concerned with two countries, China and Germany. Its aims were not identical in both. In China the purpose of the party was to make trouble for Britain, which it did as the ally of the Kuomintang. It was held that Britain (regarded as the chief capitalist power after the United States had withdrawn into isolation) would be economically damaged by the loss of her commercial empire in China. This might lead to unemployment and revolution in Britain: it might lead to an advantageous trade treaty made on condition that the Soviet Union restrained the Chinese Communist party. But in Germany the objective was revolution. The German Communist party was the most powerful outside Russia. 'Objective conditions' were ripe for a proletarian revolution. If the Russian party could do it alone, surely the K.P.D. (German party) could seize power with the aid of the Soviet Union? This is what Trotsky thought. In 1923 the other leaders (except Stalin) agreed with him. A German revolution was possible; its success would break the back of the capitalist opposition to the Soviet Union. The abject failure of the attempted revolution (November 1923) sealed Trotsky's fate. Never again would there be such a favourable opportunity. Germany was seething with anger against reparations and against the French invasion of the Ruhr. Rapid inflation added to the chaos. Yet the K.P.D. failed miserably. The German army remained absolutely loyal to the Weimar republic. After this fiasco the Comintern ceased to have so much importance. Its chief use was to ensure the loyalty of foreign communist parties in which it was highly successful. The loyalty of foreign Communists became legendary. At the drop of a hat they would reverse a policy which they had supported for years. Only Yugoslavia produced leaders who were not pathetically subservient.

The long-continued hope of a German revolution was all the stranger since it was with the German government that the Soviet Union secured its most notable diplomatic success since it had seized power. This was embodied in the Treaty of Rapallo (1922). For the first time the Soviet government received full diplomatic recognition from a Great Power. Russia and Germany pledged each other economic co-operation, and in addition Russia promised to allow the German army to have bases, training areas and factories for armament production in Russia. For Russia, the advantages of this treaty were great. She resumed normal commercial relations with the power which had been her most important trading partner in 1914. She destroyed a possible capitalist alliance against her. She established a precedent for recognition by the other Great Powers. She broke down the isolation in which she had lived since 1917. German participation in this treaty is perhaps more surprising. Like Turkey and Russia, Germany smarted under the peace treaties of 1919. The Rapallo treaty enabled her to build up a modern army, to create an economic empire in eastern Europe which would be out of reach of the Western powers. Russia and Germany were potentially the most powerful nations in Europe, but in the early twenties they were both weak and they both had an interest in becoming stronger with the aid of the other. The Rapallo line was not universally popular in Germany. It pleased Seekt, the commander-in-chief of the Reichswehr, and the General Staff. It pleased the Prussian landowning conservatives who saw in a Russian alliance the possibility of regaining the east German territory now incorporated in Poland. It pleased certain German industrial interests, particularly Krupps and Stinnes, who foresaw much profit from German exploitation of Russian iron and coal. But ranged against these 'easterners' (who were all of the German right wing) was a formidable coalition of German interests. The Roman Catholic Church was opposed to the Russian alliance. German Social Democracy, frightened by the competition of the K.P.D. for the allegiance of the workers, looked west rather than east. The anti-militarist opinion of Weimar Germany was against the Rapallo line. Large segments of German industry looked to Britain, France and the United States for markets. It was this 'westerner' anti-Russian line which was played up by Hitler. For these reasons, the Rapallo treaty was a rather brittle affair. Too many Germans had good cause to dislike it.

Rapallo horrified the Western powers. It was a step towards the accomplishment of two things which they greatly feared: the restoration of the German army and the economic recovery of Russia. Its effects were immediate. Encouraged by Rapallo the Germans refused to pay reparations and the French invaded the Ruhr (1923). This created the revolutionary situation which the Comintern failed to exploit. Clearly, Germany and Russia would have to be prised apart again. Whereas before 1914 this separation had been secured by a French alliance with Russia, in the twenties it was done by satisfying Germany. Such was the origin of the Locarno treaties (1925), a system created by the initiative of a British Conservative government.

The essence of Locarno was a British guarantee of the frontiers between France and Germany. France undertook to protect Poland and Czechoslovakia against German aggression but Britain was not involved in this. Germany was invited to join the League of Nations; fresh arrangements were made about reparations and the Germans encouraged to hope for massive American investment. From the Soviet viewpoint two features of this arrangement were salient. First, a general system of European security had been created without Soviet participation. The men of Locarno ignored the Soviet Union as if she were an extra-European power. Secondly, the absence of a British guarantee of Germany's eastern frontier aroused Soviet suspicion. It could be supposed (probably without any justification) that Britain was inviting Germany to compensate herself for the humiliations of Versailles by eastward expansion. In fact such an idea was not entertained by Austen Chamberlain, the British Foreign Secretary. For him Locarno was intended to secure peace throughout Europe. But Soviet diplomats were aware of the powerful anti-Bolshevik feelings in the British governing class. Austen Chamberlain's idealism might be turned into an anti-Soviet capitalist front. Such a fear obviously applied only to the remote future. The level of armaments maintained by Britain, France and Germany in the post-Locarno era demonstrated that the West wanted peace.

Locarno had a profound effect upon the Soviet Union. It broke the back of Rapallo; it brought the three capitalist powers together; it finally ended the hope of substantial foreign aid. It helped Stalin's rise to power for it left no possible alternative to his policy of 'socialism in one country'. Only forced industrialisation could meet the supposed long-range threat of

Locarno. For the time being Locarno meant little more than a *détente* in the West, but its logical outcome was a German threat to Russia made with the consent of the Western powers. Against this both diplomacy and the Comintern were useless. Only Soviet military strength could meet the danger, and this is what Stalinism provided.

The continued hostility of the Western powers – and particularly that of Britain – was clearly revealed by other incidents. At the Lausanne Conference (1923) where Curzon presided over a conference to decide the post-war shape of Turkey, Russia was first ignored and then humiliated. The Lausanne Convention injured Russia's interests in the Black Sea by giving foreign warships free access at all times. The Royal Navy would thus be able to threaten Russia at will. This agreement, however, was not altogether disadvantageous to the Soviet Union. It ensured the friendship of Turkey which was equally humiliated by British policy. This was a remarkable change from the old days. To secure Turkish friendship, Russian abandoned the old Tsarist drive southwards, and Russia and Turkey remained on good terms until 1939.

The last chance of obtaining useful diplomatic and trade relationships with a major capitalist power appeared and disappeared during the period of the first Labour government in Britain (1923–4). Ramsay MacDonald was hardly more friendly to the Soviet régime than were his Conservative 'rivals', but he saw the advantage of a trade agreement. Anglo-Russian trade might help to alleviate unemployment in Britain; it would satisfy the left wing of his party. Consequently, he recognised the Soviet government (1924) – a step which was immediately followed by most of the other European governments – and initiated trade talks. These were successful in one important respect. They solved the knotty problem of Soviet debts to British citizens and entitled the Soviet government to raise money on the London money market. But their very success was MacDonald's undoing. He was accused of being a Soviet sympathiser and the Liberal support upon which he relied was withdrawn. At the subsequent election (1924) the victory of the Conservative party was clinched by the publication in the Conservative press of the Zinoviev letter. This was a forgery which purported to show that the Comintern was closely linked with the British trade union movement and was encouraging it to act subversively. The new Conservative government refused to ratify the trade treaty made by its predecessor. Some of its members spoke wildly about the need to protect European civilisation against '. . . the most sinister force that has arisen . . . in European history'. Soviet leaders spoke equally wildly about the general strike (1926) and Soviet miners sent money to aid their British colleagues. After Locarno, the British Conservatives were looking for an excuse to break with the Soviet Union. In 1927 the British alleged that Soviet agents were making trouble in India. In London the police raided the headquarters of the Soviet trade delegation, Arcos. No incriminating documents were found but this was an excuse for ending diplomatic relations. Russia was again pushed out into the wilderness whence she had been briefly rescued by the Labour party.

In a longer perspective, the failure of Soviet diplomacy during the N.E.P. period was inevitable. This was the only decade since 1870 in which Germany had been weak. The weakness of Germany meant that the alliance of Russia was not worth buying. The Rapallo line had promised well for Russia but could not lead far. Russia could not afford to aid German rearmament too much: Germany could not afford to forego the profits of a Western alignment. Stalin drew the correct conclusion from German weakness: it gave Russia time to become strong. As it happened, if industrialisation had been much longer delayed Russia could not have resisted when German strength had been restored. The long perspective also shows the tragedy of Anglo-Soviet relations. British policy was based upon the myth of French military strength, but in the end only Russia was strong enough to resist Germany. Anglo-Soviet friendship could have prevented the rise of Hitler; British economic aid could have mitigated the harsh aspects of Stalinism. The lack of Anglo-Soviet understanding (for which both sides were partly responsible) was largely responsible for the insecurity of the inter-war period.

STALIN, CHAMBERLAIN AND HITLER, 1934–9

The Old Bolsheviks had looked to a German revolution for salvation; in fact out of Germany came the greatest threat to the social order established in 1917. The Russian and German peoples were locked in a national conflict which, on the surface at least, bore little relation to the class conflicts which Marxists had predicted would be the political pattern of the twentieth century. Yet it would be an error not to see in the twists and turns of Stalin's foreign policy the profound influence of ideological analysis of the power structure. On some occasions Stalin was led into error by his failure to take a completely realistic attitude towards the problems of diplomacy. But in this error he was in the company of those Western politicians contemptuously labelled 'appeasers'. Between the dreamers on either side of him, Hitler, like another Bismarck, was able to raise Germany's fortunes to an unprecedented height. Both East and West completely misunderstood fascism. Stalin showed a greater sense of reality by recovering from his earlier errors more rapidly.

During most of the twenties Soviet-German relations were governed by the Rapallo agreement. Both powers had lost by the First World War and both were intent upon altering the treaties which ended it. Germany supplied the Soviet Union with more imports than any other power, and in return the German army received secret training facilities in Russia which enabled it to elude the Versailles limitations upon its size and weapons. The Rapallo relationship cooled off slightly as Germany moved towards an accommodation with the Versailles powers. When Stresemann concluded the Locarno pacts (1925) and entered the League of Nations (1926), the Rapallo pattern was altered but not broken. The Soviet Union could no longer look forward to a fresh outbreak of hostilities between Germany and the Western powers, but Stresemann kept the line to Moscow open. While he and the Social Democrats remained in power in Germany Russia was not threatened. On the contrary, the Stresemann line was a good reason for retiring into isolation. With Germany successfully moving into friendly relations with the West there was nothing to be gained from the Rapallo line of fomenting discord between the former antagonists of the war. In longer perspective, Stresemann's success raised the prospect of a general coalition of all the capitalist powers against the Soviet Union. Such a prospect provided a good reason for Stalin to press the policy of very rapid industrialisation, for Stresemann's successors might take the next step forward from Locarno: that Germany should compensate herself for the losses imposed at Versailles by helping herself to those eastern lands which she had tried but failed to get at Brest-Litovsk.

'Socialism in one country' consequently required a radical break with the foreign policy of the N.E.P. period. The underlying reality of the new line – that the Soviet Union intended to pursue no foreign policy at all – could hardly be announced. But the fact that there was a new line could be judged from the 6th Congress of the Comintern (1928). Stalin announced that the capitalist world was just about to pass into a period of acute economic crisis; one of the more accurate Marxist predictions. In the face of this crisis, the working masses would be betrayed by the non-communist Socialist parties and in particular by the German Social Democrats. Consequently, the foreign parties were commanded to break off all relations with the Socialist parties, and to fight hard against them for the loyalty of the masses. In brief, a return was to be made to Lenin's line in 1917: no co-operation with the non-Bolshevik left. The German Social Democrats were singled out for particular abuse in this Congress. They were labelled 'social fascists' and the faithful were warned that there was no difference between the Social Democrats and all the other political parties in the Weimar Republic. All were equally enemies of the masses. It made no difference whether the Social Democrats or the Nazis were in power.

In this way the foreign Communists were launched upon a course based on a wildly mistaken analysis of fascism. The major error was consolidated by the accuracy of the less important prediction made in 1928: in 1929 the great depression did throw millions out of work and did strengthen all the parties of the left, including the Communists. These social conditions also favoured the rise of Hitler and the Nazis. In 1928, the Nazis polled 800,000 votes; in 1932, 6,400,000. The Nazi view about Soviet Russia was well known since it had been published in Hitler's *Mein Kampf* (1926). Hitler described communism as a Jewish plot to oust the Germans from their historical mission of dominating the inferior Slavs. With a wealth of historical illustration he demonstrated the crucial part played by the Germans in the building of the Russian State. If Europe were to be saved from a great Jewish plot, if the purity of Aryan racial blood were to be preserved, the Jewish-Bolshevik State must be destroyed. Such destruction would not only deliver the Russian people over to their rightful masters, it would also give Germany *Lebensraum* in the east. The task would anyway be quite simple because the Russian people were already groaning under their Jewish masters and would welcome the liberating appearance upon the scene of the master race. The whole line of pre-war German policy must be altered: like the Teutonic knights, the German people must concentrate all its energies upon the drive towards the east. Such was the heady, romantic, inaccurate but very appealing ideology of the Nazis. In the end it led the German people into a misunderstanding of Russia as profoundly mistaken as the communist analysis of contemporary German politics. Both these ideologies were based upon more or less spurious historical theories, thus demonstrating the great danger of superficial historical study.

Hitler's rise to power owed something – but not everything – to the shackles placed upon the German Communist party by the Comintern. In some cases, German Communists were actually ordered to assist the Nazis against the Social Democrats. Stalin apparently believed that between Nazis and Communists the whole German Socialist movement would be destroyed. In the final confrontation between extreme right and extreme left the Communists would hold the trump card – the universal support of the working masses. What he failed to see – and his ideology prevented him from seeing – was that the Nazi doctrine was itself revolutionary and that it would enable Hitler to build up a party which cut right across the familiar Marxist lines of class conflict: that is to say, a party which contained both the extreme right-wing groups and the industrial masses.

By 1933 Hitler was in power and was busy with the destruction of the most powerful Communist party in Europe outside the Soviet Union – a development which probably caused more sorrow to the Old Bolsheviks than to Stalin, who throughout his career looked with a jaundiced eye upon powerful foreign parties. The continued strength of the Rapallo line (and perhaps the initial uncertainty of Hitler) was demonstrated by the maintenance of friendly relations until 1934. Russia badly needed German trade (which thanks to the world slump it could have on very easy terms) and Hitler was unwilling immediately to offend those German industrial and army groups which had reasons for keeping up good relations with Russia. But within a few months Hitler had taken a step which announced his intention of following the anti-Soviet line of *Mein Kampf*. He concluded a treaty with Poland which, so Russia feared, encouraged the Poles to seek compensation in the east for the loss of their corridor, which Hitler thus early proposed to take from them.

At this point Soviet foreign policy began again. Japan already threatened from the Far East; the Versailles powers might at any moment save themselves by encouraging Hitler to turn against Russia. A new threat could only be met by a new policy, that of reviving Russia's traditional alliances with the West. But it should be noted that Stalin's motive was neither to save the West from Hitler (a hope entertained by some Western statesmen) nor was he chiefly concerned with the elimination of Hitler from the scene. His purpose was the advantage of the Soviet Union, and that could be achieved just as well by turning Hitler against the West as by turning the West against Hitler. In other words, Stalin's main purpose was to recreate the short-lived situation of 1918 in which the Western powers crippled each other in war while Russia remained neutral. But this time Russia would be strong enough to take advantage of the situation.

The new line of foreign policy was opened in 1934. Litvinov, a Jew, a 'westerner', profoundly anti-Nazi, was in charge of foreign policy. Russia joined the League of Nations and began to play a leading part in the formation of collective security. The French alliance was renewed in 1935 and linked with a similar mutual assistance pact with Czechoslovakia. The second arm of Soviet diplomacy, the Comintern, was also realigned (7th Congress, 1935). By one of the swift reversals which punctuated its dishonourable life, the Comintern jettisoned its policy of 1928. Instead, foreign Communists were to ally with any left-wing groups which were willing to combat fascism. In order to assist this 'popular front' or coalition of the left, Stalin produced his 1936 constitution, in the hope that it would persuade Western democrats that the Soviet Union was also democratic. The popular front policy had some success in the West, especially in France where it possibly discouraged the victory of domestic Fascism. But on the other hand the success of the French Communist party caused many moderate Frenchmen to make a calculation which was to be of great disservice to Stalin in 1940: namely, that while the defeat of France by Germany merely meant the imposition of fascism (which allowed one to keep one's property) the defeat of Germany by Russia would be followed by the victory of communism throughout Europe. But the immediate prospect of a left-wing, Communist-dominated government taking power in France attracted Stalin. Such a government might well decide to stand up to Hitler.

The first reactions to the new Soviet line cast serious doubts upon the possibility of stiffening the Versailles powers into common action against Hitler. In Britain, Ambassador Maisky found the governing classes deeply committed to the policy of appeasement. Without British backing, France would not act decisively. Hitler's reoccupation of the Rhineland (March 1936), in flagrant contravention of the Versailles treaties, passed without Anglo-French reaction. Similarly, Britain and France failed, at the end of 1936, to intervene on the anti-fascist side in the Spanish civil war. With extraordinary speed, Soviet war material and personnel were in action against the Spanish Fascists. At first this aid was given within the framework of popular front strategy: that is to say, Soviet aid was directed towards the victory of the non-communist, leftish Spanish Republican government. But in spite of great anti-fascist enthusiasm on the left, the British and French governments maintained a correct neutrality. By the beginning of 1937, it must have seemed clear to Stalin that his hope of involving the Western powers in war with Hitler had failed. On the contrary, it was becoming more likely that his Western twins, the appeasers, would persuade Hitler to turn east against Russia. With this possibility looming large, Stalin himself had to take steps to appease Hitler. This was a change of policy which seemed treachery in Western eyes but which was obviously perfectly justifiable in Stalin's. Much ink has been wasted on this topic by writers who have made the unjustifiable assumption that it was the duty of the Soviet Union to save the West from Hitler.

Secret negotiations between the Soviet Union and Germany were opened in 1937 by the Soviet trade delegate in Berlin, Kandelski. They were ignored, but Stalin's initiative at least informed Hitler that, in his projected revision of the Versailles treaty, Russia would be unlikely to support the Western powers. A more public demonstration of a change of Soviet policy was given in Spain. By a sudden switch (1937), the popular front policy was abandoned and a narrow communist line taken. Those

Communists who had been most prominent in the popular front were summoned home and annihilated in the purges. The implication was clear. Stalin was no longer trying to build up an anti-fascist force in the West: he was preparing to leave the West to its fate.

His reasoning seemed to be justified by the events of 1938. The appeasers allowed Hitler to take Austria and, by the Munich agreement, much of Czechoslovakia. During this crisis Stalin behaved with exemplary correctness. He stood by his treaties with France and Czechoslovakia. Soviet historians continue to assert that he would have fought Hitler in 1938 had he been permitted to by Britain and France. This is a bitterly disputed question. It is certainly true that the British and French governments were unwilling to fight alongside Russia in 1938. But in defence of the Western appeasers, it should be noted that Russia had no common frontier with either Czechoslovakia or Germany and that the Red Army was in the midst of being drastically purged. It is possible that, if war had broken out, Russia would have failed to intervene, claiming that it was impossible to move her troops across Poland and Romania. Was Stalin not behaving 'correctly' in order to lure the Western powers into a war in which he proposed to remain neutral until the pickings were to be had? Such considerations weighed heavily with the Western appeasers in 1938. Even thirty years later it is impossible to be certain that they were wrong. There is no reliable documentary basis for a study of Soviet foreign policy. The motives behind it can only be surmised.

The Munich agreement alarmed the Soviet Union. Stalin had offered to stand by his promise to defend Czechoslovakia: the Western powers had ignored him and had forced the Czechs to appease Hitler by giving up their most valuable territory. The sacrifice was in vain. In March 1939 Hitler broke the promises which he had made in the previous autumn. The German army marched into Prague and what remained of Czechoslovakia was dismembered.

This was a disaster both for Russia and for the Western powers. For five years Stalin had tried and failed to get Britain and France to join the Soviet Union in an alliance against Germany. After his success in Czechoslovakia, Hitler turned his attention to Poland. He demanded the restoration to Germany of Danzig

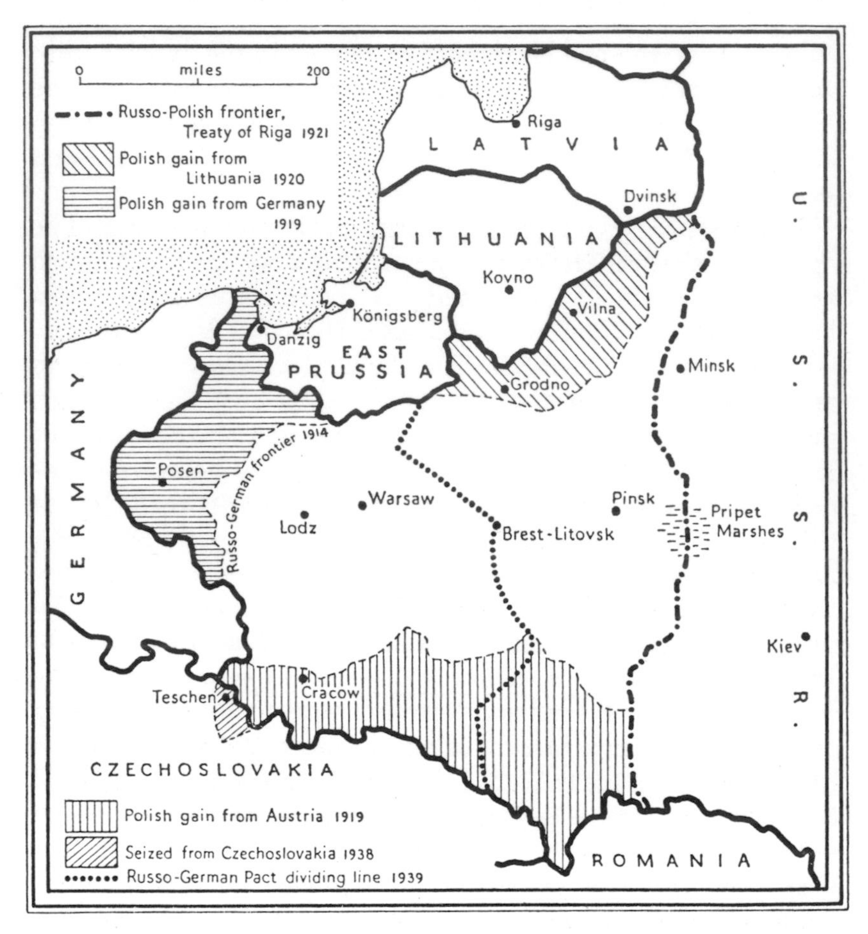

The Polish frontier, 1919–39

and the Polish corridor. The menace of a German attack on the Soviet Union was appreciably closer. Its proximity aroused in an acute form the basic dilemma of Soviet policy: should Russia continue the traditional Tsarist policy of finding Western allies against the German threat, or should she attempt to divide a potential capitalist threat by doing a separate deal with Germany. In spite of all the evidence that Britain and France were intent upon appeasing Hitler rather than fighting him, Stalin continued to try to get an alliance up to the last moment.

Chamberlain's reaction to the destruction of Czechoslovakia was swift and foolish. Without consulting the Soviet Union, he gave Poland a military guarantee against German aggression. It is (to say the least) doubtful whether he intended to keep the promise made to Poland. He meant either to frighten Hitler off or to put himself into a position from which he could persuade the Polish dictator, Colonel Beck, to accept another Munich agreement. But to Chamberlain's dismay it was assumed in Britain that he meant to keep his promise to Poland. Influential politicians like Lloyd George demanded to know why he did not take the steps which the Polish treaty obviously demanded. Neither Britain nor France could give Poland any direct military aid. That could only be provided by Russia. Prodded forward by these unwelcome pressures in Britain, Chamberlain grudgingly opened negotiations with Russia. He probably did not intend them to succeed. That they happened at all is a clear indication that the relations of the Great Powers had reached a crisis. For Chamberlain had a low opinion of the military value of the Russian alliance. He still clung to the spirit of Locarno. He continued to believe that Russia should be excluded from Europe, that her alliance was not necessary to the maintenance of the European balance, that it was positively dangerous to draw her in against Germany. Seen from the perspective of 1945 his fears had some justification. If Germany were to be eliminated as a great power both Britain and the British Empire would be at the mercy of a dangerously enlarged Russia. Similarly, if Germany were to be given a free hand to expand eastward (an interpretation of appeasement current in Soviet thought) Germany would become overwhelmingly powerful. Chamberlain's main purpose was to prevent any European war. He feared that the likelihood of such a war would be increased if Britain made a firm alliance with Russia. It is no wonder that he entered into negotiations unwillingly and pursued them half-heartedly.

They dragged on for six months. The details of the diplomatic exchanges show that the Soviet government was more eager for their successful conclusion than the British. British notes were promptly answered; Soviet replies went for weeks without response. Molotov replaced Litvinov as Commissar for Foreign Affairs (May 1939); he was a member of the Politburo and much closer to Stalin than his predecessor was. On the British side no person of consequence was ever sent to Moscow. In spite of Russian requests for a visit from the British Foreign Secretary only Strang (a Foreign Office official) was sent. When, in an attempt to break the political impasse, it was decided to hold military conversations, the Soviet government expected to receive the Chief of the Imperial General Staff. Instead they got a deputation of elderly third-rank service officers led by a retired admiral, Sir Reginald Plunkett Ernle-Erle-Drax. Even then he was not conveyed to Moscow in a style befitting the envoy of one Great Power to another. He travelled by passenger steamer to Leningrad. Stalin drew what was for him the obvious conclusion from the British disinclination to negotiate: Britain was going to make a bargain with Hitler to remain neutral while Germany drove eastward. There was only one way of preventing this: for Stalin to make his own bargain with Germany before Chamberlain gave away Poland.

The Anglo-Soviet talks revealed a profound but not insurmountable difference between the two sides. Russia wanted a general military alliance to operate automatically whenever and wherever German aggression occurred. This might be in Poland, Romania or the Baltic States. Britain was concerned with Poland only. She would not guarantee to go to the aid of Russia were Germany to attack Russia through the Baltic States. Stalin was particularly sensitive about this area. He knew that Leningrad was menaced by the corridor provided by the Baltic States. The British refusal on this point seemed to show that Britain was still trying to keep a door open in the east for the Germans. Another difficulty concerned Soviet troop movements. If the Soviet Union was to defend Poland, it must have the right to move its troops into Polish territory. Colonel Beck (heir to the centuries of Polish pride and prejudice) would not hear of Russian troops moving into Poland. Chamberlain did not press him hard. He was still angling for a deal with Hitler, and Russian troops in Poland would make this impossible.

Perhaps Stalin carried on these negotiations for so long because they raised the price of his friendship for Hitler. The latter did not want an Anglo-Soviet pact. Twice during the abortive negotiations he offered Stalin an attractive counter-proposal. Both offers were rejected. Only on 15 August (three days after the arrival in Moscow of the futile British military mission) did Stalin respond to a third German overture. Within eight days the Nazi-Soviet pact had been concluded. This speed was one of the advantages of negotiating with a dictator. In public it was declared that the two nations had signed a non-aggression pact: in private, a delimitation of spheres of influence was made. Russia was to get the Baltic States and Poland east of the rivers Narev, Vistula and San. Thus prepared, Hitler attacked Poland on 1 September.

In Britain and France the Nazi-Soviet pact marked the end of an era. For twenty years Soviet Russia had been reviled and ignored. The rise of Hitler with his avowed anti-Russian policy had fitted in well with the Locarno line. It was assumed that Russia needed the West far more than the West needed Russia. Chamberlain and the appeasers completely failed to understand that the Russia of 1939 was a great military power; with Russia and Germany allied they could not think what to do. Chamberlain tried to appease Hitler with dazzling offers of British capital. Even after a bewildered House of Commons had forced him to declare war on Germany, he still hoped to be on the German side. The 'phoney war' of 1939–40 is the best evidence that Stalin was right when he supposed that Chamberlain did not mean to fight Hitler.

While Chamberlain was active about doing nothing, the Soviet Union recovered most of what had been lost at Brest-Litovsk in 1918. This was a brilliant gain achieved with hardly any military effort, but it was not the most valuable result of the pact. By making an agreement with Hitler, Stalin gave himself a chance to deal with a problem which, in the summer of 1939, was almost as menacing as that of Germany. While negotiating with the British he had been fighting a full-scale (although undeclared) war with Japan in the Far East.

THE SOVIET UNION IN THE FAR EAST, 1919–41

Not until the defeat and death of Kolchak (February 1920) was the Soviet government able to take up the Tsarist patrimony in Siberia. The province was in complete anarchy. Japanese troops were still in control of the maritime provinces, and Japanese mercenaries terrorised the area from the coast up to Lake Baikal. Lenin was anxious not to become involved in war with Japan. He realised that Japanese ambitions in the Far East had already aroused the antagonism of Britain and the United States, while in Japan itself there was a powerful group, backed by the

Admiralty, which wanted to eliminate the Siberian adventure. With considerable skill, Lenin helped to foment the rivalries of his capitalist foes. To strengthen his hand in the Far East he adapted a method which had already proved valuable in all the borderlands reconquered at the end of the civil war. Elsewhere in Asia and Transcaucasia, the Bolshevik party had declared itself in favour of local autonomy. Behind this facade it had asserted its centralising power. In Siberia, however, Lenin went a step further. He assisted the creation of a fully independent Far Eastern Republic. Its government was democratic, and during its brief lifetime (1920–2) it entered into relations with the neighbouring powers. Its delegates were even received, in an unofficial capacity, at the Washington Conference (1922). This democratic 'front' acted as a buffer between Soviet and Japanese Siberia. It helped to arouse American feeling against Japan. Lenin stimulated American business interests by offering economic concessions. The Japanese were known to be interested in the oil of the northern half of Sakhalin, an area which the United States was determined to keep out of their hands. Lenin's patience bore fruit. Peaceful influences gained control in Japan for the time being. By the Washington Treaties she agreed to withdraw from Siberia and the maritime provinces. Lenin then hastened to wind up the Far Eastern Republic. Soviet power had reached the Pacific.

After the establishment of stability in the Far East the Bolsheviks played down their earlier promises to China to the effect that the Tsarist conquests would be restored and the 'unequal treaties' torn up. On the contrary, Lenin and later Stalin took up the Tsarist game of power politics in China. The price was even more tempting than it had been before 1914. In 1921 the Red Army entered Mongolia at the request of a 'provisional revolutionary government', and Peking was forced once more to accept Mongolian 'independence'. Again, this was a cloak for real Russian control. In 1924 Peking was forced to accept the continuation of Soviet control over the Chinese Eastern Railway. The Soviet Union thus inherited one of the most flagrant pieces of Tsarist imperialism. The motives for a forward policy in China were a complex blend of the old and the new. It was a means by which British economic interests might be threatened. It was important to prevent any other power from gaining control of China. It was tempting to strengthen the Soviet position in the Far East by linking populous and wealthy Manchuria to the thinly peopled regions of eastern Siberia. Success in China would encourage the spread of revolution in Asia, which was an important consideration at a time when the prospect of revolution in Europe was receding. In 1920 the party had held the Baku Congress of the Peoples of the East. It had proved to be a curious mixture of representatives of races in every stage of development (or lack of it), but they all had one thing in common – even Enver Pasha, who was far from being a proletarian. They all hated imperialism and particularly British imperialism. With considerable skill Lenin and Stalin persuaded the oppressed Asians that the Russian Bolshevik party was their friend. They pointed out that Russia too had been a semi-colonial territory in 1917, her economy had been in part like that of Britain and in part like that of India. Soviet Russia, half-way between Europe and Asia, was the natural friend of the anti-imperialist aspirations of the Asian masses. From the first the Soviet leaders realised that it would be futile to attempt to push all the anti-imperialist forces into the narrow mould of Bolshevism. The desire to end colonial rule united many otherwise conflicting groups. Among them were middle-class intellectuals, romantic nationalists, ambitious industrialists, landlords and peasants and a tiny urban proletariat. Although the revolutionary nationalism of these groups was directed towards the establishment of a social order incompatible with the ideals of the Soviet leaders, the Asian leaders were at least revolutionary. Their success had the immediate likelihood of undermining the prosperity of the imperialist powers. The Soviet leaders believed that British prosperity depended largely upon the exploitation of China and India. If China and India were lost it seemed likely that Britain would face an economic crisis possibly of revolutionary dimensions.

The situation in China was temptingly chaotic. The Peking government ruled only a portion of the nation. Independent warlords defied an authority already badly shaken by its evident reliance upon American and British support against Japanese aggression. Chinese nationalism had taken refuge in Canton under the idealistic but impractical leadership of Sun Yet-sen, a scholarly Confucian who had captured the mind of young China but completely lacked organisational ability. His nationalist movement (the Kuomintang) seemed the perfect instrument for Soviet ambitions. Both Sun and the Soviet leaders thought that they could exploit each other's weakness.

From 1922 to 1927 Soviet influence in the Kuomintang was very strong. Advisers like Joffe and Borodin organised its political structure along Communist party lines. Its young soldiers were sent to Moscow for training and with Russian aid a military training school was created at Whampoa. Although members of the Chinese Communist party were admitted to the Kuomintang they never dominated it. Especially after the death of Sun in 1925 the right-wing elements of the Kuomintang gained control of the movement. In 1927 the anti-communist Kuomintang leader Chiang Kai-shek scored a double victory. He marched north from Canton and gained control of the heart of China. Simultaneously he broke with the Chinese Communists and drove them out of the cities. It happened that this disaster to international communism suited Stalin quite well. He wanted to withdraw from the Chinese adventure in order to concentrate upon the reconstruction of Russia. It has even been suggested that he helped Chiang to decimate the Chinese Communist party. Whatever the truth of this, he certainly refrained from sending any further help after the disasters of 1927. Curiously enough this was of immense service to Mao Tse-tung, the Chinese Communist leader. He learned that he must stand on his own feet and that he had to base his movement on the Chinese peasantry. His independence from foreign aid eventually brought him success in a country where hatred of the foreigner was the strongest political emotion.

Although Chiang Kai-shek had annihilated Soviet prospects in China, his nationalist government was not a threat to Soviet security in the Far East. Even at the moment of triumph it was too weak; it relied too much upon British and French financial help and the goodwill of the warlords. Chiang attempted to force the Soviet Union to give up its rights to the Manchurian railways. There was a brisk Soviet military reaction (1929) against Chiang's Manchurian warlord, Chang Tso-liu. Frustrated in this direction, Chang Tso-liu began to harry Japanese interests in southern Manchuria. This also provoked a violent response. Japan had been badly injured by the world slump. The military party had long wanted an excuse to turn Manchuria into a Japanese colony, for its resources could be developed to provide the industrial products necessary for further Japanese imperialism. The weakness and isolation of Russia were an invitation to the British-dominated Kuomintang to assert itself in this area. In 1931 Japan conquered Manchuria, defied the League of Nations and created a colony masquerading as the State of Manchukuo. The Soviet Far East was again threatened. Manchukuo was rapidly industrialised by Japan at a rate which was only surpassed by the industrialisation of Russia itself. Unlike his Tsarist predecessors at the beginning of the century, Stalin met the challenge with a policy of cautious appeasement. He refrained from associating the Soviet Union with the admittedly feeble League of Nations attempt to deter Japanese aggression; he sold Japan the Russian interest in the Manchurian railways (1933); he attempted to heal the rift between the Kuomintang and the Chinese Communist party.

Japanese aggression had one profitable result for the Soviet Union: it helped to improve relations with the United States. Since 1917 the American government had refused to open diplomatic relations with the Soviet Union. This had not prevented the growth of profitable trade relations between the two nations, relations which became more important to the United States after the world slump and depression. The emergence of the Japanese threat combined with the prospect of reviving trade persuaded Roosevelt that the time had come to forget America's ideological distaste for the Communist régime. In 1933 diplomatic relations were opened. The commercial results were rather important for the Soviet Union, but this was as far as the resumption of relations went. Russia hoped that the Japanese would turn their attention against American naval power in the Pacific. The United States and Britain hoped that Japan would leave them alone and engage herself against Soviet Siberia.

In spite of Soviet efforts to foster good relations, the threat from Japan mounted during the thirties. Germany and Japan came to an understanding in 1936. They agreed to co-ordinate their policies towards Russia and China, which was very dangerous for Stalin. It raised the prospect of a war on two fronts, a war in which the Soviet Union would receive no help from Britain or America. The existence of this actute danger does much to explain the prolonged enthusiasm with which Stalin attempted to secure a British alliance. The Japanese threat also helps to explain why Britain was so unwilling to make a military agreement with the Soviet Union. Such a step would put her decisively on the anti-Japanese side in the Far East. Bad relations with Germany would cause bad relations with Japan.

Thoroughly convinced of the paralysis of the opposition, Japan moved against China again in 1937. Numerous border incidents provoked by the Japanese along the frontier between Manchukuo and the Soviet Union showed that they were trying to operate the 1936 pact with Germany. Border incidents turned into a full-scale war by 1939. The Japanese suffered 18,000 casualties at the battle of Nomunhan (August 1939). The existence of this undeclared war was a powerful motive for Stalin to make the pact with Hitler. It reduced pressure on both sides of the Soviet Union. It weakened the attraction of the German alliance for Japan. Hostilities ended immediately after the Nazi-Soviet pact. Japan realised that she would have to act alone in the Far East. The direction of her attack was settled by the spectacle of British weakness in 1940. Britain could just manage to defend herself, but she could not defend her enormous and wealthy Asian empire. This was a far more attractive bait than the wastes of Soviet Siberia. Soviet neutrality was now worth having. In April 1941 Stalin garnered one of the most valuable results of his deal with Hitler. He signed a non-aggression pact with Japan. A few months later Sorge, his excellent spy in Tokyo, was able to assure him that Japan was preparing for a Pacific war. Stalin could move his Far Eastern army to Europe where, in December 1941, it enabled him to hold Moscow at the crisis of the German invasion.

STALIN AND HITLER, 1939–41: THE SOVIET UNION PREPARES

The relations between the two dictators during the twenty-one months which elapsed between the Nazi-Soviet pact and the German attack on Russia in June 1941 were rather unsatisfactory to both. The Germans surprised Stalin by the speed with which they conquered first Poland in the autumn of 1939 and then France in the summer of 1940. Stalin had counted on a

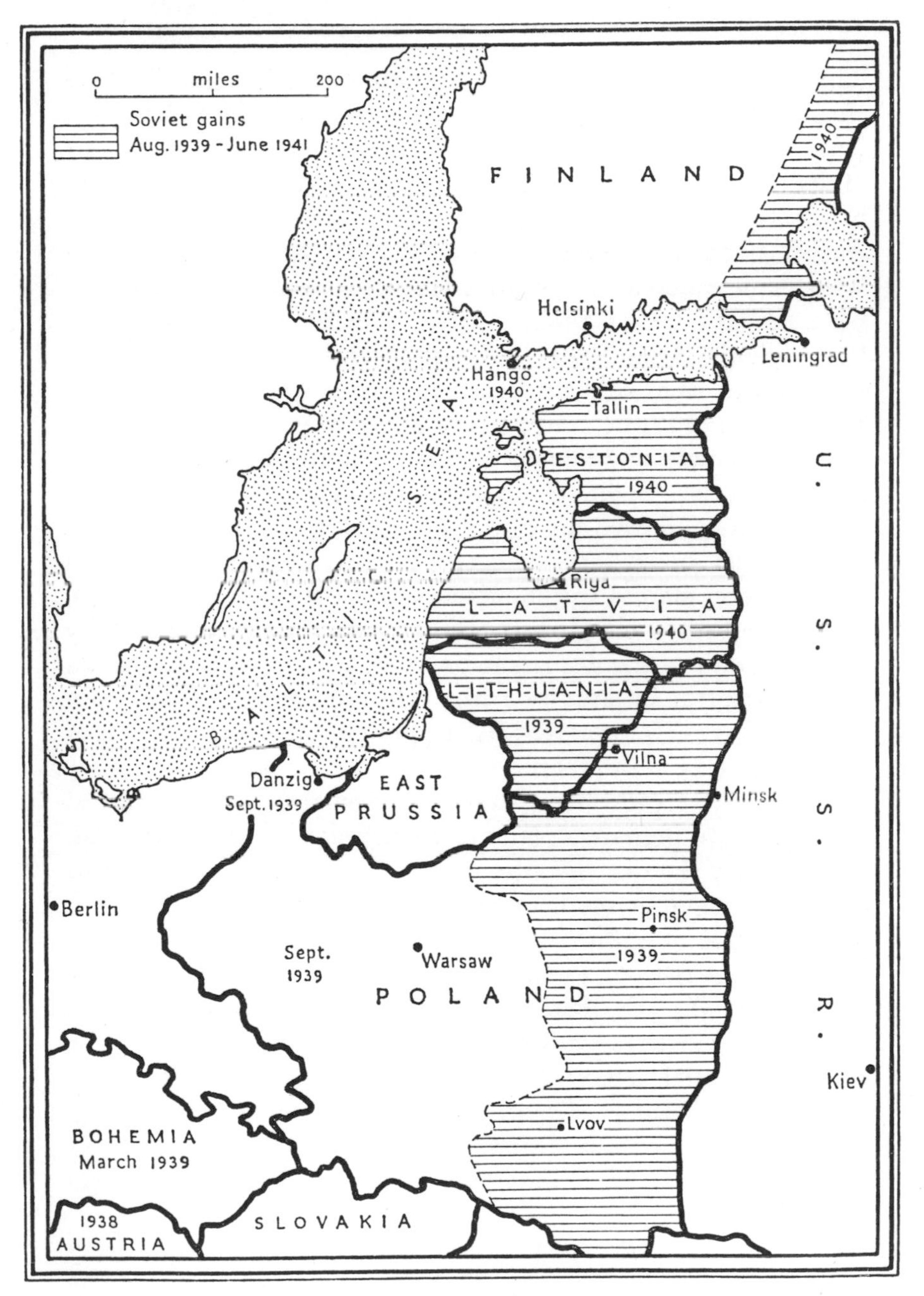

Soviet territorial gains, 1939–40

long war in the West: instead he nearly found himself in the situation which he most feared, alone with a victorious Germany. Only the unexpected resistance of Britain remained to divert Hitler from his eastern ambitions. Hitler for his part was annoyed by the quite literal manner in which Stalin interpreted the August pact. The Soviet Union was not content with mere 'spheres of influence'. Within a few weeks eastern Poland had been annexed to the Soviet Union. Finland refused to accept Soviet demands and Stalin, without consulting his German ally, invaded and eventually crushed the Finns in a winter war (1939–1940) which did little credit to the Red Army. Stalin's reactions to the extremely unwelcome German victories in the West also annoyed Hitler. Stalin incorporated the Baltic States into the Soviet Union, allowed the Soviet press to express admiration for the gallantry of embattled Britain and, most serious of all in Hitler's eyes, started to pursue an active policy in the Balkans. Bessarabia was seized (June 1940); and the pact overstepped by the annexation of the Romanian province of Bukovina. Romanian oil was of great interest to Hitler. He was enraged that Stalin should pursue this impudent Balkan policy at the very moment of German victory in the West.

Shortly after the Soviet moves in Romania, Hitler ordered Brauchitsch to prepare a plan for the invasion of Russia – a plan subsequently known by the name 'Barbarossa'. Stalin had chosen a dangerous moment to tamper with the Balkans. For if Germany was to eliminate Britain from the war without taking the enormous risks of a seaborne invasion – a risk which became unthinkable after the Battle of Britain – there was one obvious way of doing it: to drive Britain out of the Mediterranean and to cut the imperial artery at Suez. Franco refused to attack the British at Gibraltar. Consequently the German army must launch an attack on the Near East through the Balkans. But this could not be done while the Soviet Union menaced the German flank from Romania.

Before Barbarossa was put into operational planning, Hitler made an effort to wean the Russians out of the Balkans by diplomacy. Molotov went to Berlin (November 1940). He was presented with tempting offers in return for Russian neutrality and withdrawal from the Balkans. Ribbentrop invited Russia to help herself to the British Empire which, he claimed, was finished anyway. At that very moment, British bombers appeared over Berlin and the scheming diplomats had to take refuge in a cellar. Molotov did not refuse the offer of the British Empire, but he insisted upon Russian domination of the Balkans. Russia must control Bulgaria and have bases on the Bosporus. Evidently Stalin calculated that Hitler would pay almost any price for Russian neutrality. In this he was wrong, for shortly after Molotov's departure the order was given for Barbarossa to be put into planning operation.

The breakdown of an important element of the August pact was signalised during the early months of 1941 by the German invasion of the Balkans. Yugoslavia, Bulgaria and Greece were occupied by German troops without arousing any complaint from Moscow. The Communist parties in these countries were not ordered to resist the invaders and Russia continued to supply Germany with substantial amounts of oil, chrome and manganese – vital to Germany and for which the Germans had, in effect, stopped paying. Appeasement of Hitler was apparently to continue in spite of the growing evidence of Barbarossa's existence. It is true that Stalin's timid policy at this time may be explained by his fear of Japanese intentions, but this must have been considerably reduced by the signature of the non-aggression pact (April 1941). Up to the very last moment before the German attack (22 June) Stalin seems to have believed that Hitler was not serious. He would turn south against Suez; he could be bought off by the cession of Romania; he could be mollified by the delivery of yet more war supplies. Whatever the reason, whether as the result of either military or diplomatic miscalculation, Russia was not prepared for Barbarossa.

The accepted Soviet evaluation of Stalin's policy during the period August 1939–June 1941 is: that he avoided an immediate German attack; that he ensured that when Russia did go to war she was not without allies; that he gained space, which was of vital significance in 1941; and that he gained time in which to build up Soviet strength.

Some of these claims (particularly the second and third) have some foundation; the other two, however, are open to question. As for the first, it now appears that Germany was in no position to attack Russia in 1939. There was no military plan and the military and economic means were lacking. Hitler could not have attacked before the summer of 1940 and consequently the August pact gained a year's respite at the most. But it is the last claim which is the most misleading, for it is certain that Germany gained far more from the respite than did the Soviet Union. In 1939 Nazi Germany had a labour force of about 10 million: in 1941, of about 28 million. Coal production increased in the same period from 250 to 400 million tons: oil production from 1 to 7 million tons. In brief, by 1941 Hitler had behind him the economic resources of Europe, all of which could be mobilised against Russia. Stalin's gamble upon a long war in the West had gained him nothing.

But it is true that the respite was put to good use in Russia. Stalin's belief in Hitler's good intentions did not prevent him from making energetic military preparations. In industry efforts were made to increase production. The eight-hour day was introduced in 1940, absenteeism prosecuted and the multi-lathe system introduced. This was a scheme to ensure that mobilisation should not impede full production by arranging that one skilled worker could supervise the operation of many machines. It has been said, however, that Stalin showed his blind belief in the permanent nature of the August pact by failing to move important industries eastward. Such a view fails to take into account the fact that it was only a few weeks before the German attack that there was any reasonable certainty that the danger from Germany was any greater than that from Japan. To move industry eastward would certainly dislocate production and might merely place it at the mercy of the Japanese. A more certain indication that Stalin was well aware of the danger threatening him is the fact that in May 1941 he became chairman of the Council of People's Commissars. For the first time since the revolution, the party and state machines were openly merged.

It is, however, in the field of military preparation that the balance sheet of Stalin's 'Tilsit Peace' is the most difficult to draw up. External threats had provoked extensive rearmament and reorganisation from about 1934. The production of war material increased more rapidly than anything else during the second five-year plan (1934–8), and the Red Army, which had remained about half a million strong since the early twenties, was rapidly expanded to about three times that size. The purges (1937) may have eliminated many of the most experienced officers but many younger ones got unexpected chances of promotion. Perhaps the British and French armies would have fought better in 1940 had they been recently purged. But the advantages which could have been gained from the elimination of the senior officers were lost by the return (1937) to the system of dual control. That is to say, throughout the armed forces in every unit large and small, the political commissar, whose loyalty was directly to the political administration, had to countersign every order, even those of a strictly service nature. In addition, the N.K.V.D. supervised the loyalty of the political commissars. This system revealed the intense suspicion with which the party regarded the armed forces, the one power in the state which could make a bid for power and which could not be done without. In a way characteristic of him Stalin provided a carrot as well as a whip: during the late thirties the officer corps was loaded with new privileges. Pay was increased about 300 per cent in the period 1934–9: smart uniforms, gold braid, saluting, ferocious discipline and privileges in food and housing conditions helped to make the officers a race apart. Strenuous efforts were made to ensure that any promising officers were drawn into the party. In such ways Stalin was much more successful than Hitler in gaining the loyalty of the officers.

The August pact was followed by yet more frenzied preparations. The terms of military service (already compulsory for all politically reliable elements) were made much longer, particularly in specialist units. More emphasis was placed on the activities of Osoaviakhim (Society for Furthering Defence, Aviation and Chemical Warfare), a civilian body with twelve million members in 1939. Osoaviakhim did not train a mass partisan force. It concentrated upon politically reliable specialists who were trained as machine-gunners, drivers, parachutists and snipers who could be rapidly assimilated in the Red Army.

Up to 1939, Soviet military thinkers had envisaged a defence plan which embodied the French idea of static fortifications and also allowed for the fluid warfare at which the Red Army had proved itself so successful during the civil war. An elaborate defensive line – which German officers later pronounced very formidable – was started from the Baltic southward to the Pripet marshes. On the southern side of this barrier, mobile forces were to menace the flank of any invader advancing towards the northern cities. Tukhachevsky and the older officers had been keenly interested in the development of tank tactics but it is doubtful whether they had grasped the fundamental idea of Blitzkrieg: that is to say, the grouping of all armoured forces into single units. Hesitant movements in this direction were contradicted by the experience of the Far Eastern and Finnish wars (1939–40) in which success was gained by using balanced masses of infantry, artillery and tanks. It was only at the last moment, after the example of the German successes in Poland and France, that the Red Army started to regroup its tanks into exclusively armoured divisions.

In one respect at least the Soviet strategical position was improved by the August pact. The Finnish War allowed the northern frontier to be pushed back from the outskirts of Leningrad to the northern side of the Karelian peninsula. Such a move was essential if the birthplace of the revolution and a great industrial area was to be defended against a small but resolutely pro-German power. The strategical results of the other frontier changes of 1939–40 are, however, more difficult to assess. Until the rapid deterioration of relations with Germany in November 1940, the Tukhachevsky plan was retained: that is, the new forward areas were lightly held and the bulk of Soviet forces kept on the defensive line. As soon as Hitler started his forward moves in the Balkans, however, the Red Army and Airforce were concentrated in the forward zones. By June 1941, 170 divisions – five-sevenths of the total strength – were outside the 1939 frontiers. Why Stalin did this is not known. It may be guessed that he was unwilling to risk a war on Soviet soil, that he was proposing to threaten the German position in the Balkans as the German army moved southward, that he had failed to understand the implication of Blitzkrieg tactics. But whatever the reason, Stalin imperilled the whole military strength which he had spent so many years building up. He had placed his infantry, his tanks and his aircraft in a position in which they could be destroyed by the Germans during the first few days of the war. The lavish equipment which industrialisation had made possible was rapidly squandered. In June 1941 the Soviet Union had many more tanks, guns and aircraft than the Germans. In some cases Soviet equipment was notably better than its German counterpart. Guderian, the greatest German exponent of Blitzkrieg, described the Soviet T 34 as 'the best tank in any army up to 1943'. Russian artillery was good, especially a new type of multiple rocket launcher known as *katyusha*. Soviet aircraft were rather poor; there was an acute shortage of modern transport. The disasters of the winter war with Finland had led to a great improvement in the ability of Soviet troops to fight in cold conditions.

Stalin had made many solid gains from the Nazi-Soviet pact. By his military dispositions of June 1941 he threw away nearly all that he had gained. He did not adopt the traditional Russian tactics in the face of an invader – to fall back and to buy time by sacrificing space. This is what Alexander I had done in 1812 but the tactic had caused him much anxiety. He feared that the French would set his nation alight by declaring the emancipation of the serfs. Perhaps Stalin had similar fears. German victories on Russian soil would enable them to undermine the loyalty of the collectivised peasants. Stalin had good reason to know how much Communist rule was disliked, especially in the Ukraine. Why should he trust the masses when for at least a decade he had not trusted his closest friends?

14

Defeat, Victory and the Restoration of the Tsarist Hegemony: 1941-5

AT THE FRONT

'You have only to kick in the door and the whole rotten structure will come crashing down', said Hitler to Runstedt. The German generals obeyed the order but Hitler's prediction failed to come true. The history of this, the most gigantic war in human experience, centres around two large questions. First, how did the Soviet Union stand up to the initial German attack; and secondly, whence was the strength found to expel Hitler's army and finally to invade Germany? The first question was dominant until the battle of Stalingrad at the end of 1942; the second after the battle of Kursk in the summer of 1943. During the second phase of the war, in which the prospect of total defeat became the hope of complete victory, a problem reappeared which had been implicit ever since 1917, which had prevented the creation of an earlier anti-Nazi alliance, but which had been temporarily dormant during the period of German victory – that is to say, the *modus vivendi* between Russia and her new allies. Already in 1945, as the Soviet army entered Berlin and scores of Soviet, American and British units fraternised jubilantly along the front, this problem had eroded the temporary wartime alliance and the stresses of the cold war were felt. At the time it seemed a sad and unnecessary conclusion to wartime alliance. It was blamed variously upon Stalin's lack of trust, American excess of trust, Churchill's cynical realism. But in perspective it becomes obvious that the alliance was only a temporary, makeshift affair. The fear of Hitler could make some odd bedfellows but as soon as the danger was gone the underlying hostility returned. The difference in 1945 was that Russia had become a world power, no longer to be

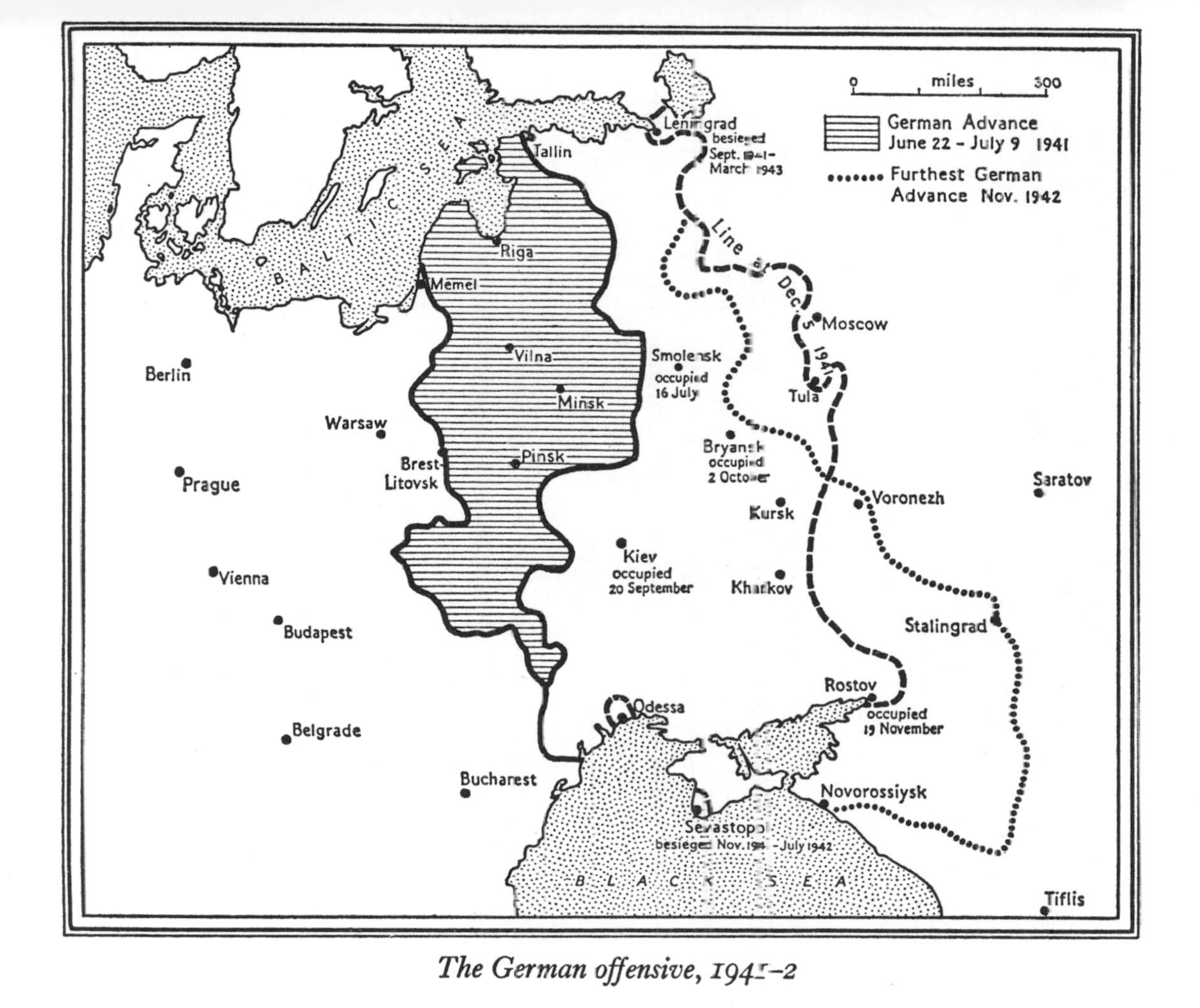

The German offensive, 1941–2

neglected or scorned. This in turn meant that the United States had to adopt the role of a counterbalancing force. In the snows and on the plains of Russia, in the critical years 1941–3, not only was the struggle for survival between two European powers fought out, but in these terrible battles the whole modern world took shape. Not in North Africa, not in the skies over Germany, not in the Pacific, not even on the Normandy beaches, but only at Los Alamos did events take place which rivalled in scale and significance the battles fought in or near three Russian cities – Leningrad, Moscow and Stalingrad.

Until the winter of 1941 the German advance moved forward almost unchecked along the 700-mile front. It has been claimed that Leningrad could have been captured at this time had Hitler concentrated his effort against it; in fact, he decided upon its investment and upon making his main effort against Moscow. Leningrad was encircled by the German army on three sides and by the Finns in the north. The determination with which its three million inhabitants withstood a two-year siege makes it very unlikely that it could have been captured in 1941. It is more likely that the German advantages in armour and aircraft would have been lost in a long street battle in which the defenders would have shown as much resolution as was later shown at Stalingrad.

The battle for Moscow began in October with the Germans in high spirits. Plans were already in being for the demobilisation of the bulk of the German army after a peace treaty which would strip Russia of the bulk of her European territories. It seemed that the Bear was already dead: since the beginning of the campaign, German intelligence calculated, the Russians had lost 2½ million men (most of them prisoners), 22,000 guns, 18,000 tanks and 14,000 aircraft – that is to say, the paper strength of the Soviet forces in June 1941. The Vyazma-Bryansk battles seemed to open the way to the capital. Only scratch forces were available to the new commander Zhukov: the remnants of broken armies and a handful of T 34s straight off the production lines. Five or six weeks remained until the winter set in, and the Panzers had only twenty to thirty miles to go. Signs of panic were seen in Moscow for the first and only time in the war. On 12 October foreign embassies and some commissariats were moved east to Kuybyshev, but Stalin remained in the Kremlin together with the main war-directing body – GOKO. On 16 October, mass flights from the city took place. Many party officials pretended that their departments had been ordered east and with their departure it seemed that the core of resistance had been broken. But on 19 October Moscow was declared to be in a state of siege: units of N.K.V.D. troops were brought in and order was restored. With his rear stabilised, Zhukov could now concentrate upon his difficult task, which was in essence to play for time. But with a confidence and skill which merit the highest rating he was not content with a merely defensive battle: he was already planning a rapid counter-offensive on the German army as it bunched up for the final assault on Moscow. His advantages were minimal. Sections of the line had to be held by hastily raised workers' battalions against triumphant Panzer forces. Heroism and ingenuity could do much, but what really slowed down the German advance were problems of supply and logistics. Wet weather preceded the snows and on the swampy ground the narrow-tracked German tanks were unable to move. The Russian roads and railways were unable to carry sufficient supplies and the massed armies in front of Moscow got in each other's way as the front narrowed. Zhukov played on the German difficulties skilfully. He used cavalry to great effect against the lines of communication; he received increasing support from the partisan movement growing up in the German rear; he carefully concealed and husbanded his trump card, the high-quality divisions which were being brought over from the Far East. These fine troops were perhaps the key factor and that they could be used in the West at all was a tribute to the Soviet espionage system. A Japanese attack had been expected since June, but during the autumn the Sorge spy rings in Tokyo had reported that the Japanese were planning a Pacific and not a Siberian war. On the night of 4–5 December Zhukov unleashed his offensive on a long front. The Siberian divisions attacked the flanks of the deep Panzer thrusts to the north and south of Moscow. For a time it seemed that the German army faced complete disaster. Its tanks were immobilised by the intense cold and the troops were weakened by an outbreak of dysentery. A recovery occurred, chiefly because Zhukov had few reserves with which to exploit his advantage. But Moscow was saved and the German army had been defeated for the first time since 1918.

By forcing his generals to hold fast in the winter of 1941–2,

Hitler gave himself what Napoleon had never had – a second chance to defeat Russia. The Red Army was exhausted by the spring of 1942 and the initiative once again passed to the Germans. Hitler's plan of campaign was simple. He realised, as his generals apparently did not, that Germany must win the war in 1942 or otherwise be defeated by the industrial strength of the United States. To defeat Russia he proposed a southern campaign which would take his armies up to the Volga. By this he would capture all the remaining Soviet industrial centres in Europe; and from the Volga he would have the choice of turning north behind Moscow or south into the Caucasus. Either way, Russia would have to seek peace or else continue to exist as only a third-rate Asiatic power. To meet this thrust the Soviet army was worse equipped than in 1941. Industrial production had been halved by the German advance in 1941, and help from the Allies was as yet unimportant. But excessive confidence had been generated by the winter campaign and Stalin decided to forestall the German attack by one of his own. It was badly mismanaged. In the Crimea an attempt to relieve Sevastopol failed with heavy losses of men and irreplaceable losses of tanks. In the north an offensive to relieve Leningrad failed and the commanding general, Vlasov, was captured. But most serious, Timoshenko's effort to recapture the fourth city of the Soviet Union, Kharkov, became a Soviet rout. The clumsy Soviet attempt to emulate Blitzkrieg tactics was a failure, and in May 1942 the Panzers scored another familiar success: two pincers closed behind a large pocket of Russian troops and nearly a quarter of a million were captured. With their morale thoroughly restored, the Germans began their drive to the Volga and the Caucasus in June. If anything, the Soviet army fought less well than in 1941. Fewer units allowed themselves to be captured as they retreated eastward across the Ukraine, but there was less individual heroism and initiative. Once again, the onward rush of the Germans had to be halted in a large city; and once again it was the appointment of Zhukov to command on the Stalingrad front (September 1942) which preceded a renewal of Russian morale.

The German attack on Stalingrad began at the end of August and continued until the Russian counter-offensive on 19 November. This was the turning-point of the war, the moment at which the last ounce was put into the balance on both sides, the moment at which the rough shape of the future was decided. The Soviet army discovered a commander of high quality to direct the battle on the spot – Chuikov. Behind him Zhukov was doing just the same as he did at Moscow, building up his reserves on the flanks and feeding into Stalingrad itself only a bare minimum of high-grade infantry. The battle took place in the ruins of the city. The antagonists fought not just for streets or even houses but for individual rooms, halls, passageways, factory floors. Chuikov's H.Q. was a few yards from the front line; no part of the Russian bridgehead on the west bank of the Volga was out of German range. The key to the battle was the river crossing by which fresh supplies and men came over nightly and the wounded were evacuated. Week by week the Germans pushed the defenders a few yards further back; week by week their strength diminished as their tanks were destroyed in the rubble of the city and as the morale of their finest troops was broken by the skilful obstinacy of the *untermenschen*. The final offensive was launched on 11 November. Although small parties of German troops, filled with alcohol and benzedrine, reached the banks of the Volga, they could not break through the last Soviet positions. On this occasion there was not even the excuse of the weather. A superior German force of high-grade troops had been fought to a standstill by a race which Nazi propaganda alleged to be backward and degenerate. Germany lost the war even before the massed guns of Zhukov's counter-offensive were heard on the night of 18–19 November. The surrender of the German army at Stalingrad in January 1943 brought the initial stage of the war to a close. Fighting on one front alone, the formerly victorious German army had been halted and defeated.

After the surrender of von Paulus at Stalingrad there still remained twenty-eight months of bitter fighting before Zhukov's units entered Berlin. The German army was still a formidable opponent, even in defeat, even after the Allied landings in Normandy (June 1944), even after various attempts had been made on Hitler's life (e.g. July 1944). The Red Army offensives were conducted with caution rather than dash. Superior man-power, larger numbers of tanks, aircraft and trucks ensured eventual victory if only local defeat could be avoided. The main military threat to total victory during these twenty-eight months occurred in July 1943 when the Germans launched their last major offensive of the war in the Kursk salient. This was a Verdun-type battle. Its object was not to secure a breakthrough but rather to draw in the bulk of the Russian armour and destroy it in a massive tank battle. Hitler had ordered the production of new tanks for this occasion, only one of which (the Tiger type) proved itself markedly superior to the T 34. There was plenty of time to prepare a defence. With typical thoroughness, Zhukov organised a system which in places was a hundred miles deep. This was more than sufficient to contain the German attack and after little more than a week's fighting the vast battlefield, the scene of the greatest armoured battle ever fought, was in Russian hands. Thus Stalingrad was shown to have been no mere fluke, the result of an unlucky decision to fight in terrain unsuitable for tanks. On ground of their own choosing, in summer weather, the Panzer armies had been defeated.

By the end of 1943 cautious advances had brought the Red Army up to the line of the Dnieper. Kiev, Smolensk and Krivoy Rog had been recaptured and in January 1944 Leningrad was relieved from its 900-day siege. The first Soviet thrusts of 1944 came in the north and the centre. Within a few weeks, Rokossovky's units had reached the Vistula and the gates of Warsaw. There they halted (August 1944) within sight of the city in which Bor Komorowski and the Polish resistance forces were being liquidated by the S.S. Soviet troops were across the 1939 frontiers for the first time. Purely military considerations now gave way to diplomatic factors. The Russians could have continued their advance into Poland in 1944 and could have reached German soil before winter. Instead, the main autumn campaign came in the Balkans which Stalin was anxious to gobble up before the Balkan States made separate peace treaties with the Allies. The Balkan campaign was just as successful as that in Poland. King Michael of Romania made peace with Russia and declared war on Germany (August 1944); Bulgaria made peace in September; in October, Soviet forces made contact with Tito and the Yugoslav partisans in Belgrade. These advances placed most of the Balkans under direct Russian control while still leaving plenty of time for the occupation of the rest of Poland. In other words, the failure of the August rising in Warsaw combined with the failure of the Western Allies to reach the Rhine, gave Stalin just enough time to conquer the Balkans.

The final stages of the campaign in Poland and Germany itself were conducted by Zhukov with the same unhurried caution. He advanced his timetable slightly in January 1945 in order to relieve the German pressure on the Allies in the West. A severe battle was fought for Berlin (conducted by Chuikov) and the Russian armies swept on to agreed stop-lines on the Elbe. The most powerful army in the world took up positions shortly to become the new East–West frontier. Its commanders, unlike the Anglo-Saxon generals, were destined not to an orgy of public adulation and memoir writing but to obscurity. Zhukov, the most brilliant of them all, was quietly removed to a subordinate command in Odessa. For behind this great army was a jealous party which, as victory approached, tightened the control which it always retained in peacetime. After a brief period in which the Russian soldiers were permitted to give the German civil population a taste of the sadistic brutality which the Russian people had endured at German hands since 1941, discipline was restored. The Russians were not good haters, and Stalin was not going to repeat Hitler's mistake of alienating his new subjects by unnecessary cruelty.

BEHIND THE LINES

The war and the German occupation inflicted a monstrous degree of suffering upon the peoples of the Soviet Union. The Red Army lost about seven million men: civilian casualties brought the total number of war dead to about twenty million. These losses fell with particular severity upon the rural population, that is to say upon that section which had the strongest reasons for hating the party and the clearest motive for desiring its overthrow. The fear that the Germans might proclaim the end of collectivisation, and that this might cause the disintegration of an army which was still largely recruited from the peasantry, was the most severe political problem facing the party. The knowledge that its authority was weak in the countryside dictated the choice of large cities – Leningrad, Moscow and Stalingrad – as the scene for last-ditch resistance. The siege of Leningrad illustrates the degree to which the party was still essentially urban. For hundreds of miles the German army had roamed almost unchecked through the countryside: it then found itself halted by the fanatical courage of hastily armed urban workers backed by a sullenly determined urban population. The whole defence was conducted by the party rather than by the army. Everybody was mobilised: the women and children to dig defences, the working men in hastily armed and trained units to fight on the front. Production was maintained even in factories under German shell fire. Women workers stopped only when there was a direct hit upon the workshop. Tanks were driven straight off the production line into battle. The Komsomol, young secondary-school children, were mobilised into a wide variety of tasks from cutting wood under fire to caring for the thousands of orphaned children and burying the bodies of those who died in the streets from starvation. The hand of the party was seen not only in the successful prolongation of siege discipline but also in the clumsy errors made; the failure to evacuate the very old, the very young and the sick before the siege began; the failure to build up adequate food supplies before the German ring closed; the extreme slowness with which a supply line over Lake Ladoga was constructed. In fact the intense suffering of the siege – intense even by Russian standards – seems to have knit party and population together. There was only one hint of treachery from within, of a break in morale. That occurred during the first days of the siege (August 1941) when the German army was assaulting the hastily built defences. Voroshilov and Zhdanov set up a military Soviet for the defence of Leningrad and in their proclamations began to play heavily not upon Soviet patriotism but upon the special local loyalty owed to Leningrad. Stalin reacted immediately even at this desperate stage of the war. A military Soviet looked like an attempt to break away from party authority, to reawaken the separatist tendencies of a city which had always tried to preserve its own party structure; perhaps it was the prelude to a separate peace. The military Soviet was dissolved and Voroshilov removed to a post in Moscow.

In some ways the party's task of maintaining order behind the battle lines was simplified by its whole history both before and after 1917. It had seized power in conditions of anarchy and it had, during the thirties, created a system of controlled anarchy. Its administrators were used to a fluid society in which hordes of displaced peasants had to be fitted into a new social pattern. It was used to sharing out small quantities of the necessities of life according to the recipient's usefulness to the State. It was used to imposing terrible suffering in the hope of future happiness: it made little difference whether the goal was labelled 'socialism' or 'victory'. It was used to being treated as ruthlessly by its superiors as it treated the masses. In brief, the party had been organised for total war for fifty years. It is not surprising that it was able to provide the last ounce of discipline which turned the balance at the critical moments. Although the courage of the millions at the front may have been born of patriotism, the will of the directing group was forged by adherence to a political ideology. This collective will was the more malleable because party membership had already been thoroughly purged. What the party had lost in intelligence, experience and initiative, it had gained in obedience and devotion.

The central body for the direction of the war, GOKO, was set up in June 1941. Its original members, appointed without regard to rank or precise shade of opinion, were at first Stalin, Molotov, Voroshilov, Malenkov and Beria. More senior members of the Politburo, Mikoyan and Kaganovich, were added in 1942. Within this body, which had absolute control over all party, Soviet and army organisations, Stalin had the controlling voice. Few hints of disagreement have been made public. The army G.H.Q., the Stavka, was directly under Stalin's command: for this reason it is futile to attempt to distinguish between those military decisions which were Stalin's and those which were forced on him by the professional soldiers. To an extraordinary degree, Stalin's will was dominant in every sphere of the war effort. He rarely left the Kremlin. It is not known that he ever visited any fighting front. But from his office came political, military and diplomatic decisions which were unfailingly put into execution by his subordinates. Every detail of official business, even down to the quantity and type of rifles to be supplied by the Americans, was known to him and discussed by him in the nocturnal sessions described by many wartime visitors to the Kremlin. He combined in a unique way the patient industriousness of a lifelong bureaucrat with a politician's sensitivity to currents of public feeling and a diplomat's long view of the peace which would follow the war. Stalin was an asset of incalculable value to wartime Russia.

Some aspects of the party control of Russia behind the front were obviously well managed. The evacuation of industrial plant from the Ukraine to the Urals is a case in point. The movement was largely improvised, it was done hurriedly and the new sites were prepared with great rapidity. The results were of great value. In spite of the losses of 1941–2, Russia was able to keep up an adequate supply of war material of high quality. Intelligent choice of a few good designs also helped. Where other armies were hindered by an embarrassing variety of tanks and guns, the Soviet forces enjoyed all the practical advantages of simple designs turned out in huge quantities. The Western Allies were allowed to provide only trucks, clothing and food; Russia could have continued the war even had the Allies made a separate peace with Germany.

But success in the ruthless choice of economic ends was not altogether surprising; the party already had a decade's experience in this field. New problems were posed by the control of partisan forces and the political direction of a gigantic army. The partisan war behind the German lines has been the subject of much romantic and propagandising mythology. During the first months after the invasion there was little spontaneous anti-German movement in the conquered zones. Large bands of Red Army soldiers cut off from their units lived in the forests of White Russia and pillaged anyone they could. Such a life was evidently preferable to the treatment they would have received upon rejoining their units. The local inhabitants stayed at home, waiting to see whether their new masters would be less oppressive than the old. Not until May 1942 did the party create a Central Partisan Staff: its task was as much political as military. Arms, equipment and experts could easily be provided for the bands which ontrolled large forest areas. More important for the party was to ensure political reliability; the partisans must not be in a position to oppose the Soviet régime when the Germans were defeated. Even at its height the partisan movement never contained more than 200,000 people; there was a good deal of friction between local leaders and infiltrated party men. The exact degree of local enthusiasm is impossible to judge because the widespread and senseless cruelty of the Germans ensured that the peasants reacted violently against the invaders. This does not mean, as official party histories assert, that the party initially commanded the loyalty of the people in the occupied zones. The military effect of partisan warfare is difficult to judge, but it was probably less than the claims made for it in the official Soviet histories.

Party control over the army followed the same pattern as in pre-war days. The problem was to ensure political reliability without destroying the professional pride of the officers. Officer morale was bolstered by fine uniforms, gold braid (an important item of foreign aid) and harsh discipline. The Red Army in wartime provided rapid promotion for men of talent. All the successful generals started the war in obscure positions: Cherniakovsky for example, one of the heroes of the battle of Kursk, began the war a major and finished it a general. But the political indoctrination of a huge peasant army could not be neglected even in wartime. From July 1941 to October 1942 the party restored tight political control. The political commissars were made the equals of the unit commanders. Later, the *zampolits*, or assistant political commissars, were restored. The chain of party control, however, still extended right down to company level. Frequent political lectures, even in front-line conditions, reminded the recruit that his ultimate loyalty was obedience to the party rather than to the officer. Watching and reporting upon both the unit commander and the *zampolit*, even in small units, were career officers of the N.K.V.D. In spite of its elaborate nature – which illustrates the extreme suspicion with which the party regarded the army – the system seemed to work well: there was only one notable case of defection in wartime, that of General Vlasov. He had been a party member since 1930 and was captured in 1942. He then lent his name to a Liberation Movement which was joined by many other captured officers and party members. Its programme was for a socialist Russia without the Communist party. Vlasov was sufficiently naïve to believe the German promise of such a political settlement after the war although anybody merely had to look at what the Germans were doing to occupied Russia to see what would happen to the Liberation Movement. The party, however, took Vlasov seriously. Elaborate efforts were made to counteract his propaganda and the N.K.V.D. tried to murder him.

Although the party was the real directing force behind the war effort, attempts were made, especially during the critical early months, to conceal the extent of party control behind a mask of patriotism. Stalin's early speeches were filled with references to the Russian heroes – sometimes the Christian heroes – of the past.

Russian patriotism was flattered by the evocation of the ghosts of Alexander Nevsky and Suvorov. Russia was said to be waging a Fatherland war, a great patriotic struggle. Stalin appeared in public wearing a jewel-bedecked marshal's uniform as if he were another Alexander I. The movement of conciliation between party and nation, between the present and the past, was brought most sharply into focus by the changed relations of the party with the Orthodox Church. A largely peasant army could still be deeply influenced by the priests; the consolations of religion could be of service to a population enduring extremes of suffering. Besides, there was an acute danger that 'Christian' Germany would use the Orthodox Church as a fifth column against the godless Communists. Official attacks on the Church ceased; more buildings were opened for worship; and in 1943 Stalin received the three Metropolitans in the Kremlin, was photographed with them and permitted them to elect a Patriarch. It seemed that soon Lenin's heir would be seen at his predecessor's tomb in Red Square accompanied by a procession of priests intoning a service for the dead. In fact, the new Holy Synod (like that of pre-revolutionary days) was little more than a department of state for religious affairs. It was answerable to a Soviet Commissariat headed by a party official, ironically known in Moscow as *Narkomop* or 'People's Commissar for Opium'. In return for these concessions, Stalin drew immense benefits. The Church put itself whole-heartedly behind the war effort: Stalin was prayed for; money was collected; schismatic Orthodox were anathematised. The Allies, and particularly the British Anglicans, were favourably impressed; the Slav peoples of the Balkans could be fed the old Tsarist line that 'Holy Russia' was their natural friend amid the encircling gloom of Christians of the wrong sort. The new Church policy proved to be a good substitute for the Comintern (abolished in 1943). This was one of Stalin's great political coups – so successful that he carried it on until his death. It seems a pity that Church leaders, no doubt as harmless as doves, should have lacked so completely the serpentine wisdom of the ex-seminarist who outwitted them.

But although it must be allowed that the party was successful in its conduct of the war effort, in its use of ideas and emotions foreign to its own tradition without in the least compromising its ultimate authority, it should be remembered that throughout the war party control was unceasingly and powerfully aided by the Germans themselves. Hitler had invaded Russia without any very clear ideas except that he meant to enslave the *untermenschen* and use them to maintain the wealth of Germany. As Russian resistance grew stronger the German propensity for senseless cruelty extended. The Jews were everywhere exterminated – a fact which did not much pain the anti-semitic inhabitants of the Ukraine. Millions of Russians were deported to work as slaves in German factories in conditions of incredible barbarity. The fight against the partisans was carried on by the S.S. with a cruelty bound to create lasting hatred. Even regular army units, in spite of subsequent protestations to the contrary, treated local populations as the less-than-human beings depicted by Nazi propaganda. All this bloodshed and cruelty, perpetrated by the race that chose to boast of its higher civilization, made it impossible for more intelligent Germans to utilise the great opportunities of weaning the conquered peoples from loyalty to the Soviet régime. Agriculture was not de-collectivised and little attempt was made to appeal to the significant nationalism of, for example, the Ukrainians. Instead, in the sickening bloodbath of Nazi-occupied Russia, in which each leading Nazi tried to carve himself out an 'empire' at the expense of the others, the survivors looked to the Reds, just as they had during the civil war, as slightly the lesser of two evils. While the German army was winning the war at the front, the Nazi party was losing it in the rear areas. This uncontrolled anarchy is in sharp contrast to the control exercised on the other side of the battle line. The comparison brings out clearly the cool, rational, purposeful character of party rule. The Germans lost the war in Russia not because they were outnumbered, frozen or stabbed in the back, but because their political system prevented them from taking advantage of their military success.

RUSSIA AND HER ALLIES

Russia did not stand alone against the Germans. She was a member of a wartime coalition which contained the leading imperialist and capitalist powers in the world. Only a common fear of Germany held the Allies together. Even before the last shot was fired in Europe the alliance was falling apart. But while it lasted Stalin extracted from it great advantages for his country. This stage of his foreign policy reveals him most clearly as a Russian patriot, grasping what he could for his country. Some would blame him for this. They argue that mankind was the loser because Stalin chose to pursue narrow nationalist aims at a moment when more generous policies of international co-operation would have been eagerly accepted by a war-torn world. Others congratulate Stalin upon his realism. They point out that when the war ended Russia had equipped herself with a line of defence against Europe, a string of satellite states which could take the shock of any future attack.

Russia had many advantages over her wartime allies. From 1941 to 1945 she engaged the bulk of the German forces. Even at the end of 1944, when the British and American war effort had developed its greatest momentum, two million Germans were fighting in the east as compared with one million in France and Italy. It was essential for the Western Allies to keep the Russians fighting hard. Without a heavy German involvement in the east the invasion of Western Europe would have been either impossible or so costly in life as to make it unacceptable to democratic nations. If Russia had collapsed or made a separate peace, German domination of Europe would have been permanent. By this means Stalin persuaded his allies to make a heavy material contribution to the Russian war effort. The United States alone sent nine billion dollars' worth of aid under Lend-Lease – that is, free of charge. Britain sent a smaller amount but stretched her naval resources to the limit to ensure that cargoes reached the north Russian ports. Russia's value to the Allies was not restricted to her ability to soak up the brunt of the German war effort. In December 1941 she had remained neutral when Japan attacked the United States. The prospect of Russian help in the Pacific War, which was to be concluded only after the defeat of Germany, was a bait with which Stalin angled most skilfully. It helped to ensure American goodwill, and the acceptance by Roosevelt of Russian political objectives which were quite the contrary of those professed by the United States. This was another of Stalin's advantages. From the beginning of the alliance he was openly following Russian national interests. Even in the dark days of 1941–2 he insisted that his allies must accept the Soviet right to the frontiers of 1941 – those frontiers which had been gained as a

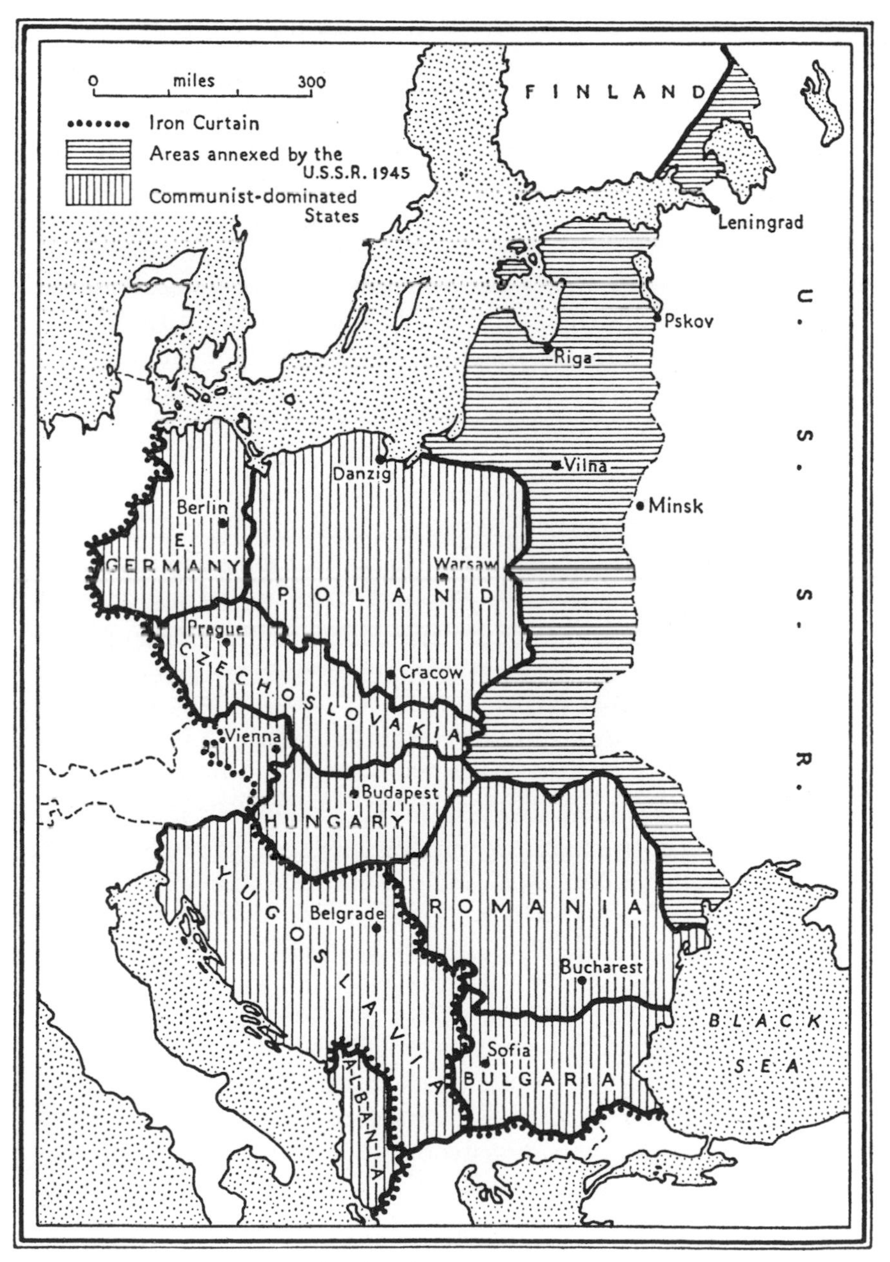

Soviet Europe, 1940–5

result of his treaty with Hitler. Roosevelt, on the other hand followed a policy which some would call naïve, others idealistic. He proposed that the war should not be fought for the advantage of any one nation and that the post-war world should be dominated by an international organisation, the United Nations. He either failed or refused to see that Stalin followed a policy of national advantage incompatible with the American ideals. Roosevelt was convinced that the United States needed Soviet friendship as much in the peace which was to follow the war as in the war itself. In order to gain and keep this friendship he was willing to make concessions which later seemed excessively generous.

At the beginning of the alliance Stalin's demands upon the Anglo-Saxons fell under three heads. First, he wanted a second front in Western Europe, a cross-channel invasion; secondly, he wanted a guarantee that his Western allies would not make a separate peace; and thirdly he wanted all the material supplies available. On the third point he was mostly satisfied. He grumbled occasionally when the Royal Navy found it impossible to protect the northern convoys. But as soon as the supply route through Persia was opened in 1942 he was mollified by the massive volume of American food, clothing and lorries. On the second point he was satisfied by the public statement made by Roosevelt at the Casablanca conference (January 1943). On that occasion – regarded by many as a turning-point in world history – Roosevelt announced a policy of unconditional surrender as the minimum acceptable basis for peace with Germany. This made it virtually impossible for the Western Allies to make peace with (for example) a German government which had overthrown Hitler. It meant that the war would have to be carried on right into Germany; this in turn virtually guaranteed that much of Germany would fall under Soviet military control. After Roosevelt's declaration not even a man as suspicious as Stalin could have believed that his allies were going to desert him before Germany was defeated.

But on the first point – the opening of a western front – Stalin had a legitimate grievance. In spite of the most urgent pleas, the Anglo-Saxons were not able to land in France until 1944. Both in 1942 and in 1943 their failure elicited Stalin's sour irony. In August 1942 Churchill made his first visit to Moscow in order to explain to Stalin why the invasion of France could not take place.

He found little sympathy for his difficulties. The Germans were at Stalingrad; the Russians had already endured enormous casualties; Stalin was angry when Churchill explained that an invasion of France would cost the Anglo-Saxons heavy casualties and would have little effect upon the course of the war in Russia. Stalin suspected that the Anglo-Saxons were fighting the war to the last drop of Russian blood. Churchill restrained his great gift for retort. In spite of some impassioned exchanges the two men parted on good terms. Both had decided that, whatever the shortcomings of the other, each could be guaranteed to fight the Germans. This did not prevent a further outbreak of accusation when Stalin was informed in May 1943 that the invasion of France had been postponed yet again. 'Scorching' and 'super-heated' messages flew between London and Moscow; but the exchanges remained private and no open sign of disagreement was allowed to encourage the Germans. By this time Stalin had learned that the main opponent of the cross-channel plan was Churchill. The Americans had always been eager for it – possibly on military grounds alone, possibly because it was obviously what their Russian ally most wanted. They much regretted that they had been drawn into a Mediterranean campaign by Churchill's enthusiastic advocacy. At Teheran (November 1943), where the three Allied leaders met for the first time, Roosevelt was delighted to be able to tell Stalin that the cross-channel invasion would certainly take place in the next summer. Stalin and Roosevelt together overcame Churchill's last plea for the Mediterranean theatre. As an indication of the increasing weight of Russia in the alliance by this date, it is significant that Roosevelt at Teheran anxiously asked Stalin to ensure that the Soviet offensives in 1944 would be so timed as to coincide with the Anglo-American landings in France. In 1942 and 1943 Stalin had asked for a second front so that German pressure in the east might be reduced. But by the end of 1943 he was the senior military partner. He promised to enable the Anglo-Saxons to land by ensuring that the Germans were fully engaged against the Red Army. The promise was kept.

As the prospect of victory came nearer the Allies naturally began to think about post-war arrangements in Europe. So far as final arrangements were concerned, all three were willing to leave most of the work to the United Nations. Each ally was to make armistice terms on its own fronts. It was agreed that the military commanders on the spot should have complete control in conquered areas since effective armistice terms were essential to the continuation of the battle against the Germans. This agreement worked quite well in Italy, the first of Germany's allies to surrender. The Russians had no forces in this theatre and accepted without a qualm the right of the Anglo-Saxons to make terms even with a former Fascist like Badoglio. Stalin recognised the supremacy of wartime needs. But the Italian precedent served him well. When his own armies entered the Balkans in 1944, he was able to answer the Anglo-Saxon demands for a share of the control of these nations by pointing to Soviet forbearance in Italy.

But there were certain points connected with the post-war settlement which Stalin was not willing to leave to the United Nations. The most vexing of these were connected with Poland. The Polish question poisoned inter-allied relations, gave comfort to the Germans and dramatically showed that Stalin's aims were as nationalist as those of Alexander I. There were two divisive issues. First the delimitation of frontiers, and secondly the problem of who was to govern Poland after the Germans had been driven out.

After their defeat in 1939 many Polish leaders had taken refuge in London. There they had been recognised as the legal Polish government. The exiled Poles had played an important part in the defence of Britain in 1940 and Polish units fought with the British forces in Africa and Italy. Naturally these Poles thought that at the end of the war they would be restored to a Poland at least as big as it had been in 1939. They expected their demands to be supported by the British government which had originally gone to war to defend Poland against Germany. But from the beginning of the alliance Stalin had insisted that he would retain the frontier which he had gained as a result of his deal with Hitler in 1939. He asserted that this frontier was based upon the Curzon line, that the British had agreed to it at the end of the First World War and that the Soviet Union had been unjustly deprived of it at the Peace of Riga. He suggested that Poland should be compensated for her losses in the east by taking German territory in the west. Eventually he wanted Poland to take not only a slice of East Prussia but also a great swathe of German territory up to the river Neisse. The Poles obstinately refused this compromise. They wanted both their old eastern frontiers and westward expansion too. They hoped that they would be compensated for their sufferings by being made a Great Power. First under Sikorski, then after his death under Mikolajczyk, they obstinately refused to compromise. Churchill did his best for them; he frequently endangered Anglo-Russian relations by pleading their cause with Stalin. But until it was too late they refused to accept the Curzon line as their eastern frontier. Roosevelt remained clear of these discussions. He claimed that he could not take the risk of losing the Polish vote in the American elections, but it is more likely that he feared the consequences of trying to thwart Stalin over what could be regarded as a purely Russian interest.

The Polish question was made still worse by the existence of a wide ideological gap between the London Poles and Stalin. The former were mostly right-wing, militaristic, anti-Russian nationalists. They were the men who had allied with Hitler in 1934 and whose obstinacy had helped to prevent the Anglo-Soviet alliance in 1939. Stalin feared the results of restoring such a clique to power in Warsaw. Would this not negate the effects of the war? Warsaw might once again move into the German orbit and the Soviet frontiers would be as vulnerable as ever. Stalin was determined to ensure a friendly government in Warsaw. This meant a Communist government – desirable not because it was Communist but because it was friendly.

The London Poles made Stalin's task fairly simple for him. In 1943 the Germans discovered a mass grave of Polish officers in the Katyn forest. They asserted that these men had been murdered by the N.K.V.D. The Russians replied that they were innocent and that the Germans were trying to create discord among the Allies. Rather too hastily – as they themselves later admitted – the London Poles called for an international investigation. Infuriated (or pretending to be infuriated) by this lack of trust in an ally, the Soviet Union broke off diplomatic relations with the London Poles. This was the only major, public breach in the wartime alliance. It was never healed, partly because of the obstinacy of the London Poles but chiefly because the Russians did not want it healed. Stalin now turned to the formation of a more friendly Polish group in Russia. This (eventually known as the Lublin Committee) took over the provisional government of

Poland as the Red Army moved in. Churchill did his best to create a coalition between the London and Lublin Poles. He might have succeeded had it not been for the tragedy of the Warsaw rising (August 1944). The Red Army had reached the Vistula, a few miles from Warsaw. The London Poles were anxious to show their strength in Warsaw. If they could recapture it from the Germans with their underground army this would compensate for past humiliations and give them a valuable bargaining counter against the Soviet Union. General Bor Komorowski ordered his 35,000 men out into the streets; the Germans brought in reinforcements and a terrible battle began. The Red Army made little or no attempt to cover the few miles between the Vistula and Warsaw. Stalin claimed that his northern offensive had been concluded and that he did not want to get involved in street fighting in a large city. He had good reason to fear a battle of Stalingrad in reverse. But the London Poles naturally put another interpretation upon his inaction. The Germans were doing his job for him. In the course of the Warsaw battle they killed off most of the underground leaders loyal to London, and destroyed London's last chance of influencing post-war Poland.

In the end, Stalin imposed his will upon Poland. Russian national aims were achieved without subjecting the Polish question to a peace conference. A Communist-dominated government was set up in Warsaw. More important, the western frontier changes made the new Poland wholly reliant upon the Soviet Union. Millions of Germans had to be expelled from their homes in the Oder-Neisse territory. This ensured that Poland would be permanently faced by the German demand for revenge and only the Soviet Union could defend her against the German threat. Between the wars Poland was supposed to defend Europe against Russia: after 1945 Poland became Russia's bulwark against Europe.

The Polish model was followed in the Russian dealings with the conquered Balkan countries. As the Red Army overran these areas (1944–5) only the fiction of Allied control was retained. In fact, the Soviet government alone imposed armistice terms, set up friendly provisional governments and made political arrangements of a permanent nature. In Romania, for example, the Soviet Union without consulting its allies overthrew one provisional government in order to instal another more obedient to its will. To Anglo-Saxon objections Stalin always offered two excuses. First, that the Anglo-Saxons had done the same in Italy; and secondly that it was essential to ensure safe lines of communication while the war continued against the Germans. But there was another side to Stalin's 'spheres of influence' policy. He recognised the point at which he had to stop. Churchill discovered this when he went to Moscow in October 1944. With dramatic ease he made a verbal agreement with Stalin to the effect that Russia would not give aid to the Greek communists. The latter were causing grave concern to the British government. British troops were already in action against them and the commitment seemed intolerable while the fighting against Germany continued. Stalin had no plans for a Russian invasion of Greece. His political aims stretched no further than the Red Army could march. In France, too, he was willing to accept a political settlement highly disadvantageous to the powerful French Communist party. He recognised the right of de Gaulle to head the French provisional government and he backed de Gaulle's demands for an equal voice in the alliance. He could not have more clearly demonstrated the main impulse behind his foreign policy. None knew better than Stalin that the aftermath of war is revolution. But he showed that he did not intend to take advantage of the revolutionary situation in Europe as a revolutionary leader. He meant only to build an empire out of the conquests of the Red Army.

Although the development of Stalin's policy caused much anxiety among his allies – an anxiety most acutely felt by Churchill – the second meeting of the Big Three at Yalta (February 1945) passed off very cordially. Roosevelt was determined to prevent the Polish problem from poisoning relations. He was delighted to find that Stalin seemed much more enthusiastic about the future rôle of the United Nations. He was relieved to get from Stalin a firm undertaking to enter the war against Japan three months after the surrender of Germany. Both he and Churchill were pleased by the recent evidence of Soviet military co-operation. When the western front had been temporarily endangered by the German Ardennes offensive Stalin had willingly altered the timetable of Soviet offensives in order to take the weight off his allies. Roosevelt has been accused of excessive generosity towards Stalin at Yalta. It is true that some

American officials, like Kennan, had already perceived the menace of Russian ambitions. But Roosevelt was convinced that any post-war settlement must rely upon goodwill between the Soviet Union and the United States. He thought that it would be impossible for his government to keep troops in Europe after the end of the war. Only by good relations could Russia be kept from dominating the whole continent. Since none of the governments had any fixed policy about a permanent German settlement they merely agreed on provisional arrangements to follow the German surrender. Military zones were allocated and an inter-allied control council created for the settlement of questions which concerned the whole of Germany. The prospect of these temporary arrangements becoming permanent was raised by the Soviet insistence upon enormous reparations. In fact as the Russian armies advanced the Soviet Union helped itself to the economic resources of the areas under its control. In the long run this was bound to lead to the creation of an economic structure in eastern Germany quite different from that of the West. But at Yalta it seemed only common justice that Russia, which had suffered so terribly from the German invasion should recoup some of her losses. The one thing which might have prevented Stalin from doing this – a firm promise of American post-war economic aid – even Roosevelt was unable to give.

After Yalta relations between the Allies deteriorated rapidly. Roosevelt's successor, Truman, did not share his predecessor's sense of common partnership in a great cause. He was more influenced by the growing opinion among his officials that Russia was actually a menace to American security. American military opinion was less certain of the necessity of Russian aid against Japan. Reports indicated the revolutionary nature of the atomic bomb which had recently been successfully exploded for the first time. At the Potsdam conference (July 1945) Truman took a harsher line than his predecessor. He showed Stalin that Russia could not expect an American loan; he told Stalin about the bomb; and he said that Russia could not expect to take part in the post-war occupation of Japan.

At Yalta Stalin had gained an extravagant price for Soviet entry into the war against Japan. From Japan he was to get the Kuriles and southern Sakhalin. From China he was to get control over the Manchurian railways and special rights over Port

Arthur and Dairen. In fact, Russia was to be restored to the position which she had occupied before 1905. In order to gain these national territorial objectives Stalin abandoned the Chinese Communist party. He accepted Chiang Kai-shek as the rightful ruler of China – although it should be observed that he weakened and humiliated the Nationalist government of China by making these territorial demands. The Soviet acceptance of the Chinese Nationalists was a great relief to Roosevelt. Chiang had not had a successful war. In spite of massive American aid he had been consistently defeated by the Japanese. Even in victory his government would be weak. If the Soviet Union had chosen to back the Chinese Communists the whole basis of American policy in the Pacific would have been undermined.

In spite of Truman's coolness, Stalin hurried to gain the Far Eastern territories promised at Yalta. On 6 August the first bomb was dropped on Hiroshima. On 8 August Russia declared war against Japan. Soviet forces in Manchuria indirectly, and perhaps unwittingly, helped the Chinese Communists to get the lion's share of surrendered Japanese war material. This helped them to seize power four years later – an event almost as unwelcome to Stalin as to the American government. Russia got what she had been promised at Yalta but nothing more. From Moscow's viewpoint it could be argued that Japan was the American sphere of influence just as the Balkans were the Russian.

Many have wondered whether the destruction of the wartime alliance could have been prevented by any one person or by any single act of policy. The world gained so much from the strange alliance and has lost so much by its dissolution into the Cold War. An approach to the problem is offered by an analysis of the possibilities open to Stalin at the end of the war. As has been seen, he adopted a policy of national expansion. Two other courses were possible. First, he could have exploited the revolutionary potential of international communism. It is unlikely that this course would have caused less tension between Russia and her allies. Secondly, he could have fallen in with Roosevelt's plan for an international settlement based upon the continuation of the wartime alliance. But as a Marxist Stalin could have felt no confidence in the continuation of an alliance between a socialist and capitalist powers. The United States and her allies would dominate the United Nations; they would insist upon peace terms which would deny Russia such security as her position in 1945 allowed her to take. American wealth and the American bomb endangered Russian interests in the long run just as much as Germany and Japan had threatened her previously. Stalin thought in terms of inevitable conflict. The elimination of old enmities merely made room for new ones and a divided Germany symbolised the new order. Could Stalin possibly have hoped for such an advantageous settlement from any peace conference? The Anglo-Saxons were bound to want to restore German prosperity and economic unity in the end. Only by putting an iron curtain round his conquests could Stalin ensure that Germany would be too weak to repeat for a third time the eastern conquests which had twice in his own lifetime devastated Russia.

Although the decisive factors had changed enormously since the time when he took over the conduct of foreign policy, Stalin's style remained constant until his death in 1953. 'Socialism in one country' had meant a cautious, defensive policy in which Soviet interests had been placed a long way ahead of revolutionary ideals. Yet these ideals have never been openly repudiated and the historical analysis upon which they are based remains the driving force of the ruling class of the Soviet Union. In the party view, there remains an irreconcilable conflict between capitalism and communism and, however long it may be necessary to wait, History is on the communist side. In the short view Stalin may be the man who betrayed the revolution; but in a longer perspective he may be seen as the man who preserved it during the days of its weakness.

The position of the Soviet Union in 1945 is comparable with that of the Russian Empire in 1815. A terrible defeat had in both cases been turned into a swift and overwhelming victory. In both cases, suspicious allies resented the extent of Russian power. These suspicions were immediately transformed into a quarrel about the political frontiers of central Europe. But here the similarities end. In 1815, Alexander I had secured the consent of conservative Europe for the peace settlement which he helped to create. The rulers of Prussia and Austria, sometimes the rulers of France, were the natural allies of the Russian Tsar against any attempt to overthrow the 1815 settlement. Stalin was far more powerful than Alexander I. Only the United States was in the same class of military and economic power. But the ruler of

Communist Russia could not hope to secure consent for a European peace treaty as Alexander had done. For Metternich, Alexander had been the bulwark of a conservative social system. For Truman and Dulles, Stalin represented a threat subversive to their ideology. If Russia in 1945 had merely been another Great Power, no doubt some accommodation with the United States could have been reached. But Russia was much more than a rival for world power. Her social system was an affront to American ideologists, just as that of Tsarist Russia had been to British Russophobes in the nineteenth century. Russia had no right to flaunt such a different way of life. The fact that Soviet Communism was also quite successful made it all the more dangerous. When religious enthusiasm has waned, political systems become all the more important. Human nature was not so diverse as to admit the simultaneous existence of political systems so different as 'the American way of life' and Soviet Communism. One must be right and the other wrong. This was the basis of the cold war. It was intensified by military and strategic factors but it was at bottom an ideological conflict.

It has been asserted that mid-twentieth-century technological developments will make both the capitalist and the communist systems identical. This seems doubtful in the light of past Russian history. Tsarist and Stalinist Russia followed the West through the Industrial Revolution but Russia still remained distinct from the West. It will be difficult (even if it were desirable) to eradicate the effects of centuries of autocracy. Russia has continued to move along her own path of development. It has been a path which has led her to emulate many of the achievements of the West, to try to become more powerful than the West, to beat the West at its own game, but few Russians have ever thought it desirable to copy the social and economic forms of Western Europe and the United States. Alexander II's attempt to graft Western liberalism was a total failure. The Russian views about the nature of man, the State and political power have been and remain different from those of the West.

Select Bibliography

BIBLIOGRAPHIES, ENCYCLOPAEDIAS AND ATLASES

Adams, Matley and McCagg, *An Atlas of Russian and East European History* (1966).
Harkins, W. E., *Dictionary of Russian Literature* (1957).
Shapiro, D., *A Select Bibliography of Works in English on Russian History 1801–1917* (1962).
Taaffe and Kingsbury, *An Atlas of Soviet Affairs* (1965).
Utechin, S. V., *Everyman's Concise Encyclopaedia of Russia* (1961).

GENERAL HISTORIES

Charques, R., *A Short History of Russia* (1959).
Florinsky, M. T., *Russia, A History and an Interpretation* (1953).
Karpovich, Michael, *Imperial Russia 1801–1917* (1932).
Klyuchevsky, V. O., *A History of Russia 1926–31* (5 vols. 1911–31).
Kornilov, A. A., *Modern Russian History* (1916).
Pares, B., *A History of Russia* (1955).
Riasanovsky, N. V., *A History of Russia* (1963).
Seton-Watson, H., *The Russian Empire 1801–1917* (1967).
Sumner, B. H., *Survey of Russian History* (1948).

CHAPTER I

Anderson, M. S., *Britain's Discovery of Russia 1553–1815* (1958).
Clausewitz, C. von, *The Campaign of 1812 in Russia* (1843).
Rogger, Hans, *National Consciousness in 18th Century Russia* (1960).
Tarle, E. V., *Napoleon's Invasion of Russia* (1942).
Wilson, Sir Robert, *The Invasion of Russia* (1860).

CHAPTER 2

Leslie, R. F., *Polish Politics and the Revolution of November 1830* (1956).
Mazour, A. G., *The First Russian Revolution, 1825* (1937).
Monas, S. L., *The Third Section: Police and Society in Russia under Nicholas I* (Cambridge, Mass., 1961).
Pipes, R., *Karamzin's Memoir on Ancient and Modern Russia* (1959).
Raeff, M., *Michael Speransky, Statesman of Imperial Russia 1772–1839* (The Hague, 1958).
Riasanovsky, N. V., *Nicholas I and Official Nationality in Russia 1825–1855* (Berkeley, Calif., 1959).
Zetlin, M., *The Decembrists* (1958).

CHAPTER 3

Blum, J., *Lord and Peasant in Russia from the Ninth to the Nineteenth Century* (Princeton, N.J., 1961).
Maynard, J., *The Russian Peasant and other Studies* (1962).
Maynard, J., *Russia in Flux* (New York, 1948).
Pavlovsky, G., *Agricultural Russia on the Eve of Revolution* (1930).
Robinson, G. T., *Rural Russia under the Old Régime* (1949).
Treadgold, D. W., *The Great Siberian Migration* (1957).

CHAPTER 4

Greene, F. V., *Sketches of Army Life in Russia* (1881).
Greene, F. V., *The Russian Army and its Campaigns in Turkey 1877–78* (1879).
Kucherov, S., *Courts, Lawyers and Trials under the Last Three Tsars* (New York, 1953).
Leslie, R. F., *Reform and Insurrection in Russian Poland, 1863* (1963).
Leroy-Beaulieu, A., *Un homme d'état russe* (1884).
Leroy-Beaulieu, A., *The Empire of the Tsars and the Russians* (1893–6).
Mosse, W. E., *Alexander II and the Modernisation of Russia* (1959).

Seton-Watson, H., *The Decline of Imperial Russia 1855–1914* (1964).
Wallace, D. M., *Russia* (1905).

CHAPTER 5

Berlin, I., *The Marvellous Decade* (*Encounter*, 1953).
Billington, J. H., *Mikhailoysky and Russian Populism* (1958).
Bowman, H. E., *Visarion Belinsky* (1955).
Carr, E. H., *Michael Bakunin* (1937).
Carr, E. H., *The Romantic Exiles* (1933).
Footman, D., *Red Prelude* (1944).
Hare, R., *Pioneers of Russian Social Thought* (1951).
Herzen, A., *My Past and Thoughts* (1968).
Kropotkin, P., *Memoirs of a Revolutionist* (1899).
Lampert, E., *Studies in Rebellion* (1957).
Lampert, E., *Sons Against Fathers: Studies in Russian Radicalism and Revolution* (1964).
Malia, M., *Alexander Herzen and the Birth of Russian Socialism* (1961).
Riasanovsky, N. V., *Russia and the West in the Teaching of the Slavophils* (1952).
Stepnyak, S., *Underground Russia* (1883).
Venturi, F., *Roots of Revolution* (1960).
Yarmolinsky, A., *Road to Revolution* (1957).

CHAPTER 6

Black, C. E. (ed.), *The Transformation of Russian Society: Aspects of Social Change since 1861* (Cambridge, Mass., 1960).
Charques, R., *The Twilight of Imperial Russia* (1965).
Fischer, G., *Russian Liberalism, from Gentry to Intelligentsia* (1958).
Gurko, V. I., *Features and Figures of the Past: Government and Opinion in the Reign of Nicholas II* (1939).
Keep, J. L. H., *The Rise of Social Democracy in Russia* (1963).
Kochan, L., *Russia in Revolution 1890–1918* (1966).
Laue, T. H. von, *Sergei Witte and the Industrialisation of Russia* (1963).

Lyashchenko, P. I., *A History of the National Economy of Russia* (New York, 1949).
Pipes, R., *Social Democracy and the St Petersburg Labour Movement* (Cambridge, Mass., 1963).
Pobedonostsev, K. P., *Reflections of a Russian Statesman* (1966).
Troyat, H., *Daily Life in Russia under the last Tsar* (New York, 1962).
Witte, S. Y., *The Memoirs of Count Witte* (1921).

CHAPTER 7

Florinsky, M. T., *The End of the Russian Empire* (1961).
Golovin, N. N., *The Russian Army in the World War* (New York, 1931).
Katkov, G., *Russia 1917: the February Revolution* (1967).
Kokovtsov, V. N., *Out of my Past* (1935).
Pares, B., *The Fall of the Russian Monarchy* (1939).
Pares, B., *My Russian Memoirs* (1931).
Pares, B. (ed.), *The Letters of the Tsaritsa to the Tsar 1914–1916* (1924).

CHAPTER 8

Anderson, M. S., *The Eastern Question* (1966).
Gleason, J. H., *The Genesis of Russophobia in Great Britain* (1951).
Kukiel, M., *Czartoryski and European Unity* (1955).
Langer, W. L., *The Franco-Russian Alliance* (1929).
Medlicott, W. N., *The Congress of Berlin and After* (1963).
Mosely, P. E., *Russian Diplomacy and the Opening of the Eastern Question in 1838 and 1839* (1934).
Mosse, W. E., *The European Powers and the German Question* (1958).
Puryear, V. J., *England, Russia and the Straits Question 1844–1856* (1931).
Stavrianos, L. S., *The Balkans since 1453* (New York, 1958).
Sumner, B. H., *Anglo-Russian Relations* (1948).
Sumner, B. H., *Russia and the Balkans 1870–1880* (1962).
Taylor, A. J. P., *Struggle for Mastery in Europe 1848–1918* (1954).
Temperley, H., *England and the Near East: The Crimea* (1964).

CHAPTER 9

Baddeley, J. F., *The Russian Conquest of the Caucasus* (1908).
Churchill, R. P., *The Anglo-Russian Convention of 1907* (Iowa, 1939).
Curzon, G. N., *Russia in Central Asia in 1889 and the Anglo-Russian Question* (1967).
Dallin, D. J., *The Rise of Russia in Asia* (1950).
Kennan, G., *Siberia and the Exile System* (1891).
Lensen, G. A., *The Russian Push towards Japan: Russo-Japanese Relations 1697–1875* (1959).
Malozemoff, A., *Russian Far Eastern Policy 1881–1904* (Univ. of California, 1958).
Pahlen, K. K., *Mission to Turkestan* (1964).
Pierce, R. A., *Russian Central Asia 1857–1917* (Berkeley, Calif., 1960).
Romanov, B. A., *Russia in Manchuria 1892–1906* (Ann Arbor, 1952).
Sumner, B. H., *Tsarism and Imperialism in the Middle and Far East 1880–1914* (Brit.) Acad. Proc. 1941 (1944).
Wheeler, G., *The Modern History of Soviet Central Asia* (1965).

CHAPTER 10

Berdyaev, N., *The Origin of Russian Communism* (1937).
Carr, E. H., *The Bolshevik Revolution* (1966).
Chamberlin, W. H., *The Russian Revolution* (1952).
Deutscher, I., *The Prophet Armed; Trotsky 1879–1921* (1954).
Fischer, L., *The Life of Lenin* (1965).
Futrell, M., *Northern Underground: Episodes of Russian Revolutionary Transport and Communications through Scandinavia and Finland 1863–1917* (1963).
Kennan, G. F., *Soviet-American Relations 1917–20* (Princeton, 1956).
Grey, I., *The First Fifty Years Soviet Russia 1917–67* (1967).
Haimson, L. H., *The Russian Marxists and the Origins of Bolshevism* (1955).
Radkey, O. H., *Agrarian Foes of Bolshevism* (New York, 1958).
Reed, J., *Ten Days that Shook the World* (1962).
Shub, D., *Lenin* (1966).
Schapiro, L., *The Communist Party of the Soviet Union* (1960).

Sukhanov, N. N., *The Russian Revolution 1917* (1955).
Treadgold, D. W., *Lenin and his Rivals* (1955).
Treadgold, D. W., *Twentieth Century Russia* (Chicago, 1959).
Trotsky, L., *The History of the Russian Revolution* (1965).
Wheeler-Bennett, J., *Brest-Litovsk: The Forgotten Peace March 1918* (1938).
Wilson, E., *To the Finland Station* (1960).
Wolfe, B. D., *Three who Made a Revolution* (1966).
Zeman and Scharlau, *The Merchant of Revolution: The Life of Alexander Israel Helphand (Parvus) 1867–1924* (1965).

CHAPTER 11

Baykov, A., *Development of the Soviet Economic System* (1946).
Carr, E. H., *The Interregnum 1923–1924* (1954).
Carr, E. H., *A History of Soviet Russia: Socialism in One Country* (1950–64).
Deutscher, I., *The Prophet Unarmed, Trotsky 1921–1929* (1959).
Fainsod, M., *How Russia is Ruled* (1963).
Hare, R., *Maxim Gorky: Romantic Realist and Conservative Revolutionary* (1962).
Kolarz, W., *Religion in the Soviet Union* (1961).
Pipes, R., *The Formation of the Soviet Union* (1965).
Simmons, E. J. (ed.), *Continuity and Change in Russian and Soviet Thought* (1955).
Slonim, M., *Modern Russian Literature* (1953).
Struve, G., *Soviet Russian Literature 1917–1950* (1951).
Trotsky, L., *My Life* (1930).
Trotsky, L., *Literature and Revolution* (Tr. 1925).

CHAPTER 12

Russian Institute, *The Anti-Stalin Campaign and International Communism* (Columbia, 1956).
Chamberlin, W. H., *Russia's Iron Age* (1935).
Crankshaw, E., *Khrushchev* (1966).
Dallin and Nicolaevsky, *Forced Labour in Soviet Russia* (1947).
Deutscher, I., *Stalin: A Political Biography* (1966).
Deutscher, I., *The Unfinished Revolution 1917–1967* (1967).
Dobb, M., *Soviet Economic Development since 1917* (1966).

Fainsod, M., *Smolensk under Soviet Rule* (1959).
Frankland, M., *Khrushchev* (1966).
Ginzburg, E. S., *Into the Whirlwind* (1968).
Jasny, N., *The Socialised Agriculture of the U.S.S.R.* (1949).
Jasny, N., *Soviet Industrialisation 1928–1952* (New York, 1961).
Koestler, A., *Darkness at Noon* (1941).
Nikolaevsky, B. I., *Power and the Soviet Elite* (1966).
Nove, A., *The Soviet Economy: an introduction* (New York, 1961).
Souvarine, B., *Stalin* (1939).
Trotsky, L., *The Revolution Betrayed* (1937).

CHAPTER 13

Beloff, M., *The Foreign Policy of Soviet Russia* (1947–9).
Dallin, D. J., *Soviet Russia and the Far East* (1948).
Kennan, G. F., *Russia and the West under Lenin and Stalin* (1961).
Laqueur, W. Z., *Russia and Germany* (1965).
Maisky, I., *Who helped Hitler?* (1964).
Taylor, A. J. P., *The Origins of the Second World War* (1963).
Toynbee, A. J. (ed.), *The Impact of the Russian Revolution 1917–1967. The Influence of Bolshevism on the World outside Russia* (1967).

CHAPTER 14

Clark, A., *Barbarossa* (1963).
Clark, D., *Three Days to Catastrophe* (1966).
Conquest, R., *The Soviet Deportation of Nationalities* (1960).
Deane, J. R., *The Strange Alliance* (1947).
Dallin, A., *German Rule in Russia* (1957).
Erikson, J., *The Soviet High Command: A Military-Political History 1918–1941* (1962).
Feis, H., *Churchill, Roosevelt and Stalin* (1957).
Fischer, G., *Soviet Opposition to Stalin* (Harvard, 1952).
Liddell Hart, B. H. (ed.), *The Soviet Army* (1956).
Pavlov, D. V., *Leningrad 1941: The Blockade* (New York, 1966).
Stettinius, E., *Roosevelt and the Russians* (1950).
Werth, A., *Russia at War 1941–1945* (1964).

Index